CYBER-MARX

NICK DYER-WITHEFORD

CYBER-MARX

Cycles and Circuits of Struggle in High-Technology Capitalism

UNIVERSITY OF ILLINOIS PRESS

Urbana and Chicago

Library of Congress Cataloging-in-Publication Data
Dyer-Witheford, Nick, 1951–
Cyber-Marx : cycles and circuits of struggle in high-technology
capitalism / Nick Dyer-Witheford.
       p.   cm.
Includes bibliographical references and index.
ISBN 0-252-02479-6 (acid-free paper)
ISBN 0-252-06795-9 (pbk. : acid-free paper)
1. High technology industries. 2. Technological innovations—
Economic aspects. 3. Capitalism. 4. Information technology—
Economic aspects. 5. Socialism. 6. Business cycles. I. Title.
HC79.H53D94 1999
338.4'762—ddc21   98-58056
              CIP

1 2 3 4 5 C P 5 4 3 2 1

To my daughters, Adrienne and Miranda

CONTENTS

ACKNOWLEDGMENTS

Brecht somewhere remarks that "the formation of intellectuals is a long and difficult process, and sorely tries the patience of the masses." The writing of this book, integral to my formation as an intellectual, may have tried the patience of many individuals and collectives, who have, nevertheless, unstintingly contributed to its completion.

Foremost among these is Rick Gruneau, who, as my doctoral supervisor, allowed me the latitude to follow my interests, offered insightful criticism, and had the generosity of spirit to support a dissertation some of whose specific arguments he disagreed with. Other faculty members at Simon Fraser University also strongly influenced the project, without bearing responsibility for its outcome. Chin Banerjee's course on Lukács, interrupted as it was by a campus strike, set the political ball rolling. Michael Lebowitz's profound lectures on *Capital* and *Grundrisse* showed me what Marxism could, and should, be.

Two great teachers deeply marked my thinking about the issues discussed here. Margaret Benston introduced me to cyberspace and, much more important, gave me an enduring example of warmth and intelligence on the left before her tragically early death. Dallas Smythe, whose writings I had long admired, did me the honor of reading a draft chapter and discussing it one memorable afternoon very shortly before he passed away. If these pages contain anything worthy of such mentors, scholars and revolutionaries both, then the writing time was well spent.

My friends Dorothy Kidd and Santiago Valles played a very special role in the production of this book, reading and commenting on nearly every chapter. They also allowed me to read their own work-in-process: since we all, while following different lines of inquiry, drew on a common set of sources and traditions, this was enormously enriching. We discussed what was at stake in scores of conversations. Dorothy and Santiago challenged me creatively on crucial issues—Dorothy on the history of radio and video and on gender, Santiago on slavery, the international division of labor, and war. In different ways, they made inroads on the characteristic thought-patterns of an author who remains irredeemably white and male. I cannot pretend to have answered all the questions they raised. But I can thank them for a rare experience of academic and political comradeship.

There are many others in Vancouver and at Simon Fraser University who contributed their time, their insight, or their experience to this work. Salutations to Colleen Wood, who survived the early stages of its composition. Thanks to the members of the Canadian Union of Public Employees, Local 2396, especially Laurine Harrison, Steven Howard, and Rhonda Spence, with whom I learned, through trial and much error, political lessons indirectly reflected in the text. Thanks also to Bob Everton, for friendship when it was most needed, and for an inspiring example of political commitment; to Sue Cox, for several excellent conversations about technology theory; and to Sid Shniad, director of research for the Telecommunications Workers' Union of British Columbia, whose ceaseless circulation of electronic information not only was practically useful but also gratifyingly exemplified some of the arguments made here about the subversive potential of virtual labor.

I owe a great debt to many in the international network of autonomist Marxists: to Ed Emery, for keeping an extraordinary archive, and for his invaluable translation work; to Harry Cleaver, for so many insights and provocations; to Massimo De Angelis, for several rewarding exchanges. Thanks also to those with whom I have had discussions online, but not yet in person, especially Steve Wright and Michael Hardt. Thanks too to the various members of what was once the Vancouver InfraReds Collective, now gradually dispersing to the four corners of the world, especially to Conrad and Jennifer Herold, Julian Prior, and T.J.

Finally, in a class of their own are the thanks due to my wife, Anne, whose composure and expertise have so often saved me from recalcitrant computers; whose combination of spiritual insight and social activism has so deeply instructed me; and whose sweet companionship is a constant incentive to struggle for the shortening of the working day.

CYBER-MARX

1

DIFFERENCES

In William Gibson and Bruce Sterling's novel *The Difference Engine*, the year is 1855, the place is England, and the information age has arrived a century-and-a-bit ahead of schedule.[1] Charles Babbage's attempts to develop a mechanical computer, instead of petering out in an expensive failure, have triumphantly succeeded. The Industrial Radical Party, headed by Lord Byron, forges an alliance between bourgeois commerce and scientific "savantry." Ruthlessly repressing Luddite insurgency, it applies the phenomenal powers of steam-driven cybernetic engines to a convulsive transformation of society, automating factories, extending surveillance, and perfecting weapons in a global consolidation of imperial power. Across this digitalized Victorian landscape bizarre intrigues unwind, as nefarious "clackers"—the adepts of the new mechanical computing—governmental security forces, and criminal subversives all pursue a secret accidentally discovered by Babbage's co-inventor, Lady Ada Byron, "Queen of Engines," while attempting to meet her gambling debts: the secret of self-conscious artificial intelligence. Meanwhile, societal catastrophes pile up around the conspirators: ecological disasters, Gulf War–style carnage in the Crimea, mass unemployment and dispossession all converge on chaos. Yet the alliance of science and capital seems irresistible, even as it drives toward unthinkable transformations in the fate of the human species.

What interests me in this steampunk fantasy—at once historical novel and science fiction, yet so manifestly about neither past nor future, but rather a defamiliarized portrait of our own verge-of-the-twenty-first-century present— is one little detail, tangential to the main plot, a mere corner of the canvas. For in the world of *The Difference Engine*, Karl Marx is alive and well. His employment by the *New York Daily Tribune* (for whom the actual Marx worked during the 1850s as a foreign correspondent in the biggest information industry of his day) has clearly resulted in migration to the United States, a visit yielding momentous consequence. For, in a North America wracked by regional separatism and civil war, revolutionaries have seized the "means of information and production" of the largest city of the New World.[2] And the Manhattan Communards now provide a nucleus for an international ferment of dissidence that, combining re-emerged Luddites, renegade clackers, anarcho-feminists, Blakean-situationist artists, and immiserated proletarians, boils beneath the surface of the bourgeois universe, waiting for the next calamity to burst into revolt.

In what follows, I propose a Marxism for the Marx of *The Difference Engine*. That is to say, I analyze how the information age, far from transcending the historic conflict between capital and its laboring subjects, constitutes the latest battleground in their encounter; how the new high technologies— computers, telecommunications, and genetic engineering—are shaped and deployed as instruments of an unprecedented, worldwide order of general commodification; and how, paradoxically, arising out of this process appear forces that could produce a different future based on the common sharing of wealth—a twenty-first-century communism.

Marx and Babbage

To establish some of the issues and conflicts central to this study it may be useful for a moment to look back in the past, to the "actual" Babbage and Marx. In fact, the opposition between Babbage—capitalist-computer-savant— and Marx—insurrectionary revolutionary—that Gibson and Sterling propose is well founded in the historical archive. Although Babbage's pioneer attempts to develop machine intelligence collapsed, partly because of the limits of nineteenth-century engineering, partly because of his managerial conflicts with the craft-workers crucial to the production of the "engines," his influence was far in excess of that normally associated with a failed inventor. As Simon Schaffer has shown, Babbage was an eminent member of a coterie of radical utilitarian thinkers, including such figures as the political economist Andrew Ure, the philosopher Jeremy Bentham and his brother Samuel, and industrialists such as Marc Brunel and Henry Maudsley, all dedicated to the scientific organization of a nascent industrial capitalism.[3]

Babbage himself wrote a book in the tradition of Ricardian political economy—*On the Economy of Machinery and Manufactures*—which in its argument for the deskilling and fragmentation of labor is now recognized as anticipating Frederick Winslow Taylor's system of "scientific management."[4] Babbage's search for mechanical means to automate labor, both manual and mental, was the logical extension of the desire to reduce and eventually eliminate from production a human factor whose presence could appear to the new industrialists only as a source of constant indiscipline, error, and menace. And this in turn was part of a wider project of industrial planning that foresaw the societywide mobilization of theoretical knowledge in the service of manufacture, overseen by a "new class of managerial analysts," such as Babbage himself, who would become "the supreme legislators of social welfare" and be rewarded with "newfangled life peerages and political power."[5] In such schemes, the mechanical maximization of capitalist profit mercifully coincided with the highest theological aspirations, for Babbage believed that "machine intelligence was all that was needed to understand and model the rule of God, whether based on the miraculous works of the Supreme Intelligence or on his promise of an afterlife."[6]

Marx, Babbage's contemporary, read his work. And what he found in its pages was not evidence of the ineluctable march of progress, or an approach to divine wisdom, but a strategy of class war. Writing in London, within living memory of the Luddite revolts that had seen hundreds hanged or transported and vast sections of England subject to martial law, Marx analyzed the introduction of machinofacture as a means by which the bourgeoisie strove to subjugate a recalcitrant proletariat. He alludes to Babbage's writings in the great chapter of *Capital*—"Machinery and Large Scale Industry"—where he describes how the factory owners' relentless transfer of workers' skills into technological systems gives class conflict the form of a "struggle between worker and machine."[7] He cites, as evidence of the political economist's technological strategy, the work of Babbage's colleague, Ure, who in the conclusion to his 1835 *The Philosophy of Manufacture* declared, "when capital enlists science into her service, the refractory hand of labour will always be taught docility."[8] "It would be possible," Marx observes, "to write a whole history of the inventions made since 1830 for the sole purpose of providing capital with weapons against working class revolt."[9]

Later, in a section of volume 3 of *Capital* entitled "Economy Through Inventions," Marx again footnotes Babbage. Commenting on capital's ever-increasing use of machines, he notes that "mechanical and chemical discoveries" are actually the result of a social cooperative process that he calls "universal labour": "Universal labour is all scientific work, all discovery and invention. It is brought about partly by the co-operation of men now living, but partly also by building on earlier work."[10] The fruits of this collective project are, Marx argues, generally appropriated by the "most worthless and wretched kind of money-capitalists."[11] But the ultimate source of their profit is the "new developments of the universal labour of the human spirit and their social applications by combined labour."[12]

Marx had already discussed this tension between the social nature of technoscientific development and its private expropriation by capital—in the final pages of the notebooks for *Capital,* the *Grundrisse.* Here, he again makes passing reference to Babbage as, in some of the most volcanically brilliant of all Marx's writing, he foretells the future technological trajectory of capitalism.[13] At a certain point, Marx predicts, capital's drive to dominate living labor through machinery will mean that "the creation of real wealth comes to depend less on labour time and on the amount of labour employed" than on "the general state of science and on the progress of technology."[14] The key factor in production will become the social knowledge necessary for technoscientific innovation—"general intellect."[15]

Marx points in particular to two technological systems whose full development will mark the era of "general intellect"—automatic machinery, which, he predicts, will all but eliminate workers from the factory floor, and the global networks of transport and communication binding together the world market. With these innovations, Marx says, capital will appear to attain an unassailable pinnacle of technoscientific power. However—and this is the whole point of Marx's analysis—inside this bourgeois dream lie the seeds of a bourgeois nightmare. For by setting in motion the powers of scientific knowledge and social cooperation, capital undermines the basis of its own rule. Automation, by massively reducing the need for labor, will subvert the wage relation, the basic institution of capitalist society. And the profoundly social qualities of the new technoscientific systems—so dependent for their invention and operation on forms of collective, communicative, cooperation—will overflow the parameters of private property. The more technoscience is applied to production, the less sustainable will become the attachment of income to jobs and the containment of creativity within the commodity form. In the era of general intellect "capital thus works towards its own dissolution as the form dominating production."[16]

Babbage and Marx were both prophets of today's information society. But their prophecies are radically opposed, one promising the technoscientific consolidation of market relations, the other the dissolution of that rule. Both spoke, as befits nineteenth-century men of science, in tones of certainty. After the catastrophes and surprises of the twentieth century, such teleological confidence should no longer be available to anyone. Nevertheless, the predictions of both Babbage and Marx are alive and well today, present as vectors of struggle, antagonistic potentialities meeting in a collision that I term "the contest for general intellect."

But surely this must be a joke? Are not Marx and Marxism now so thoroughly discredited, so fatally consigned to the dustbin of a history that has itself been dispatched to postmodernist on-screen trash cans, that any attempt to invoke their memory can only be an exercise in speculative dreaming or historical nostalgia? Since Marxism, assailed from all quarters, is generally deemed to have died the death of a thousand cuts, it is important, at the very outset, to take difference with this prevailing view.

Deaths of Marxism I: The Neoliberal Critique

In the eyes of many, the fate of Marxism has been sealed by the collapse of state socialism—by the disintegration of the ex-USSR and its East European bloc and the absorption of China into the world market. Unfolding through a progression of scenes—intensifying economic crisis, the people in the streets, confrontation with security forces, bloody repression, or flight of demoralized leaders—which seemed in every respect to fulfill the revolutionary anticipations of the Left, only with the diabolic twist that it all culminated not in the collapse of capital but in the fall of socialism, these events have shattered the long-flagging confidence of Marxist militants and intellectuals everywhere.

In the many jubilant postmortems conducted by neoliberal intellectuals over the corpse of Marxism, a wide variety of reasons have been invoked for its demise: the inherent imperfectability of humanity, the innate superiority of markets over state planning, the inevitable transformation of revolutionary aspiration into despotic tyranny, and so on. Not the least important of these is the alleged incapacity of Marxism to comprehend the "information revolution." Many analysts suggest that the evident failure of the Soviet regime to deal successfully with new technoscientific conditions of production—computerization, telecommunication, mass media—is traceable to intrinsic flaws and anachronisms in the legacy of Marxian theory. This argument is, for example, fundamental to that most pompous of neoliberal self-congratulations, the "end of history" announced by Francis Fukuyama, for whom the innate superiority of liberal capitalism in developing the "mechanism" of modern technoscience determines its role as the *summum bonum* of human development.[17]

Fukuyama's work has, however, provoked the emergence of a surprising champion of Marx. In a scathing critique of the "end of history" thesis, Jacques Derrida has questioned the fashionable assumption that the end of state socialism has exorcised the revolutionary "specter" that has haunted capital for so long.[18] Reviving the recognition—long standing in some quarters—that Marxism is not a monolithic body of thought but comprises a multiplicity of intertwined and radically contradictory strands, Derrida challenges any belief that the Bolshevik tradition exhausts this legacy. He further argues that, rather than Marxism being rendered obsolete by the information age, it is only in the light of certain "informational" developments—globalization, the preeminence of the media, telework—that we can see the full importance of certain themes within the texts of Marx, for example, their emphasis on the internationalization and automation of production. Marxism, Derrida insists, will manifest a continuing "spectrality," an uncanny refusal to stay dead and buried, that is profoundly linked to the increasingly "spectral," immaterial, virtual nature of contemporary techno-capitalism.

Derrida's points are important ones, even if his insights into the multistranded nature of Marx's legacy are not original. Marxism is a diversity—so much so that it would be possible to speak, in exemplary postmodern fashion, not so much of Marxism as of "the Marxisms." This heterogeneity goes right back to the oeuvre of Marx himself. For Marx said and wrote different things at different times, not all of which are consistent, or—more important—all of which can be arranged to form different consistencies. In the historical development of Marxism these statements have been selected, permuted, and refracted into an array of very different, and sometimes fiercely antagonistic, forms.

The Leninist strand was only one of these. Its historical preeminence over the last century has to be seen as resulting from a mutational process inherent in the relation of communist movements to the capitalism they struggle against. For in the war between capital and anticapital the combatants are each constantly transforming themselves in order to answer or preempt the strategies of their opponent, spiraling in a "bad infinity" of reciprocal reshaping that can be broken only if one finally extinguishes the other. Inherent in this process is an evident problem, for both sides, of mirroring and introjection—of becoming that which is opposed. Seen in this light, Leninism should be understood as a Marxism highly adapted—indeed, fatally overadapted—to a particular moment of capitalist development, namely, that of Fordist capitalism, with its characteristic Taylorist division of labor, industrial mechanization, and emphasis on "mass organization."

As Karl Heinz Roth has argued, the Leninist party in its division of party managers from proletarian masses uncannily emulated the Taylorist division of labor.[19] The Soviet state carried this mirroring yet further in its concept of socialism as "soviets plus electrification," its embrace of scientific management, the adoption of the stopwatch and the assembly line, its gigantism of industrial factories and standardization of social life.[20] Ultimately, this led to a path of modernization and forced industrialization that under Stalin constituted nothing so much as a version—hideously enlarged to Russian, rather than English, and twentieth-, rather than eighteenth-century, scale—of capitalism's era of so-called "primitive accumulation."[21]

As several commentators have pointed out, this process was, by capitalist standards, a great success—producing the fears, current in the 1950s and now long forgotten, that Russia and China would overtake the West in economic growth.[22] The other side of the coin, which I would emphasize, is that this introjection of capitalist norms of efficiency, labor discipline, industrialism, and accumulation was, in communist terms, a catastrophic defeat—entailing the suppression of workers' self-organization and the bloody annihilation of every different form of Marxism that remembered this aspiration. State socialism thus became a competitor with, but not an alternative to, capitalism.

The eventual collapse of this regime (as opposed to its much earlier abnegation of revolutionary goals) was, as neoliberals claim, intimately related to the new information technologies and post-Fordist production techniques. For these reduced to global irrelevance the industrial, Fordist methods to which Bolshevism had so tightly bound itself. In this respect, the arms race resulted in a victory for the West, not in the anticipated apocalyptic form of a nuclear exchange, but rather because military expenditures provided a superstimulus to the development of the high technologies that formed the basis for a whole new stage of capitalist restructuring. Blinded by a deeply embedded "factory-ism," unable to adjust an authoritarian regime of labor discipline suitable for digging canals or running assembly lines to what was needed for making computer software, and vainly trying to impose central state command on ever-proliferating international and domestic media channels, the Soviet state could not adapt to these new conditions. It disintegrated under the pressure of movements that, in their dissident use of *samizdat* and computer networks, manifested a quintessentially "informational" subjectivity.

The reader will find no apologies or laments for "actually existing socialism" here, no debate as to whether Stalin, Trotsky, Lenin, or Engels should be blamed for its failures, nor even any attempt to absolutely exonerate Marx from all the stain of its catastrophe. The question is rather whether there is anything *else* in the Marxist legacy with which to confront our own informational commissars. For rather than identifying this disintegration of Bolshevism with the end of Marxism, it can be seen as opening a space within which other, repressed branches of the Marxist genealogy can emerge and blossom.

What makes this probable is that post-Fordist, informational capital exhibits tendencies to catastrophe and conflict perhaps even wider and deeper than those of the Fordist, industrial predecessor that beckoned Bolshevism into being. The unleashing of computerization, telecommunications, and genetic engineering within a context of general commodification is bringing massive crises of technological unemployment, corporate monopolization of culture, privatization of bodies of knowledge vital for human well-being and survival, and, ultimately, market-driven transformations of humanity's very species-being. In response to these developments are emerging new forms of resistance and counterinitiative. And insofar as the force with which these movements collide is capitalism—perhaps a post-Fordist, postmodern, informational capitalism, but capitalism nonetheless, and not some postindustrial society that has transcended commodification—Marx's work can continue to provide participants in these struggles a vital source of insights. As Fredric Jameson has said in a slightly different context, "whatever its other vicissitudes, a postmodern capitalism necessarily calls a postmodern Marxism over against itself."[23]

In the last twenty-five years, over the very period of the post-Fordist, post-modern restructuring of capitalism, the theoretical elements of such a meta-morphosed Marxism have, slowly, painfully, out of the experience of defeat and disintegration, been recomposing themselves. It is a Marxism that, learning from the failure of the Bolshevik experiment, draws from the multiplic-ity of Marx's writings threads different from those out of which the Leninist flag was woven, and, moreover, transforms what it takes in the light of the new "informational" conditions of exploitation and revolt. But this Marxism will mark a reappearance of the very specter that capital has fled so fast into the future to avoid.

Deaths of Marxism II: The Post-Marxist Critique

Any such reconstruction of Marxism has to confront another line of criticism, coming not so much from the free-market neoliberals but from the so-called new social movements—feminism, green movements, antiracist groups, gay and lesbian rights activists, and others. It is generally claimed that since at least the 1960s, these "new" movements have displaced the "old" working-class struggles—with which Marxism was so closely identified—as the ma-jor source of social dissent in advanced capitalist societies.

This phenomenon, too, is often related to the new informational conditions of automation, computerization, and media saturation. For many "social movement theorists," from Alain Touraine through Alberto Melucci to Timo-thy Luke, the new forms of social upheaval are specifically linked to the ad-vent of a postindustrial order, in which manual labor plays a diminishing role, and the emergence of unprecedented forms of technocratic power elicits novel forms of struggle beyond the ken of conventional class analysis.[24] Such "antitechnocratic" interpretations may not reflect the self-understanding of many feminist, antiracist, environmental, or peace activists. But what is cer-tain is that from these movements, and their academic interpreters, has come a devastating indictment of Marxism's claims to be in forefront of social struggle.

In Marxism, these critics say, people are understood *reductively*, solely in terms of class identity—that is, their position within an economic system of production. But this view strips them of gender, race, culture, or significant relation to nature. This reductionism is reinforced by the *totalizing* nature of Marxist theory—its claim to map and account for the entirety of social relations. This totalizing, reductive perspective generates a series of disastrous theoretical omissions and repressions: blindness to patriarchy and racism, denial of cultural diversity, scientific triumphalism. From these theoretical flaws flow the often catastrophic record of actual Marxist regimes and parties in terms of sexism, ecological despoliation, and totalitarian repression.

The result of this critique has been the increasing fashionability among the Left of a "post-Marxist" position of the sort most famously theorized by Ernesto Laclau and Chantal Mouffe.[25] This decisively rejects the centrality Marx ascribes to issues of capital and class, now dismissed as the result of a crude, mechanistic economic determinism. In its place is proposed a new lexicon of difference and discourse. Class relations are no longer "privileged" but rather seen as only one among a diversity of semiotically constructed identities. The extraction of surplus value is simply included within a range of dominations and oppressions (sexism, racism, homophobia, industrialism) none of which can be accorded any priority over the other. Progressive politics has to be rethought on a more plural and populist basis, as a series of variegated struggles against numerous relations of subordination, all of which, though distinct, may be related in a project not of revolution but of "radical democracy."[26]

Although my differences with theorists such as Laclau and Mouffe will rapidly become evident, it should be said at once that I find many of the criticisms leveled by social movement activists against Marxism telling. In the pages of Marx himself there are major blind spots to issues of gender, ethnicity, and the destruction of nature. That these are characteristic of his age does not diminish the seriousness of their consequences. In many respects such problems have been magnified, rather than corrected, in the later development of the Marxist tradition. Why not then just say goodbye to all that? Or, at the very least, why not adopt the sort of post-Marxist position in which analysis of class and exploitation, rather than occupying a crucial position, is deployed eclectically alongside other approaches?

To this the short answer is: because of capitalism—unfinished business of a serious magnitude. Post-Marxists have seriously mistaken the target of their attack. The major source of practical, brutally effective reductionism and totalization at work on the planet today is not Marxism, but the world market, now enabled by computer networks, satellite broadcasts, just-in-time production, and high-tech weaponry. This is a system based on the imposition of universal commodification, including, centrally, the buying and selling of human life-time. Its tendency is to subordinate all activity to the law of value—the socially imposed law of exchange. It relates a monological master-narrative in which only money talks. Such a system operates by process of massive reduction—Marx called it "abstraction"—that perceives and processes the world solely as an array of economic factors. Under this classificatory grid—this "classing" of the world—human subjects figure only as so much labor power and consumption capacity, and their natural surroundings as so much raw material. This reductionism—the reductionism of capital—

has today a totalizing grip on the planet. Other dominations, too, are reductive—sexism reduces women to objects for men, racism negates the humanity of people of color. But neither patriarchy nor racism has succeeded in knitting the planet together into an integrated, coordinated system of interdependencies. This is what capital is doing today, as, with the aid of new technologies, it globally maps the availability of female labor, ethno-markets, migrancy flows, human gene pools, and entire animal, plant, and insect species onto its coordinates of value.

In doing so, it is subsuming every other form of oppression to its logic. Contrary to the post-Marxist belief that different kinds of domination politely arrange themselves in a nonhierarchical, pluralistic way the better not to offend anyone's political sensibilities, capitalism is a domination that really dominates. This is not to say—as Marx and many later Marxists sometimes suggest—that the corrosive power of commodification necessarily abolishes patriarchy or sexism (although it can sometimes work in that direction). Indeed, it is possible now to see much better than Marx in his day could how the capitalist international division of labor often incorporates, and largely depends on discrimination by gender or ethnicity to establish its hierarchies of control.

Nevertheless, sexism and racism do not in and of themselves act as the main organizing principle for the worldwide production and distribution of goods. Patriarchal and racist logics are older than capital, mobilize fears and hatreds beyond its utilitarian economic understanding, and are virulently active today. But they are now compelled to manifest themselves within and mediated through capital's larger, overarching structure of domination: as market-racism, commodity-sexism. Class—capital's classification of its human resources—*does* tend to assert itself as definitive of social power. It is "privileged" in all senses of the world—not because of any essential, ontological priority of economics over gender, ethnic, or ecological relations, but because of society's subordination to a system that compels key issues of sexuality, race, and nature to revolve around a hub of profit.

Looked at in this way, the conventional division between "old" class politics and "new" social movements seems profoundly mistaken. Capital is a system inimical not only to movements for higher wages, more free time, or better working conditions—classic labor movement objectives—but also to movements for equality-in-difference, peace, and the preservation of nature. This is not because it creates racism, sexism, militarism, or ecological despoliation, phenomena whose existence handsomely predates its appearance, but rather because it treats them only as opportunities for or impediments to accumulation. Because capital's *a priori* is profit (its own expanded replication), its logic in regard to the emancipation of women, racial justice, or the preservation of the environment is purely instrumental. The prevention of male violence toward women, the saving of rain forests, or the eradication of racism is a matter of bottom line calculus: tolerated or even benignly supported when costless, enthusiastically promoted when profitable, but ruthlessly opposed as soon as they demand any substantial diversion of social surplus. Hence capitalism is antithetical to any movements for whom these goals are affirmed as fundamental, indispensable values.

In this respect, the 1980s and 1990s have been perversely illuminating. Any belief that the advent of the new social movements marked a transition from the "old" struggle over social surplus must crumble away in the face of neoliberalism's doctrinaire reaffirmation of the market, attack on the welfare state, and unconstrained expansions of commodity exchange. Over this period virtually every objective of social movements—wilderness preservation, equal pay for women, funding for day-care, battered-women's shelters, or AIDS education—has had to be fought for, often lost, in the teeth of governmental and corporate insistence on the primacy of austerity, restraint, cutbacks required by global competition, and the reestablishment of wavering profit rates. Insofar as there have been victories, cracks in the reductive logic of capital, it is usually only because movements have been prepared to challenge the overriding priorities of corporate growth in the name of other, differing visions of societal good.

In a bold metaphor, John McMurtry has referred to this era as "the cancer stage of capitalism."[27] Previously restricted by the "communist threat" and workers' movements, capital has now, he argues, entered into a phase of uncontrolled expansion marked by global mobility and the explosion of financial speculation divorced from any productive function. This process is attacking the social institutions that maintain public health and life in a way analogous to the metastasizing encroachments of tumorous cells on a human body. Capital, McMurtry says, is engaged in a systematic subversion of the "social immune system." Environmental despoliation, unemployment, the redistribution of income from poor to rich, and the dismantling of public forms of life-provision are the symptoms of a malignancy that diverts more and more social resources to fuel its own growth:

> Indicative of the classic pattern of cancer mutation and spread are the synergistic effects of money capital's cumulative destruction of the planet's basic conditions of life (air, sunlight, water, soil, and biodiversity), its increasingly aggressive invasions and assaults on social infrastructures and self-protective systems of life sustenance and circulation, its systemic intolerance of bearing the costs of maintaining social and environmental carrying and defence capacities, and its rapidly escalating, autonomous self-multiplication that is no longer subordinated to any requirement of life-organisation.

McMurtry remarks that the essential problem of such a cancerous form of growth is that "the host body's immune system does not effectively recog-

nize or respond to the cancer's challenge and advance." In the case of capitalism, this occurs because the surveillance and communication systems of
host social bodies across the world—i.e., their mass communication and education systems—are themselves subordinate to transnational capital and
largely reject and refuse to disseminate messages that identify the source of
the disease.

The academic fashionability of post-Marxism is an aspect of this failure of
recognition and response. In its refusal to acknowledge the full depth of
capitalism's subsumption of the planet, and in its dismissal of the very political and intellectual tradition that has consistently applied itself to this
issue, it is part of a problem of globally life-threatening dimensions. But a
reinvented Marxism, one that learns from the new social movements without forgoing its focus on the contradictions specific to capitalism, could be
part of the solution.

Back to the Laboratory

This book aims to assist such a reinvention. I imagine it as a laboratory investigation, disinterring seemingly long-dead strands of theoretical DNA coiled
within Marx's texts and exposing them to new mutations. This metaphor of
course betrays the influence of biotechnological science-fiction movies such
as *Jurassic Park* or *Alien Resurrection*. But the story line made familiar by
these films has to be significantly altered. Hollywood's reanimation fantasies
tell of inhuman terrors brought back from the past, or from extraterrestrial
origins. To grasp the situation of late capitalism, however, we must imagine
a planet, our planet, on which the dinosaurs (obvious metaphorical figures for
the gigantic, alienated powers of global corporations and financial institutions)
have lived on, well beyond their appointed time of extinction. The hominid
population has by some catastrophic evolutionary detour (in terms of our trope,
the failure of early socialisms) been diverted from attaining its full development. It now endures a stunted and terrorized existence, scurrying around the
feet of these monsters. The emergence of a truly human form of life, free from
chaotic violence and arbitrary predation, becomes conceivable only by genetic
experimentation aimed at reviving certain near-extinguished lines of species-
being—or, to translate again from the biological to the political, by rediscov-
ering the possibility of a collective, communist transformation of society.

In attempting to recover some theoretical cell-matter for such a transformation, I proceed as follows. In the second chapter I review the work of the
heirs of Babbage—today's information revolutionaries. Looking at a line of
social theorists that runs from Daniel Bell to Francis Fukuyama, I show how
these thinkers conceive of informatics as a high-technological "fix" for the
conflicts and crises of capitalism—and how their theories have developed in
an antagonistic dialogue with the specter of Marxism.

In the third chapter I turn to the Marxist reply to such theories. Starting with an examination of tensions and contradictions around the technology issue in the work of Marx himself, I investigate how these have been developed in very different directions by various Marxian schools and tendencies— "scientific socialists," "neo-Luddites," and "post-Fordists"—and suggest why none of these represent an adequate answer to the challenge of the information revolutionaries.

Having taken foes and friends alike to task, it is clearly now time for me to show my own hand. In chapter 4 I therefore introduce the perspective that has substantially shaped my thinking on these issues—that of "autonomist Marxism." After briefly explaining why this theoretico-political current offers a way beyond the impasse depicted in the previous two chapters, I use the autonomist concept of "cycles of struggle" to offer a historical analysis that locates the origins of the information society in the conflict between labor and capital, and examines current controversies about class composition in a digitalized era.

I adopt a more synchronic approach in chapter 5, proceeding around the "circuit of capital," examining the conflicts that attend the informationalization of production, consumption, social and ecological reproduction, finishing with a look at the cyberspatial realm, which increasingly provides a medium both for capitalist control and for the "circulation of struggles."

In chapter 6 I expand the territorial scope of the study, so far focused principally on conflicts within the so-called advanced or developed world and take up the international dimensions of resistance to high-technology capital: I examine "globalization" and argue that this process, in which new communication technologies obviously play a central role, can be understood only in terms of two conflicting vectors: the expansion of the world-market and countervailing, oppositional movements increasingly linked in what I term "the other globalization."

I shift register from the technological to the cultural in chapter 7 and take up the issue of "the postmodern." Building on the analysis of others who suggest that postmodernist thought can be seen as a response to a world radically restructured by high-technology capital, I suggest that a new critical analysis of the "postmodern proletariat" opens horizons beyond the traditional theoretical polemics between postmodernists and Marxists.

In chapter 8 I raise the issue of how computers and other information technologies might play a part in the constitution of a postcapitalist society. Here I consciously break with many other autonomist analysts—who have often been reticent about utopian speculations—and make some futuristic proposals of my own about the possible form of an information-age communism.

Finally, I return in chapter 9 to Marx's category of "general intellect." I examine more closely his formulation of this concept in the *Grundrisse* and then turn to recent reworkings of it by intellectual-activists associated with the French journal *Futur Antérieur*. Drawing on their work, I review the over-all dynamic of conflict in high-technology capitalism. I then look at the situation of universities, which, along all other forms of educational and knowledge-transmitting institutions, are being rapidly transformed by the capital's information revolution, and conclude by assessing the possibilities for academics, such as myself, to intervene in the "contest for general intellect."

2
REVOLUTIONS

On the eve of the twenty-first century the only revolution spoken of in advanced capitalism is the information revolution. Few other ideas have proven so compelling for people attempting to comprehend incessant and accelerating technological change in their daily lives. Along with a number of synonymous or associated terms—"postindustrialism," "superindustrialism," "the technetronic society," "the wired society," "the control revolution," "high-technology society," "the second industrial divide," "post-Fordism," "the globalization of technology"—the phrase "information revolution" has come to define contemporary anxieties and hopes about the future. For, according to the theorists of this revolution, the technoscientific knowledge crystallized in computers, telecommunications, and biotechnologies is now unleashing an ongoing and irresistible transformation of civilization, dramatic in its consequences, unavoidably traumatic in the short term, but opening onto horizons nothing short of utopian.

The development and content of the doctrine of information revolution have already been given extensive critical analysis.[1] But I want here to relate it to a different body of revolutionary theory—one whose star has fallen, even as that of the information revolution has risen: Marxism. Marxists have shared information revolutionaries' belief in the profound social consequences of

technoscientific change. But they have differed from them in relating the dominative and liberatory potential of machines to the struggle between labor and
capital and to another kind of revolution—communist revolution. No propositions could today appear more fatally archaic. In the age of cyberspace, Lenin
lies in ruins. And many would say that the inverse trajectories of Marxism
and the information revolution—one ascending as the other declines—are
causally connected. Marxism, information revolutionaries claim, was unfit for
the information age, doomed by allegiance to a labor theory of value in an era
of intelligent machines; by a base/superstructure model of society blind to the
significance of symbolic data; by a despotic statism that tried in vain to repress irresistibly proliferating channels of communications; and by a concept
of revolution made obsolete by technological progress.

But if information revolutionaries have polemicized against Marxism, they
have also themselves claimed many characteristically Marxist themes—notions of "progress," of "materialism," of "liberation," and, of course, of "revolution" itself. This common vocabulary in part goes back to the Enlightenment heritage that the insurrectionary Marx shared with technocratic
utopians such as Babbage, men whose schemes for a perfected industrialism
overseen by scientific experts are the forerunners of information society
theory.[2] But it also has a more recent basis. Some of today's most prominent
information revolutionaries are themselves one-time Marxists, apostates who
have drawn heavily on their former beliefs even while developing a new creed.
This chapter therefore examines the information revolutionaries' hostile
annexation of Marxism, showing how they turn Marx against Marx in pursuit of a technologically altered world where communism is neither possible
nor necessary.

From the End of Ideology to Postindustrialism

Although it is only recently that the idea of "information revolution" has
become widely current, it is the immediate descendant of a concept of the
late 1960s—postindustrial society. But to understand the relation of both these
theories to Marxism it is necessary to look further back and glimpse behind
the shoulder of postindustrialism the shape of a yet earlier concept—that of
the "end of ideology."[3]

In the late 1950s and early 1960s a number of intellectuals, surveying the
apparently calm and prosperous conditions of North American and European
"industrial" societies, suggested that these had reached a plateau of more or
less permanent stabilization. Postwar affluence, the institutionalization of
collective bargaining, and the welfare state had banished the class conflicts
of an earlier era from the scene. Such societies presented *the* successful so-
cioeconomic model, toward which other experiments, including those in the
"underdeveloped" and "socialist" world, would gradually converge. This was

the condition of the "end of ideology"—which meant, in general, an end of alternatives to liberal capitalism, and, more specifically and pointedly, an end to Marxism as a revolutionary force. Among the most eloquent spokespersons for this thesis was Daniel Bell, a rising young intellectual rapidly departing early Trotskyite flirtations on a rightward trajectory that would eventually deliver him as a founding figure of American neoconservatism.[4]

Few social theories have, however, had the misfortune to be as swiftly discredited as the "end of ideology" thesis. Within a matter of years the appearance of peaceful, passionless capitalist stability was spectacularly contradicted by the upsurge of domestic and international dissent in the late 1960s and early 1970s. Industrial society—the unsurpassable pinnacle of modernity, prosperity, and technological advance—went into paroxysm, its military machine stalled in the jungles of Vietnam; its urban ghettos burning through successive summers; its huge automobile factories paralyzed by labor conflict; its university campuses in rebellion; its culture subverted by the music, drugs, and politics of youth revolt; its domestic arrangements and relation to nature shaken by nascent feminist and ecological movements.

Bell's "second coming" as a prophet of postindustrialism can be understood as a reaction to these events.[5] Faced with the unexpected convulsions of "industrial society," many intellectuals sought explanations in the possibility that these tumults marked nothing less than the growing pains associated with the emergence of a radically new social order. Such notions were variously inflected, embracing both right and left variants. But the most influential version, the one from which a direct line to today's concept of the information revolution can be traced, arose among the think tanks and sponsored research projects offering futurological guidance for U.S. state policy and corporate strategy.

From this context emerged ideas such as that of the "technological society" fostered in Harvard's IBM-sponsored Program on Science and Technology (1971), the "knowledge society" predicted by management guru Peter Drucker (1968), the "technetronic era" described by soon-to-be U.S. national security adviser Zbigniew Brzezinski (1970), the "year 2000" scenarios elaborated by Herman Kahn and Anthony Wiener out of the RAND Corporation and the Hudson Institute (1967), and, most famously, the work of Bell, whose *The Coming of Post-Industrial Society*, published in 1973 but expressing ideas that its author had been developing since at least 1968, was to prove definitive of the entire genre.[6]

Taking the United States as the exemplar of future global developments, Bell argued that out of the crises of his day was appearing a new type of "postindustrial" society, to be fully visible "in the next thirty to fifty years."[7] The principal motor of this postindustrial transition was the increasingly systematized relationship between scientific discovery and technological application, which was making theoretical knowledge society's central wealth-producing resource. Around this axis of change were grouped loosely associated transformations: a shift from a goods-producing to a service economy; a move in occupational distribution away from manual labor to the preeminence of professional and technical work; increasing capacities of assessment and forecasting; and a new "intellectual technology" of games theory and systems analysis, materially embedded in computer systems.[8]

The result would be a society "organized around knowledge for the purpose of social control and the directing of innovation and change."[9] The most important agents in this postindustrial society would be scientists, engineers, and administrators, a new "knowledge class" lodged primarily within government and academia, bearers of the rationalist skills and virtues required by increasing organizational and technological complexity.[10] Bell argued that the endeavors of this new class could create an epoch of rationalized integration and prosperity, which, while not without its own problems, would finally escape from the material want, economic crisis, and class conflict of the industrial era.

As he advanced this new position, Bell had firmly in mind the adversarial presence of Marx. For although the upheavals of the late sixties challenged socialist parties and governments as well as capitalist ones, they were undeniably shot through with the spirit of the very revolutionary tradition that the "end of ideology" thesis had pronounced defunct. Marx was present in the support for Vietnamese and Cuban guerrillas, in the theories of the New Left, and in the slogans of workers and students in Paris, Turin, and Detroit. *The Coming of Post-Industrial Society* in fact opens with the image of Marx in the British Museum hearing "in every faint sound of riot or each creaking downturn of the business cycle the rumblings of revolution and the abrupt transformation of society."[11] Saluting Marx's work, Bell situates his own efforts in the same tradition of "social forecasting"—and then launches into a sustained attack on Marxist claims that capitalist societies must violently succumb to their internal contradictions.[12]

This rebuttal proceeds not by a simple rejection of Marx, but by an ingenious recuperation.[13] Bell proposes that there are actually two contradictory "schemas" in Marx's analysis of capitalism. The first, best-known, is the "revolutionary" prediction of sharpening class contradictions, market anarchy, and deepening crisis contained in volume 1 of *Capital*. The second, Bell claims, is suggested in the later volumes, and it envisages a quite different "rationalizing" tendency, glimpsed by Marx but better understood by theorists such as Max Weber, a tendency apparent in the separation of professional management from capitalist ownership, the rise of a "middle" class, the bureaucratization of enterprise, and the spread of stockholding. This latter trend, Bell says, blurs and softens class conflict. The history of the twentieth century is the story of the cancellation of the former revolutionary prediction by the latter rationalizing one—culminating in the advent of post-industrial society.

Knowledge, says Bell in one of his most widely repeated formulations, will replace both labor and capital as the main factor of production. Between the opposition of capitalist and worker emerges a new class—"a professional class, based on knowledge rather than property."[14] The rise of this new class follows a quasi-Marxian logic that relates the emergence of new historical subjects to new forces of production but effectively negates its revolutionary force.[15] Capital will be transformed by technical and administrative experts, abandoning fixation with profit, becoming more socially responsible, and giving "moral issues" equal priority with balance sheets.[16] Labor too will be transfigured. Technological development will raise living standards, automate manual toil, and thereby liquidate Marx's subject of history—the immiserated industrial proletariat. "If there is an erosion of the working class in post-industrial society," Bell asks, posing the question all information society theorists will subsequently hurl at Marxism, "how can Marx's vision of social change be maintained?"[17]

Ultimately, in an ambivalence that persists throughout information society theory, Bell equivocates as to whether this regime of scientific expertise peacefully *transcends* capitalism or simply *elevates* it to a new level of stability and organization.[18] He toys with the idea that the "knowledge class" will become a new ruling class, only to retreat regretfully from this suggestion. But in any case its appearance is sufficient to nullify Marx's prediction of war between capital and labor, smoothing the sharp edges of bipolar class antagonism so as to make the idea of communist revolution a quaint anachronism.

The postindustrial prophecy thus projects into an imminent future the very conditions of stabilization that the "end of ideology" thesis had mistakenly declared already achieved. As Krishan Kumar has pointed out, Bell and his colleagues, faced by the revelation that contemporary society was not fully pacified, responded by proposing an extra stage to the march of progress.[19] With the suitable application of expertise and technology, the lingering problems would be cleared up once and for all around the year 2000.

Often, Bell speaks of this outcome with oracular certainty. Yet this tone is at odds with another, more urgent and combative element in his writings—condemnation, polemic, warning. Rational progress—embodied in the technocratic state and its knowledge elite—is under siege by the irrational protest by the New Left, student revolt, affirmative action groups, and an "adversary culture."[20] Only if the pilotage of society is entrusted to the cadres of technical experts, scientists, engineers, and administrators will chaos be avoided and the dawning era safely ushered in. No mere extrapolation from predetermined trends, but a determined assertion of what those trends will be, postindustrial futurology foresees the future it intends to make.

From Postindustrialism to the Information Society

In the late 1960s and early 1970s such postindustrial theory enjoyed wide popularity among academics, government experts, and corporate managers. Nowhere was it more avidly received than in Japan. There, translated texts by North American futurists were reworked by authors such as Tadeo Umesao, Kenichi Kohyama, Yujiro Hayashi, and Yoneji Masuda to produce the concepts of *johoka shakai* or *joho shakai*—"informational society" or "information society."[21] According to Tessa Morris-Suzuki's study of Japanese information society theory, *joho shakai* gave particular emphasis to computers' potential for changing industrial production methods by introducing unprecedented levels of automation and of integration between office, factory, and consumer.[22] At the same time, the content of production was envisaged as becoming more "information intensive," in the sense that innovation, planning, design, and marketing would represent an integral and increasing share in the value of goods and services.

In the work of futurists such as Masuda these transformations were linked to an idealistic vision of an emergent society in which increased availability of information and free time resulted in declining materialism, improved self-actualization, voluntary civic participation, enhanced global and ecological consciousness, and, ultimately, a revival of spirituality—in short "computopia."[23] But this concept of extensive computerization also entered the domain of public policy, sponsored by Japan's powerful Ministry of International Trade and Industry, as a hard-headed development strategy aimed at overcoming shortages in labor and natural resources, securing international markets, and remedying the widespread social disaffection of the 1960s. The creation of an "information society" became a centerpiece of Japanese economic planning.

In North America and Europe, interest in these ideas was accelerated by economic recession, whose first tremors had appeared in the late sixties. Bell and his colleagues had assumed an uninterrupted continuation of postwar rates of economic growth. But by the mid-1970s this prediction was abruptly confounded as social disorder was met by austerity, recession, and economic crisis. However, as the West's leaders searched for solutions to social economic malaise, their eyes turned to the "Japanese miracle"—only to discover *joho shakai* as a strategy for computerization, robotization, workplace reorganization, and systematic "softening" of the economy. Under this guise, postindustrialism earned a new lease of life. In 1978, a conference of Japanese and U.S. communications scholars resulted in the publication of the first North American book to use the term "information society" in its title.[24]

At the same time, related ideas were independently gaining currency on both sides of the Atlantic. In 1977, the U.S. government's Office of Telecommunication published Marc Porat's influential study of the "information

economy," which suggested that an increasing portion of the GNP depended
on "information activity" and a growing proportion of jobs on "information
work."[25] In Europe, a broadly similar effect was produced by the publication
in 1978 of a French government report on computerization, *L'Informatisation
de la Société*, by Simon Nora and Alain Minc.[26] Nora and Minc argued that
the convergence of computers and telecommunications—which they termed
"telematics"—would alter "the entire nervous system of social organization."[27] In the light of this transformation, national well-being depended on
the fostering of domestically based high-technology industries and the computerization of government operations.

By the late 1970s, the "information revolution" was emerging as a central
category in government and corporate planning. In 1979 Bell recast his original postindustrial thesis in the new, fashionable terms, emphasizing the
importance of computer and telecommunication networks and speaking of
an "information explosion" constituted by:

> a set of reciprocal relations between the expansion of science, the hitch-
> ing of that science to a new technology, and the growing demand for news,
> entertainments and instrumental knowledge, all in the context of rap-
> idly increasing population, more literate and more educated, living in a
> vastly enlarged world that is now tied together, almost in real time, by
> cable, telephone and international satellite, whose inhabitants are made
> aware of each other by the vivid pictorial imagery of television, and that
> has at its disposal large data banks of computerized information.[28]

This statement was simultaneous with and succeeded by a spate of similar
academic studies; by best-selling popularizations such as Alvin Toffler's *The
Third Wave* and John Naisbett's *Megatrends*; by a burgeoning business literature devoted to managing in the information age; and by journalistic coverage of the type that made the microcomputer *Time*'s "Person of the Year"
for 1982.[29] All of this translated theories of the information revolution into
a popular idiom of the 1980s.

These theories revamped the postindustrial vision of epochal transition,
giving it glossier sheen, leaner design, and enhanced computing power.
Postindustrialism had primarily defined the new era in terms of its departure
from the crises of industrialism. Information society theory gives this shift
a more substantial content: industry is succeeded by information. The bor-
derline between eras is that dividing mechanical from digital machines, steel
mills from silicon chips, railroads from communication networks. Postindus-
trialist technocracy, moreover, had worn the mark of an attachment to governmental bureaucracy. Information revolution, more attuned to the climate
of Thatcherism and Reaganism, dispenses with this. Technocracy is replaced
by high tech, organization men by intelligent machines, experts by expert
systems, intelligentsia by artificial intelligences, mainframes by microcom-

puters, pyramidal hierarchies by distributed systems, central office by cyber-space.[30]

In this form, the idea of an information revolution—a revolution simultaneously inevitable and desirable—became a crucial intellectual and rhetorical component in a project of high-technology restructuring pursued collaboratively by state and corporate sectors throughout the advanced capitalist world.[31] For corporations, the image of an approaching information age provided a slogan to accompany the robotizing of factories, automating of offices, selling of cable television, and marketing of microcomputers, new media, and on-line services. For government, the approach of the information society was invoked to justify public subsidization of corporate high-technology research, the forging of academic-business partnerships, the deregulation of phone companies, and the privatization of telecommunications and other information utilities in the public domain.

Those who propounded its doctrine—political leaders, corporate executives, state bureaucrats, research scientists, academic theorists, journalistic popularizers—did not merely describe the future. They prescribed it. Although the arrival of the new epoch was declared inevitable, definite steps were demanded to adjust to its realities, hurry its benefits, preempt its problems, and secure positional advantage within it. These included massive investment in new machines; vast restructurings of work and unemployment; the stimulation of new markets; the inculcation of unfamiliar leisure habits and cultural forms; the reorganization of research, education, and training; the treatment of technophobia; and the crushing of "Luddism." The proffered choice was adaptation or obsolescence. And insofar as such exhortation did result in a deepening social commitment to, and dependence on, information technologies, it secured for itself the virtuous circularity of self-fulfilling prophecy—generating the reality it predicted.[32]

Revolutionary Doctrine

Theories of the information revolution are not all the same. At each stage in the unfolding of the doctrine advocates of the most recent version urge the novelty of their position and distance it from the preceding one. There are also substantial differences within each generation of the argument, as well as significant variations of tone between its various academic, popular and official registers.[33] Nonetheless, the principle claims of the information revolutionaries can be summarized in seven points of "revolutionary doctrine."[34]

1. *The world is in transition to a new stage of civilization, a transition comparable to the earlier shift from agrarian to industrial society.* In this transition computers and telecommunications play a role equivalent to that of the steam engine and railroad in the nineteenth century. Underlying this idea is a powerful technological determinism. Masuda writes: "When epoch-

making technological innovation occurs, changes take place in the existing society and a new society emerges. The steam engine precipitated the industrial revolution, bringing about the changes that led to a new economic and political system. . . . The information epoch resulting from computer-communication technology will bring about a societal transformation just as great or even greater than the industrial revolution."[35] Other accounts acknowledge that the effects of technology on society are not immediate, nor the interaction entirely unidirectional. But the overall tenor of the argument is usually that machines are the real makers of social change. The transformative effects of information technologies are usually conceived of as becoming visible in the 1960s, although originating earlier, starting in developed economies—Japan, the United States, and other OECD countries—and proceeding at an accelerating rate and with expanding scope as we approach the millennium, moving on a trajectory that is basically benign, eventually universal, and certainly unavoidable—the latest phase in the march of progress.

2. *The crucial resource of the new society is technoscientific knowledge.* While technological innovation is understood to have always been the critical factor in societal transformation, the distinguishing mark of the current epoch is generally held to be the direct harnessing of scientific research to this process. Whereas previously scientific discovery and technological application proceeded with relative independence and only sporadic intersection, now the pure knowledge of science can no longer be sharply distinguished from its practical realization in technology. Science and technology are so institutionally integrated as to fuse in a single operation, which Bell designated by the phrase "research and development" and is more recently signified as "technoscience."[36] The result is what Drucker calls a "knowledge society," or what Alvin Toffler terms a "powershift" whereby "both force and wealth themselves have come to depend on knowledge."[37]

3. *The principle manifestation and prime mover of the new era is the invention and diffusion of information technologies*—that is, technologies that transfer, process, store, and disseminate digitalized data: computers, telecommunications, and, by some accounts, biotechnology. Information revolutionaries point to the extraordinarily swift and broad development each of these fields of informatics has undergone since 1945—computers passing through successive generations, each of smaller size, larger capacity, and higher speed; telecommunications moving from analog to digital signals, and adopting new switching and transmission methods that dramatically improve performance, reliability, and costs; biotechnology advancing from the initial discoveries of DNA and RNA to everyday in-vitro fertilization and transgenic species creation. Information revolutionaries anticipate that this pace of innovation will not only continue but accelerate at an exponential rate.

Moreover, they point out, the real power of information technologies lies not so much in their independent capacities but rather in the fact that their

common digital language permits the convergence of their discrete capabilities into increasingly powerful, combined, synergistic technological systems.
Thus the full potential of communications and computer technologies emerges
at their confluence as a single stream of "compunications," "telematics,"
"computer mediated communication," or "intelligent networks," enabling the
creation of on-line data banks, email services, and global computer connec-
tivity. There are signs of similar fusions between biotechnology and microelectronics.[38] This process of convergence is seen as eventually culminating
in the creation of a generalized digital medium within whose networks an
enormous range of transactions and operations—from manufacturing through
messaging to medicine—will be conducted. The information revolution is thus
perceived as a technological change that does not just alter individual products but pervades the fundamental processes of an entire culture.[39]

4. *The generation of wealth increasingly depends on an "information economy" in which the exchange and manipulation of symbolic data matches,
exceeds, or subsumes the importance of material processing.* Since Porat's
study of the "information economy" the idea that information technologies
are provoking a qualitative change in the nature of employment and the
sources of wealth has been variously interpreted but widely accepted.[40] The
prevailing view now declares that information is a central "economic resource" of the twenty-first century.[41] Jorge Schement has aptly characterized
this creed as "informational materialism."[42] Its main tenets are summarized
in Toffler's account of the contemporary "super-symbolic economy"—a "new
system of accelerated wealth creation" increasingly dependent on "the exchange of data, information and knowledge," where land, labor, financing, and
raw materials become less important than the symbolic knowledge that can
increasingly discover substitutes for them; where technological and organizational innovation are at a premium; where faster decision making and better internal communication are a central commercial objective; where mass
production is replaced with flexible production systems synchronized to detailed customer feedback about market conditions and preferences; where
electronic transfers replace metal or paper money as the major medium of
exchange; where goods and services are modularized and configured into systems requiring a constant multiplication and revision of standards; where new
abstract and intellectual skills demanding high levels of education and training become the crucial attributes of the labor force; where computerized
monitoring governs the profitable recycling of wastes; and where global news
and data flows are an essential strategic asset.[43] Although other information
revolutionaries might dispute the details of this portrait, it embodies most
of the conventional wisdom about the economic importance of technological knowledge.

5. *These techno-economic changes are accompanied by far-reaching and
fundamentally positive social transformations.* Here information revolution-

aries display their most enthusiastic optimism. The undesirable features of industrial society—meaningless work, huge impersonal organizations, rigid routines and hierarchies, anonymous and alienating urban existences—are seen dissolving. In their place, the information age holds out the hope of diversification, localism, flexibility, creativity, and equality. Promises include the computer-aided recovery of craft skills and artisanal traditions; the convenience of universal teleshopping, telebanking, and interactive entertainment; the assistance of expert systems for education, health care, psychotherapy, and home security; the revivification of domestic life in an electronic cottage; the participatory democracy of electronic town halls; and a historically unprecedented diffusion of every sort of knowledge—"all information in all places at all times." A brilliant culture of individual and collective self-actualization is seen arising from the matrix of the networks.

This is not to say that information revolutionaries deny potential problems. Technological unemployment, intrusive surveillance, electronic crime, and "future shock" are all duly acknowledged. But they are represented as problems of adjustment—temporary setbacks or avoidable hazards on what remains in essence an ascending path. Bell, no facile utopian, recognizes anxieties about technological domination and dehumanization, especially in the cultural realm, but he nevertheless insists that the tendency of information systems is toward "the freeing of technology from its 'imperative' nature," and the creation of "alternative modes of achieving individuality and variety within a vastly increased output of goods."[44] Others have been less restrained: W. P. Dizard, for example, speaks of the information society as one where the "search for a new Eden through the melding of nature and machine" eventually yields "social salvation through better communication and information."[45]

6. *The information revolution is planetary in scale.* Although early postindustrialists focused on changes in the developed world, they quickly identified a tendency toward a unified world economy as one major consequence of enhanced communication technologies.[46] Recognizing the disparity between advanced economies and the Third World, they nevertheless believed in the overall trajectory of "development" by which Western societies pioneered advances that would eventually, given suitable aid, expert direction, and trading connections, be adopted and emulated by other regions. Later information society theorists followed this logic. Some, strongly influenced by Marshall McLuhan's notion of an electronic "global village," amplified on this one-world theme in a very optimistic manner.[47] Some have argued that rapid computerization would enable Third World countries to spring from a preindustrial to a postindustrial society—leapfrogging over the industrial stage. Others suggest that computer and telecommunications would open up possibilities for decentralized, de-urbanized, village-based industry bringing material prosperity to the Third World without destroying cultural autonomy and tradition—what Toffler calls "Gandhi with satellites."[48] Even those who

don't share these high hopes tend to see global disparities being rectified by a trickle-down economics in which huge technologically generated increases in productivity, initially concentrated in the developed world, will eventually be disseminated across the planet.

7. *The information revolution marks not only a new phase in human civilization but also a new stage in the development of life itself.* At the extreme limits of their prediction, many information revolutionaries see the augmenting powers of intelligent machines tending logically toward the creation of "synthetic life."[49] The steady transfer of human abilities to machines will, it is argued, lead to the production of technologies whose capacities exceed those of their creator. The roboticist Hans Moravec typifies this view. "Sooner or later," Moravec asserts, "our machines will become knowledgeable enough to handle their own maintenance, reproduction and self-improvement without help."[50] When this happens, humanity will pass away, "having lost the evolutionary race to a new kind of competition," superseded by its own "mind children."[51] Computers are thus not viewed merely as servants for humankind but also as a potential successor species—the next stage in evolution.

Tofflerism: Marx against Marx

As the thesis of postindustrial society metamorphosed into the theory of information revolution, its anti-Marxism simply remodulated itself. There was perhaps less talk of a new technocratic class mediating the tensions between capital and labor. But increasingly the direction of technological development itself was claimed to contradict Marx's analysis. The computer was discovered as the nemesis of socialism, a machine whose astounding capacities confounded class struggle.

Again these arguments appeared particularly telling because their proponents often claimed to be not so much repudiating Marx as simply updating him—following his own logic through to unanticipated conclusions. Pointing to Marxism's customary emphasis on the development of the means of production—and interpreting it as referring entirely to innovations in machinery—information society theorists said, in effect, that if "the handmill gives you society with the feudal lord; the steam-mill with the industrial capitalist," then what arrived with the microcomputer was the information society.[52] The real "historical materialists" are those who recognize the arrival of this new order rather than clinging to outdated notions of capital and class.

No one has pursued this line more energetically than the indefatigable popularizer of information revolution, Alvin Toffler. Toffler is himself a former Marxist convinced by Stalinism and American affluence that "Marxism was a misleading, obsolete tool for understanding reality in the high technology world. Using Marxism to diagnose the inner structures of high technology societies today is like limiting oneself to a magnifying glass in the age of the electron microscope."[53] But although Toffler is a relentless polemicist against "antique Marxist ideas, applicable at best to yesterday's industrialism," his own concept of history owes an obvious debt to Marx.[54]

Toffler's work hinges on a narrative, adapted from Bell's schema of preindustrial, industrial, and postindustrial societies, of civilization propelled by a series of "waves"—the First, agrarian; the Second, industrial; the Third, current, wave, informational.[55] Hendrick Hertzberg has pointed out an eerie if superficial similarity between this and Marx's story of how feudalism (the equivalent of Toffler's agrarian First Wave) gives way to capitalism (the equivalent of Toffler's Second Wave), and capitalism, in turn, is replaced by communism (the equivalent of Toffler's cybernetic Third Wave). As Hertzberg observes, "Each stage, in its time, constitutes a tremendous advance in human progress; each eventually becomes obsolete (the 'contradictions,' as the Marxists say, begin to get out of hand); and the next emerges from the collapsing ruin of its predecessor."[56] Moreover, Hertzberg notes, Toffler even sounds like Marx. The first sentence of his book *Creating a New Civilization* reads, "A new civilization is emerging in our lives, and blind men everywhere are trying to suppress it"—an obvious plagiarism of the famous opening of *The Communist Manifesto:* "A spectre is haunting Europe—the spectre of Communism. All the powers of old Europe have entered into a holy alliance to exorcise this spectre."[57]

The crucial difference is, of course, that in Toffler's account the advent of the new civilization has nothing to do with class war and everything to do with computers. Exploitation of labor, alienation, dehumanizing mechanization, centralization and concentration of wealth, immiseration—all are characteristics, not of capitalism per se, but rather of the fading Second Wave of industrial civilization—a civilization to whose premises Marxism is itself profoundly tied. The advent of the information-driven Third Wave will overcome such ills. Struggle against capital is irrelevant, because everything once (and so deceptively) signified by the red flag—the classless society, nonalienated work, the dissolution of property—will be achieved simply by the operation of the technology that capital is itself so frenetically developing. "Archaeo-Marxists" who "nurse dreams of revolution drawn from the yellow pages of yesterday's political tracts" are left standing as we "speed into a new historical zone."[58]

The inability of Marxism to respond to the realities of the new era is, Toffler argues, deeply inscribed in its theoretical tenets. Forged in reaction against the Hegelian idealist philosophy, Marx's materialism is predicated on an opposition between the physical, sensuous world of objects—the site of production—and the ethereal, abstract realm of ideas. This binary contrast underpins Marx's notorious "base/superstructure" metaphor, by which "information, art, culture, law, theories and other intangible products of the mind were merely part of a 'superstructure' which hovered, as it were, over the economic base of society. While there was, admittedly, a certain feedback between the two, it was the base that determined the superstructure, rather than the reverse."[59] Such dualism renders Marxism inherently blind to the productive power of data exchange, symbolic manipulation, and the expansion of knowledge—the very activities central to the modern economy. For Marxists, "hardware was always more important than software"; now, however, the computer revolution teaches us that the opposite is true. Today, says Toffler, "it is knowledge that drives the economy, not the economy that drives knowledge": "Marx, in arguing the primacy of the material base, stood Hegel on his head. The great irony of history today is that the new system of wealth creation, in turn, is standing Marx on his."[60] In a classic dialectical trope, historical materialism has been dematerialized.

Where Toffler finds the anachronism of Marxism most obvious is in its concept of the industrial proletariat as the agent of revolutionary change. It was, he says, not so much capitalist ownership of the means of production but rather the crude technology of the "smokestack era" that generated the drudgery against which revolutionary socialism fought. "Marxism," remarks Toffler in typical style, "glorified beefy workers straining muscles in steel mills and factories."[61] Now the legions of mass labor are vanishing: the information economy is eliminating the factory—and with it, Marxism's historical protagonist.

This farewell to the working class—an adieu bidden not only by Toffler and his colleagues but also by many left intellectuals during the 1980s—takes two forms in information revolution theory. The first, most straightforward, simply argues that automation will progressively liquidate labor. There will be less and less work—hence less and less of a "working" class. Early versions of postindustrialism were often linked to the idea of an emergent "leisure society" in which the most pressing social problem would be the overcoming of boredom. This vision has never entirely faded from information society theory. However, an obvious problem diminishes its appeal—namely, that in the context of a wage economy such a liberation from work manifests as unemployment. Anxious to refute any idea that they merely aim to replace the tedium of the assembly line with misery of the welfare queue, information revolutionaries like Toffler have often tended *not* to focus on the labor-saving consequences of automation and instead pursue a quite different argument.

In this second version, work, instead of being terminated, is transformed. Emphasis falls not on the quantitative reduction of labor but on its qualitative improvement. Automation, it is conceded, will eliminate jobs, primarily in manufacturing. But this will be compensated for by new work, appearing in high technology, information-intensive industries. However, the new jobs will be different from the ones they replace; they will be *better* jobs. Here information society theory elaborates an argument first influentially stated by the sociologist Robert Blauner during the 1960s in a critique of Marx's theory of alienation—namely, that advanced technology reverses the inhuman, estranging effects of industrial machinery on workers.[62]

Computers, it is claimed, are fundamentally different from earlier forms of mechanization. Transmuting manual drudgery into mental labor, manipulating symbols rather than objects, informatics not only frees workers from routine drudgery but also places a new premium on critical and diagnostic capacities, cooperative problem solving, and the reintegration of previously fragmented tasks. These potentials tend to reverse the Taylorist simplification and fragmentation of work. It either *permits*, in the weak form of the argument, or, in its more determinist version, *requires* dissolution of traditional hierarchies and command structures, and the introduction of new dimensions of autonomy and job-satisfaction.

Thus a crucial part of Toffler's description of the Third Wave production depends on the intellect and skills of the work force. Industrial workers owned few of the tools of production; today however "the most powerful wealth-amplifying tools are the symbols inside workers' heads."[63] Workers, therefore, "own a critical, often irreplaceable, share of the 'means of production.'"[64] The foundation for Marx's theory of class conflict thus drops away. The consequence of the high-technology, post-Taylorist workplace is the evaporation not only of the hostility but even of the distinction between management and labor; in its place emerges a shared ethos of participation and professionalism, reinforced by profit sharing, stock options, and workplace quality circles. While there will still be work, there will be no working "class," because class as a collective identity based on adversarial relations of production will have been dissolved.

At some points Toffler goes even further and suggests that the Third Wave will transform not only work but also property. This is often represented as a necessary consequence of the economic peculiarities of information-intensive goods and services. Because information is not exhausted by use, can be reproduced easily and cheaply, and often multiplies in value the more widely it is distributed, such goods and services are—supposedly—immune from ownership or commodification. Since information constitutes the central resource of the new age these property-transcendent features herald the advent of an increasingly sharing, cooperative, equalitarian society. According to Toffler, Marxists have an "obsession with ownership" that is anachronistic in an era of "info-property"—"non-material, non-tangible, and potentially infinite."[65] In the unfolding of this transformation revolutionary, overthrow of the ruling class is crudely beside the point. What will occur is rather a gentle auto-dissolution of ownership.

Here there is an interesting bifurcation in the work of information revolutionaries. Some theorists, at some moments, look to a future "beyond capitalism." This perspective is exemplified by the early work of Toffler, and by the "computopia" prophecies of Japan's most famous futurist, Masuda. It sees information technology bringing a gradual, spontaneous and nonantagonistic relaxation of capitalist relations—with corporate ownership eventually assumed by technologically participatory workers and citizens and the abundance of information-generated resources dissolving commodity exchange. What results is nothing less than an electronically created classless society.

Other information revolutionaries—or sometimes the same theorists at other moments—look only to a "better capitalism." This is the view implicit in all the governmental and corporate descriptions of the information society. It is also the perspective of Toffler's more recent work, clearly adapted to the free-market climate of the 1980s and 90s. In this perspective, information technologies still produce incredible economic and societal benefits. But these result mainly from an improved position in an ever-more-intensely competitive market society. Digitalization yields, not postcapitalism, but new investment possibilities, more efficient management techniques, better marketing opportunities—faster, swifter, more efficient commodification.[66]

Yet despite their apparent divergence, both the "beyond capital" and the "better capital" versions of the information revolution point in the same direction: to a future in which the capitalist development of technology leads to social salvation, whether through the perfection of the market or its transcendence. And in practice, information revolutionaries straddle both positions without apparent embarrassment. Masuda, who writes about the dissolution of the commodity form even while serving as an adviser to Japan's economic planning agencies, speaks of his "computopia" not only as a "classless society" but also as the fulfillment of Adam Smith's vision in *The Wealth of Nations* of a "universal opulent society."[67] Toffler hopped with ease from talking about postcapitalism to advising ultra-right-wing free-marketeer Newt Gingrich.

Indeed, in many moments of information society theory both visions merge in the synthesis of a capital without contradictions, conflict, or competition. In a typically nebulous but heartfelt panegyric, William Halal asserts that "the relentless advance of technology has become the driving force for social change," and he celebrates the emergence of a "hi-tech/hi-touch" business organization that unites enterprise and democracy. "Rising like a phoenix from the ashes of a dying epoch," the resulting "New Capitalism" will be so transformed that "it is really no longer capitalism at all" because "it is governed democratically to serve a full range of human goals rather than profit alone—yet it is still free enterprise."[68]

Both the "beyond capitalism" and the "better capitalism" versions of the information revolution see high technology reshaping society, and both see this as a good thing. Their shared technological determinism means that the radical possibilities announced by the visionaries of the "beyond capital" school are conceived of as a direct, linear consequence of the innovation directed by the pragmatists of the "better capitalism" tendency. For this reason the positions are complementary rather than antagonistic: the one is the perfect idealist counterpoint to the utilitarianism of the other. In both cases the prognosis is the same—more technology. And in both cases, what is decisively off the agenda of the future is Marx's concept of revolution as class struggle.

The End of History: Déjà Vu

The ultimate vindication of this information-age anti-Marxism was of course the end of "actually existing socialism." In the 1970s some postindustrialists had prophesied a certain convergence of capitalist and socialist systems as each resigned "ideological" attachment to notions of either the free market or world revolution in favor of a common resort to technocratic planning. But in the 1980s, the era of the Second Cold War, this argument gave way to a more aggressive line. Totalitarianism was the inevitable outcome of Marxism, but computers and telecommunications were "technologies of freedom" with an intrinsic antipathy to such statism.[69] In arguments strongly marked by the influence of Frederick Hayek, it was widely argued that the creation of a knowledge economy was inherently related to the play of the open market.[70] High-technology innovation depended on levels of enterprise and initiative antithetical to rigid state control. Moreover, application of such innovation would produce a complex and accelerated economy, dependent on data flows that elude centralized control. Any regime that attempted to restrict these flows would inevitably fall victim to the populist technological empowerment brought by the proliferation of microcomputers, video, and fax systems.

It would seem hard to imagine a more convincing vindication of such arguments than the ignominious disintegration of the Soviet bloc in 1989. As the statues of Lenin toppled across Europe, Brzezinski, one of the originators of postindustrial theory, ascribed the Soviet state's degeneration to a failure to grasp the "technetronic revolution" that made its relative achievements in the field of heavy industrialization and mass education obsolete.[71] Kenichi Ohmae, theorist of business in a "borderless world," enunciated a common verdict when he declared that information "never respected the Berlin Wall"; he wrote, "in an age of instant information, a wired-for-pictures world . . . any government that cannot offer Western-style choices of material goods, services and travel will arouse the enmity of its citizens."[72] Toffler, of course, truly knew how to twist the knife in the wounds of old comrades. Declaring that "the central failure of the great socialist experiment of the 20th century lay in its obsolete ideas about knowledge," he observed that

> Marx himself had given the classic definition of a revolutionary moment. It came, he said, when the "social relations of production" (meaning the nature of ownership and control) prevent further development of the

"means of production" (roughly speaking, the technology). That formula perfectly described the socialist world crisis. Just as feudal "social relations" once hindered industrial development, now socialist "social relations" made it all but impossible for socialist countries to take advantage of the new wealth-creation system based on computers, communication, and, above all, on open information.[73]

The most ambitious statement of such ideas was, however, that of Francis Fukuyama, a former deputy director in the U.S. State Department and consultant with the RAND corporation, who in a widely acclaimed article announced "the end of history."[74] This, he hastened to point out, did not mean a cessation of empirical events, but rather that such events could no longer be "understood as a single, coherent evolutionary process" that would culminate when "mankind had achieved a form of society that satisfied its deepest and most fundamental longings."[75] Beyond such a point no further progress in the development of underlying principles and institutions could occur, because "all the really big questions had been settled."[76] This idea of history had, Fukuyama observes, been enunciated by Hegel but "made part of our daily intellectual atmosphere by Karl Marx," who, he claims, believed that the "end of history" would be marked by the advent of communism. Now, in the aftermath of the collapse of the USSR, it was clear that, on the contrary, the "end of history" was achieved by the triumph of capitalist liberal democracy.

Fukuyama is not a conventional information society theorist. But he shares with these theorists a teleological faith in technological progress. He finds the "mechanism" that explains the directionality and coherence of history in the "logic of modern science."[77] This, he claims, "would seem to dictate a universal evolution in the direction of capitalism."[78] Because the unfolding of applied science makes possible the limitless accumulation of wealth to satisfy ever-expanding human desires, and also confers inestimable military advantages, it dictates homogenization toward the form of society best able to reap its benefits. This form is capitalist democracy, whose competitive enterprise, decentralized market decisions, and work ethic favors technological innovation. This superiority had seemed in doubt when the centrally planned economies of the USSR and China were able to rival the capitalist bloc in industrial production. But the inevitability of evolution in the direction of "decentralized decision making and markets" became apparent with the transition to a postindustrial order placing a premium on invention and information: "One might say in fact that it was in the highly complex and dynamic 'post-industrial' economic world that Marxism-Leninism as an economic system met its Waterloo."[79] With this sorry example of the failure of alternatives, the global adoption of capitalism by the countries of the developing world—Fukuyama calls it "the victory of the VCR"—is inevitable.[80]

Whatever problems the future holds will arise primarily from boredom with the universal "peace and prosperity" created by the technoscientific achievement of capitalism.[81]

This announcement may provoke an uncanny sense of déjà vu. For we have come full circle. The "end of history" Fukuyama presents is a massively enlarged version of the "end of ideology" thesis, now global in scope and engineered not by industrialism but by postindustrialism. At last, aided by the "mechanism" of information technology, the specter of Marxism has finally been laid to rest.

The Road Ahead?

Since 1989 history has, of course, refused to lie down and die. Nothing, however, has diminished the prevailing conviction that the information revolution represents the destiny of humankind. In the United States, the embrace of this doctrine by corporations and state reached a new level of intensity with the Clinton administration's announcement of the so-called "information superhighway"—a high-bandwidth, omnipurpose, digitalized telecommunications network interconnecting the nation's computers, phones, and televisions by fiber-optic strands, coaxial cables, satellites, and radio waves. In 1994, the National Information Infrastructure (NII) bill initiated construction of the "highway" as a governmentally subsidized but privately built, owned, and operated network. The immediate consequence was a frenzy of mergers by telephone, cable, and entertainment corporations positioning themselves to reap profits from video-on-demand, teleshopping, telegambling, interactive gaming, and on-line advertising.

As many commentators have pointed out, the "highway" image—with its connotations of linear movement, physical transportation, and material solidity—seems hopelessly inadequate to convey the multidirectional, telecommunicational, virtual interactions of cyberspace.[82] Yet it clearly displays the purposes of the promoters of digital infrastructures. For the metaphor invokes memories of the post–World War II golden age of capitalism—the period sometimes known as Fordism, in tribute to the central role of the auto industry as a provider of jobs, production techniques, and consumer goods.[83] In this era, road building was an essential component in the reordering of social life that integrated assembly-line labor, mass consumption of manufactured goods, suburban housing, and privatized mobility in an industrial regime that sustained three decades of extraordinary prosperity. The rhetoric of the "highway" summons up remembrances of this boom period (while conveniently forgetting about negative effects of automobile-centered growth such as pollution, congestion, alienated labor, and community destruction). By analogy, it defines information technologies—computers and telecommunications—as the twenty-first-century successors of the twentieth-century automobiles and roads, the leading technologies in what is hoped to be a new cycle of capitalist growth.

In this context, governmental and corporate leaders of all complexions have spouted the formulas of information society theorists. Vice-President Al Gore has made a stock-in-trade of promising a cornucopia of possibilities for virtual education, democratization, and self-improvement. In a speech on the "National Information Infrastructure" at the University of California in Los Angeles in 1994—a speech stirringly subtitled "Information Conduits, Providers, Appliances and Consumers"—Gore expanded on the highway metaphor by noting that if cars had advanced as rapidly as computer chips, a Rolls Royce would today go a million miles an hour, cost twenty-five cents, and be one millimeter in length. Such a rate of development, Gore declared, amounted to a "world revolution." Rhetorically brushing away any hint of conflicting interests from his picture of the cybernetic future, Gore went on to unblushingly promise business a compliant regulatory climate, in which the state would facilitate but not encroach on commercial opportunities of cyberspace, *and* simultaneously to guarantee citizens "open access" to the networks. Declaring that the economic future of the United States depended on its ability to grasp the opportunities of the digital age, he contrasted the "innovation and entrepreneurship" of capitalism with the dire example of the ex-USSR—"a country that used to put armed guards in front of copiers"—and declared his hope that "America, born in revolution, can lead the way in this new, peaceful world revolution" based on digital technology, and exhorted his audience "not to predict the future but to make firm the arrangements for its arrival."[84]

Gore's technophilia was, however, matched by that of his political rival, Newt Gingrich, until November 1998 Speaker of the House of Representatives. An aficionado of the works of Toffler—for one of whose books he wrote an introduction—Gingrich synthesized futurist revolutionary rhetoric with the most reactionary of right-wing politics, rhapsodizing freely on the need to wire every child into cyberspace while simultaneously slashing at the welfare programs. Gingrich's Progress and Freedom Foundation hosts major conferences on the confluence of capitalism with the information age.[85] In 1994 it published a document, "Cyberspace and the American Dream: A Magna Carta for the Knowledge Age," coauthored by information-age luminaries such as Toffler, George Gilder, and Esther Dyson. The document begins with a grandiloquent declaration that "the central event of the 20th century is the overthrow of matter. . . . The powers of mind are everywhere ascendant over the brute force of things." It is built around a Toffleresque contrast between the second "industrial" age and the third "informational" age, now elaborated with reflections about cyberspace as a "bioelectronic frontier," a "land of knowledge" whose "exploration can be civilization's truest, highest calling."[86]

After ruminating on "the nature of freedom" and "the essence of community" and rejoicing in the power of cyber-communications to liberate us all from "smokestack barons and bureaucrats from the past," the "Magna Carta" finally gets down to brass tacks with some policy recommendations. These are remarkably to the point: strong intellectual property rights to protect private ownership of information; a "highway" infrastructure to be owned by an unregulated private monopoly; tax breaks for information-oriented companies; and the widespread dismantling of federal government regulations. These proposals for the consolidation of information age capitalism are far from airy dreaming; much of the spirit of the "Magna Carta" proposals moves in the 1996 U.S. Telecommunications Bill, a legislative testament of faith in the power of deregulated, concentrated capital to manage the new informational environment.[87]

The corporate sector itself has been almost as fulsome as its government clients about the prospects for virtual capitalism. The ruminations of Bill Gates, owner of Microsoft Corporation and cyberspace's premier captain of industry, can be taken as exemplary. In his biography *The Road Ahead*—a title that carefully echoes the rhetoric of the information highway—Gates looks forward to what he calls "Friction-Free Capitalism." In this scenario, omnipresent digital technologies become the basis for the perfection of the market. Gates, who has the frankness to acknowledge that the driving force behind the information highway is "the race for the gold," nonetheless introduces a utopianism of his own when he suggests that the movement of business into cyberspace will produce Adam Smith's dream of a world of "perfect knowledge" or "perfect information," a prerequisite for "perfect competition." Ignoring the ironies that such words invite in the mouth of the information age's most aggressive monopolist, Gates promises us "a new world of low-friction, low-overhead capitalism, in which market information will be plentiful and transaction costs low. It will be a shopper's heaven."[88] Freed by technology from its rigidities and imperfections, the market passes into a veritable paradise of exchange, in which the global digital grids and lattices connect the whole planet in the limitless transaction of prosperity and freedom.

However, even in such utopian pictures a few shadows sometimes appear. In the context of a unified capitalist world economy, discussion of the information revolution is now inseparable from that of "globalization."[89] A harder, more anxious note replaces rhapsodies about the "global village." For it is now the pressure of a communicationally integrated and increasingly competitive world market that enforces adaptation to the information age. Techno-idealism falls to computer-age *realpolitik*. Rhetoric urging the rapid adoption of new technologies now relies not only on the utopian promises of such technologies, but, even more, on the costs—in terms of lost jobs and declining living standards—of refusing them. However, if this introduces a newly anxious note to the approach of the information revolution, it in no way diminishes its inexorability. While in an era of mounting technological unemployment and global corporate mobility there may be some qualms about the universal benevolence of the information age, there is even less doubt about its necessity.

The worldwide collapse of socialist regimes, or their clear subordination to market discipline, has meant that anti-Marxist diatribes now seem beside the point for contemporary high-tech futurists, such as George Gilder, Nicholas Negroponte, Michael Rothschild, or Kevin Kelly and the editors of *Wired* magazine. Rather, they focus on the necessary identification of technological progress and the market economy. Many commentaries endorse the views put frankly by Rothschild, whose "bionomic" analysis of an "economy derived from technical information" asserts that "capitalism is simply the way technology evolves" and is the "inevitable, natural state of human affairs"— a phenomenon that it is a "waste of time and mental energy" to oppose, because "like it or not, the sun rises in the east."[90]

Yet despite the loss of any easily identifiable ideological opponent, information capital's revolutionary intellectuals retain a messianic sense of mission. Some predictions by Hans Moravec remind us of the scope of their ambitions. Envisaging the emergence within the foreseeable future of highly advanced artificial intelligences, he asks us further to imagine that "most of the human universe has been converted to a computer network—a cyberspace—where such programs live, side by side, with downloaded human minds and accompanying simulated human minds."[91] Moravec then outlines the political economy of this world. The cyberspatial entities will all make their living "in something of a free market way," trading the products of their labor for the essentials of life in the networks—memory space and computing time. Some will convert undeveloped parts of the universe into cyberspace or improve the performance of existing patches, thus creating new wealth. Others will act as banks, storing and redistributing resources, buying and selling computing space, time, and information: "Some entities in the cyberspace will fail to produce enough value to support their requirements for existence—these eventually shrink and disappear, or merge with other ventures. Others will succeed and grow."[92] Moravec says that the closest present-day parallel to the existence of these virtual creatures would be "the growth, evolution, fragmentation, and consolidation of corporations, whose options are shaped primarily by their economic performance."[93] Noting that "a human would likely fare poorly" in such a cyberspatial market, he looks, without regret, to the necessary conclusion—our species merger with or supersession by these corporatized synthetic entities.[94] Reading such apocalyptic visions, one cannot but hear the echoes of some lines of Marx's of which McKenzie Wark has reminded us in his brilliant discussion of computerized stock markets—lines in which the young Marx speaks of the ultimate destination of capital: "finally—and this goes for the capitalists too—an inhuman power rules over everything."[95]

Appropriations and Exorcisms

The doctrine of the information revolution, as it has unfolded over the last half century, has proven to be much more than just futurist speculation or even sociological description. Rather, it has become an indispensable ingredient in a massive reorganization of advanced capitalist societies, centered on the introduction of new technologies. Formulated and promoted within the think tanks, policy institutes, laboratories, government offices, and consultancy circuits of the most powerful and prosperous centers of the capitalist world economy, the theory of an inevitable information revolution provides the rationale for this restructuring, legitimization for social dislocation, and exhortation toward a radiant future.

In its development, this idea has been propelled by competition with another revolutionary theory that aimed to become a "material force": Marxism. This was the foe that was meant to have been defeated by the "end of ideology" in the affluence of postwar industrial societies. It was in response to an unforeseen crisis of these societies, a crisis of international and domestic insurgencies permeated by the spirit of supposedly dead and buried Marxism, that Bell and his colleagues produced the concept of postindustrialism. Their annunciation of a new age was not merely a prediction, but a project, an effort both of prophecy and partisanship aimed at setting in motion the social and technological measures necessary to restore the stability of an order threatened by what they saw as chaotic and subversive forces. This is the idea that has subsequently flowered into theories of the information revolution and virtual capitalism.

The relation of these theories to Marxism is, however, not just one of antagonism but of appropriation. Produced by intellectuals who were often familiar with or had actually espoused Marxist ideas, the concept of the information society derives much of its analytic force and imaginative power from a rewriting of Marxism. That rewriting retains the notion of historical progress toward a classless society but reinscribes technological advance rather than class conflict as the driving force in this transformation. It thus annexes the idea of "revolution." The collapse of actually existing socialism in popular uprisings intimately linked to the capacities of new media to carry messages across the walls and curtains behind which Marxian regimes had sheltered from the world market is, in the eyes of information revolutionaries, the vindication of this project. It marks the final, technologically aided exorcism of the ghost of Marx. In what follows, however, I will argue that this exorcism has failed. But first we must see what other Marxists have made of "the information revolution."

MARXISMS

It is too late in the day to become in-
tensely vexed as to what Marx "really said" about technology. For Marx was,
like all of us, a multiple. He wrote variously about technology, making statements that cannot all be reconciled one with another—or, at least, that can
be reconciled in very different, sometimes radically opposed, ways. In the
historical development of Marxism this heterogeneity of utterances has
yielded a volume of interpretation that now weighs considerably on the brain
of the living and whose influence powerfully refracts every re-reading of their
source.

Marx's Machines

This chapter begins with a scan of the many representations of the machine
in Marx's texts. It then moves on to see what other Marxists have made of
and from these representations as they respond to the "information revolution." I examine three positions: scientific socialism, which sees technoscience as a central agent in a dialectical drama culminating in the inevitable
defeat of capital; neo-Luddism, which focuses on technology as instrument
of capitalist domination; and post-Fordist perspectives, which often look to
the possibility of a technologically mediated reconciliation between labor and

capital. While this is by no means a complete inventory of Marxist, or Marx-
ist-derived, thought on technological change, it does muster the major analyses with which I want to take issue. For, I argue, scientific socialists, neo-
Luddites, and post-Fordists all fall short of an adequate response to the
challenge of the information revolutionaries, though in very different ways.

As we have already seen, there is a certain Marx very close in spirit to the
information revolutionaries—the Marx of "the handmill gives you society
with the feudal lord; the steam-mill with the industrial capitalist."[1] This
technologically determinist Marx is not a negligible figure.[2] His hand has been
seen at work in the *Preface to the Contribution to a Critique of Political
Economy,* in the celebrated account of how "in the social production of their
life men enter into definite relations that are independent of their will, relations of production which correspond to a definite stage of development of
their material productive forces."[3] At a certain stage in their development,
Marx says, "the material productive forces of society come into conflict with
the existing relations of production" and "from forms of development of the
productive forces these relations turn into their fetters"—thereby initiating
social revolution.[4]

What precisely constitutes the forces of production and what the relations
of production, and the precise nature of the interaction between the two, are
among the most controversial questions in Marxist theory.[5] But what is certain is that a view that sees the forces of production as technological, and only
the relations of production as social, with the former having primacy over
the latter, seems to have taken root very close to Marx himself, in the work
of his friend Frederick Engels, who wrote that with the advancement of modern machinery "the productive forces themselves press forward with increasing power towards . . . their deliverance from their character as capital."[6] From
there extends a line of Marxist thought, along which are posted figures such
as Nikolai Bukharin, J. D. Bernal, and Gerald Cohen, which understands technological development as an autonomous force, a motor of history, whose
ever-expanding productive powers smash relentlessly through anachronistic
forms of property ownership in a trajectory heading straight to the triumph
of socialism.[7]

However, there are other passages in Marx that modify and indeed contradict this mechanistic view of history. For example, the major treatment of
factory machinery in *Capital* tells a story in which capitalism, as it deepens
its control of the workplace and society, transforms methods of production.
Marx describes this process in terms of successive degrees of "subsumption."[8]
In "formal subsumption"—roughly the early stages of the industrial revolution—capital simply imposes the form of wage labor on preexisting modes
of artisanal production. But in the subsequent phase, "real subsumption," it
undertakes a wholesale reorganization of work. Science is systematically
applied to industry; technological innovation becomes perpetual; exploitation

focuses on a "relative" intensification of productivity rather than an "absolute" extension of hours.[9]

Central to this process of subsumption is the replacement of manual methods of work by machinofacture. And the impetus for this development is, Marx says, the factory master's drive to enhance command over his labor force by deskilling craft workers and enlarging the reserve army of the unemployed. Such a narrative reverses the technologically determinist account. For it is social relations—capital's requirement for total control over the valorization process—that shapes machines, not vice versa. From the reading of such passages flows a different line of analysis the exponents of which run from Georg Lukács through to Harry Braverman and David Noble, who insist that machinery is only a moment in forces of production whose constitution is itself a matter of social power.[10]

Even if it could be agreed that Marx posits a complex interaction between "social" and "technological" factors—complex to the point where the two categories are understood as so inseparably bound up as to make him one of the first theorists of what today are termed "socio-technical systems"—there would still be space for disagreement in his writing on machines. Many readers have been impressed by his nightmare portrayal of nineteenth-century factory masters' use of technology. Throughout his work, Marx again and again tells us how machinery confronts the worker in production as the power of capital incarnate—or at least metalized. The steam engine serves as an "instrument of torture" in the hands of the factory owner. In a necrotic tyranny, the "dead labour" of automatic machinery becomes a "mechanical monster" with "demonic power" that "dominates, and pumps dry, living labour power," converting the worker into a "living appendage."[11] Or, as Marx put it in a speech to the Chartists in 1856, "At the same pace that mankind masters nature, man seems to become enslaved to other men. . . . All our invention and progress seem to result in endowing material forces with intellectual life, and stultifying human life into a material force."[12] From this and many other passages can be distilled a technophobic, dystopian, neo-Luddite Marx, a Marx who rages against the machine.

Yet the production of such a Marx depends on a considerable effort of edition and selection. For there are other moments where Marx speaks not just of the infernal effects of machines, but also of their emancipatory promise. For example, in one passage of *Capital* he discusses how "modern industry" continually transforms itself "by means of machinery, chemical processes and other methods" and in doing so "incessantly throws masses of capital and of workers from one branch of production to another," in a way that "necessitates variation of labour, fluidity of functions, and mobility of workers in all directions."[13] Under capital, Marx says, this incessant technological change is an appallingly destructive, immiserating force, which "does away with all repose, all fixity and all security as far as the worker's life situation is concerned."[14] However, he argues, such relentless innovation also has a potentially positive side. By annihilating the narrow specializations that previously characterized craft production it makes possible "the recognition of variation of labour and hence of the fitness of the worker for the maximum number of different kinds of labour."[15]

Modern industry thus opens the vision of an alternative—communist—social order in which the "monstrosity" of technological unemployment is replaced by the "possibility of varying labour."[16] The "partially developed individual who is merely the bearer of one specialised social function" will give way to "the totally developed individual, for whom the different social functions are different modes of activity he takes up in turn."[17] Citing a French worker who claimed that constantly changing trades in California made him feel "less of a mollusc and more of a man," Marx recommends the development of technical, agricultural, and vocational schools, in which "the children of the workers receive a certain amount of instruction in technology."[18] From such moments can be constructed another Marx, an enthusiast for the progressive possibilities of human-machine interaction.[19]

Although much of Marx's writing on machines concerns factory automation, a broadly similar ambivalence informs his observations about the other great technological innovations of his age—those in the sphere of communication and transport.[20] For Marx, the telegraph, the steamship, and the railway were the inseparable concomitants to the development of factory production, instruments for the creation of the world market necessary to supply the raw materials and absorb the goods produced by industrial machinery, an extension of capital's ceaseless revolutionizing of the means of production. They were the manifestations of a relentless dynamic that "chases the bourgeoisie over the whole surface of the globe," compelling it to "nestle everywhere, settle everywhere, establish connections everywhere."[21]

As such, the new channels of travel and communication were tendrils for the extension of a system of domination. In a passage that strikingly anticipates the conditions of contemporary globalization, Marx writes: "Every development in the means of new productive forces is at the same time a weapon against the workers. All improvements in the means of communication, for example, facilitate the competition of workers in different localities and turn local competition into national."[22] Elsewhere, Marx analyzes the new means of communication as an essential component in the "autonomisation of the world market," elevating monetary exchanges into a force whose impersonal and relentless processes appear to stand over and against any possibility of human intervention or transformation.[23] These observations—especially when linked to Marx's remarks on ideology and commodity fetishism—have provided planks for a Marxist political economy focused almost entirely on the dominative effects of capitalist media and information industries.

At the same time, and even more emphatically than in the case of industrial machinery, Marx also saw liberatory possibilities in the nineteenth-century communications revolution. The telegraph, fast mails, and travel broke down parochialism, localism, and narrow national interests. As such, they were potential catalysts for proletarian internationalism. *The Communist Manifesto's* famous exhortation to the "workers of the world" is prefaced by a series of enthusiastic observations on how this "ever expanding union of workers" is "helped on by the improved means of communication that are created by modern industry and that place the workers of different localities in contact with one another."[24] This is one vital aspect of a process by which the bourgeoisie forges "the weapons that bring death to itself" and also calls into existence "the men who are to wield these weapons—the modern working class-the proletarians."[25]

In his own life, Marx was eager to take advantage of such possibilities. According to James Billington, Marx and Engels on one occasion planned to penetrate the international wire agencies in Brussels, through a leftist press agency, in order to distribute their messages more widely.[26] As Peter Waterman notes, this may not be quite enough to make Marx a "hacker" *avant la lettre.* Nevertheless, the enthusiasm for the revolutionary possibilities of mass communication so evident in his texts has resonated with theorists from Bertholt Brecht and Walter Benjamin to Hans Magnus Enzensberger onward.[27]

These synoptic observations only skim the surface of Marx's machine-writings. But they are perhaps enough to establish that throughout these texts there runs an electric tension, an alternating current that oscillates between rival possibilities. At one pole, technology is an instrument of capitalist domination, a means for the intensification of exploitation and the enchaining of the world in commodity exchange. At the other, it is the basis for the freedom from want and the social intercourse that are prerequisites for a communist society. How much emphasis is given to each pole, and by what logic or narrative they are connected, is, however, a matter of huge contention. Later, like all the other interpreters, I will select my own favored points of reference, the passages where, for me, Marx's antinomies about the machine fuse at white heat into brilliant insight. But for the moment, we will see what others have made of Marx's ambiguous machines.

Scientific Socialism

I use the term "scientific socialism" to designate that form of Marxism—also variously referred to as "objectivist," "classical," or "neo-orthodox" Marxism—which, taking its direction from Marx and Engel's observations about the contradictions between forces and relations of production, sees history driven by scientifically predictable laws of motion toward a socialist destination.[28] Perhaps the most sophisticated recent example of this school of thought is to be found in the work of Ernest Mandel, the eminent theoretician of the Fourth International.

Mandel's magnum opus, *Late Capitalism*, was first published in 1968 and translated into English in 1975. At that time, many societal phenomena were being claimed by postindustrial theorists to mark the definitive supersession of Marxism—cybernetics and other new technologies, the increasing importance of planning and education, and the increasingly "knowledge based" nature of economic development. Mandel's work represents a magisterial attempt to reinsert these changes within the framework of historical materialism. For Mandel's fundamental claim is that the societies of contemporary Europe and North America, far from having transcended the features of capitalism described by Marx, in fact exhibited them in a singularly pure form.[29]

Mandel argues that there have been three fundamental moments in capitalism, each one marking a dialectical expansion over the previous stage: market capitalism, monopoly capitalism, and our own phase, late capitalism. He links the appearance of these stages to Kondratieff's famous theory of "long waves"—successive, rhythmic episodes of economic growth and stagnation that supposedly dominate the last two centuries of Western history. In each wave, surges in technological innovation are precipitated by temporary increases in the rate of profit after a protracted period of underinvestment. Corresponding to the three phases of capitalism are three "general revolutions in technology"—steam-driven machinery of the 1840s, electric and combustion motors of the 1890s, and, from the 1940s on, the "third technological revolution" of nuclear power and computerization.

The central feature of this latest phase is the increasing level of automation, and, in particular, the replacement of industrial workers by cybernetic systems and continuous flow processes based on automatic control. This brings with it a series of interrelated developments, which reverberate through the capitalist economy. These include a shift of living labor from the actual treatment of raw materials to preparatory or supervisory functions; new developments in organized research and university education; a speedup in production and a consequent pressure for more effective inventory control, market research, and demand management; and increasingly large, and increasingly quickly obsolete, investments in technological systems. These developments in turn generate a compulsion to introduce exact planning of production not only within each enterprise but also within the economy as a whole—leading to more state intervention. All of these changes, however, relate back to the overwhelming imperative of capitalism, the maintenance of the rate of profit.[30]

This analysis brings Mandel into direct confrontation with the first expressions of postindustrial theory. Categorically rejecting any idea that the new economic centrality of science and technological knowledge mark some unprecedented historical epoch, Mandel argues that "late capitalism, far from representing a 'post-industrial society,' . . . appears as the period in which all branches of the economy are fully industrialised for the first time."[31] Specifically citing Bell's work as an example of prevalent theories of "technological rationalism," he declares that "belief in the omnipotence of technology is the specific form of bourgeois ideology in late capitalism": "This ideology proclaims the ability of the existing social order gradually to eliminate all chances of crises, to find a 'technical' solution to all its contradictions, to integrate rebellious social classes and to avoid political explosions."[32] However, Mandel says, the idea that new technologies allow capitalism to transcend its perennial antagonisms and crises is spurious; on the contrary, such innovations only bring closer its inevitable collapse.

Although Mandel enumerates a wide array of factors, all of which he sees interacting to generate breakdown, the centerpiece of his argument is a traditional mainstay of "objectivist" Marxism: the falling rate of profit, consequent on the rising organic composition of capital. To understand this argument a brief technical exposition is necessary.[33]

The Marxist theory of value holds that the source of surplus value is the exploitation of living labor. Capitalist production can be represented in value by the formula $C + V + S$. C is "constant capital"—the part whose value is not increased in production but merely preserved by it—buildings, raw materials, and, especially, machines. V is "variable" capital, the part used by the capitalist to buy labor power, so termed because it is the only part of capital that lets the capitalist increase the value of his or her capital. S is the "surplus value"—the portion of the newly created value appropriated by the capitalist. The rate of profit is the ratio between surplus value and total capital, that is, $S / (C + V)$. The ratio between constant capital and variable capital, C/V, is the "organic composition of capital."

The fundamental tendency of the capitalist system is to increase the ratio of constant capital (machines and raw materials) to variable capital (wages). For Mandel—and most other objectivist Marxists—the principal impetus in this direction comes from the "whip of competition" among capitals, which compels entrepreneurs constantly to automate in order to raise productivity.[34] But if the organic composition of capital, C/V increases, other things being equal, the profit rate $S / (C + V)$ will decline. The more completely mechanization expels workers from production, the more the rate, and eventually the mass, of surplus value diminishes. This decline in profitability causes faltering investment, catalyzes class conflict, and drives irrevocably toward revolutionary crisis. Capital's profit-driven compulsion to expand the forces of production thus becomes the instrument of its self-destruction.

This formally elegant argument is a topic of immense controversy, even among Marxists. In his original account of the "falling rate of profit," Marx identified certain countertendencies—intensified exploitation of labor; cheapening of the elements of constant capital (i.e., increased efficiency in the manufacture of machines, new sources of raw materials); the opening up of industries with low organic composition; increases in foreign trade; speedup in the turnover in capital—all of which might alter the inevitability of the falling rate of profit.[35] But in neo-orthodox accounts these tend to be seen as subsidiary factors.[36] Certainly Mandel believes that "the fall in the average rate of profit is inescapable."[37]

Cybernetics, by bringing in sight the "workerless factory," drives this process to a climax, placing on the horizon what Mandel terms "the absolute inner limit of the capitalist mode of production"—the point where fully automated production no longer allows the creation of surplus value.[38] "The absolute limit . . . lies in the fact that the mass of surplus value itself necessarily diminishes as a result of the elimination of living labor from the production process in the course of the final stage of mechanisation. Capitalism is incompatible with fully automated production in the whole of industry and agriculture, because this no longer allows for the creation of surplus value or valorisation of capital."[39] To secure this prediction Mandel makes certain theoretical assumptions that rule out capital's discovering a way of lowering the average organic composition by moving outside its traditional factory base. The development of the service sector is discounted on the grounds that most work in this area, because it does not change the "bodily form" of a commodity, is "unproductive."[40] A shift of labor power to spheres of research and design is similarly rejected because such a transformation "would imply a radical suppression of the social division between manual and intellectual labour," which would "undermine the entire hierarchical structure of factory."[41] Having blocked off these boltholes, Mandel can be confident that the third technological revolution seals the fate of capital.

Mandel does not see capitalism straightforwardly automating itself into oblivion. Rather, he believes that declining profits will ultimately cause it to check automation. But the closure of this route to expansion will lead to crisis-ridden stagnation and intensified conflict over the allocation of surplus. In fact, capitalism is caught in the historical trap foreseen by Marx, where its achievement in expanding the forces of production unleashes conflicts that explode the social relations its continuance requires. While Mandel qualifies the finality of his verdict, admitting of reprieves and postponements, the teleology is inscribed in his masterwork's title—*Late Capitalism*.

In many ways, Mandel's work is a brilliant answer to Bell and the postindustrialists. By showing how so many of the allegedly new features of contemporary society cited by these theorists relate to the very old logic of accumulation, he effectively refutes the claim that the logic of capital has been replaced by some unprecedented and benign informational principle. Moreover, at the time of its publication Mandel's prediction of renewed economic crisis showed remarkable prescience compared with the postindustrialists' rosy forecasts of unimpeded economic growth.

What is striking, however, is the subterranean affinity between Mandel and his postindustrial opponents. To a remarkable degree such "automatic Marxism" mirrors the assumptions of the very theories it opposes.[42] There is disagreement about the prospects for scientific-technological innovation yielding capital a smooth, evolutionary future. But there is a common view of the forces of production—seen primarily as machines—as central instruments of inevitable social transformations. In *Late Capitalism* the dance of machines and capitalists moves like clockwork toward a foreordained conclusion that uncannily echoes the linearity of postindustrial doctrine.

Unlike more vulgar scientific socialists, Mandel is not a technological determinist who reduces revolution to a consequence of autonomous scientific progress. On the contrary, he dialectically relates capital's mechanical self-destruction to its competitive drive for innovation. But he is a social determinist for whom technology relentlessly executes a predecided verdict. The distance between this position and the "bourgeois" faith in the "omnipotence of technology" is not as great as he would like to imagine. As interpreted by Mandel, the doctrine of the falling rate of profit functions as a mirror image of the upward path of progress espoused by Bell and the postindustrialists, the one leading as surely to socialist victory as the other does to capitalist stability.[43]

There are theoretical reasons even for those who share Mandel's premises to doubt his conclusions. As I have pointed out, Marx himself noted the existence of countertendencies to the "falling rate of profit," and many Marxists see its supposed inevitability as a special case obtaining only under specific conditions.[44] Capitalism's deployment of new technologies certainly drives living labor out of production (through automation), but it can also enhance the countertendencies against the falling rate of profit by increasing the rate of exploitation (through surveillance and monitoring), cheapening machine production (robots making robots), opening new areas of exploitation with a low organic composition (tertiarization), speeding circulation (through advertising, marketing and innovation), and integrating the world market (telecommunications). Mandel rejects such possibilities with arguments whose intricacy verges on the quasi-theological. But such possibilities seem significant enough to cast doubt on his teleological certainty. This is *not* to ratify the postindustrialists' dreams of unimpeded market expansion. But it is to see crisis as contingent on the outcome of series of social struggles over the scope, scale, and velocity of commodification rather than guaranteed by capital's own internal logic.

Mandel's account is remarkable for the absence of any agency for such struggles. At the moment of crisis, of course, the working class is summoned to seize the revolutionary hour. But a striking feature in the pages of *Late Capitalism* is that this crucial protagonist, the ostensible raison d'être of the whole drama, is largely invisible—far less closely analyzed than capital and its machines. When, elsewhere, Mandel does discuss the modern proletariat, it is essentially to reaffirm the verity of Marx's description of the industrial worker, dismiss the significance of the "manipulations" of the mass media, and assert the guarantee of revolutionary commitment given by "the basic structural stability of the proletarian condition."[45] In such objectivist analysis there is little sense of labor as a living subject, animated by needs and desires; little sense that this subject might change, altering in complexity and capacity in ways at least as dynamic as that of the dead labor embodied in machines, or that capitalist development might itself be crucially shaped by its efforts to harness and contain the energies of this collective subject. Mandel's dialectic of productive forces and relations, in short, skips over class struggle. It is rhetorically prominent but analytically ancillary, the insurgencies of the laboring subject merely the predestined reflex of capitalism's autodestruction.

Moreover, this covert affinity between the determinism of Marxist scientific socialism and bourgeoisie theories of technological development extends further to touch the very concept of socialism. For if socialism is seen as a byproduct of the advance of science and technology, rather than as a result of people's rebellion and self-organization, the revolutionary task easily becomes defined as the speeding of technoscientific advance at all costs—including the suppression of any resistance or alternative offered by the very workers in whose name the revolution is undertaken. Where the consequences of this concept appeared in truly grotesque form was of course in the late Soviet regime—in which the objectivism of scientific Marxism combined with a logic of vanguardism, substitutionism, and technocratic expertise in a fatal mix.

As a student of Trotsky, Mandel necessarily maintained a highly ambiguous position toward the Soviet Union. But his notion of a "third technological revolution" has a strong similarity to the notion of a "scientific technological revolution," or "STR," embraced by Soviet officials and academicians in the 1960s and 70s.[46] Such theories, which foresaw a new historical epoch inaugurated by cybernetic automation, essentially recapitulated bourgeois theories of postindustrialism, with the caveat that the beneficiary of the STR would be not capitalism, but socialism. In the Soviet bloc the planned realization of the STR would be a vital lever for the achievement of a classless society, while in the West, the anarchy of the market would intensify contradictions, conflict, and disintegration. But the essential terms of the analysis were little different from Bell's or Brzezinski's—and the accompanying injunctions about the necessity of adjusting people's subjective attitude to the new objective realities were, if anything, even more chilling.

What links information society theory and scientific socialism is a shared, though differently inflected, determinism that subordinates the wishes of human subjects to the necessity of technoscientific advance. Each duplicates the other's linearity, scientism, and technocratic tendencies. As such, both are doctrines suitable for regimes in which the means of production have been sequestered from collective control, whether by a corporate or a bureaucratic class. This is precisely why information revolutionaries have been able to borrow so much from scientific socialists, and vice versa.[47] The former were of course more successful than the latter: the Soviet advocates of STR failed to make the innovations that the Western information revolutionaries are, with at least temporary success, effecting. But what divides the promulgators of such doctrines is the sort of distinction that differentiates carnivorous dinosaurs into tyrannosaurs—bulky but deadly—and velociraptors—fast, agile, and even more lethal. With the demise of the Bolshevik experiment, all the teleological certainties of scientific socialism have been thrown up in the air. The one thing that *is* sure, however, is the irrelevance to future struggles of a Marxism convinced of predestined triumph, fixated with the industrial factory, and carrying internally the seeds of the very dominative logic against which it contends.

Technology as Domination

From the late 1960s—in the very period postindustrial theory emerged—attitudes among many European and North American Marxists toward technoscience moved in a direction notably different from that of scientific socialism. Confronting assembly lines, napalm manufacturers, and nuclear power plants, growing numbers of theoreticians and activists rediscovered the dark, nightmarish aspects of Marx's writings on technology. Seen through the window of such writings, emergent technologies of automation and communication seemed more likely to strengthen capital than undermine it. The new forces of production appeared not as agencies automatically and autonomously bursting apart the old relations of production, but rather as themselves implacably shaped by those relations, designed and deployed at the behest of a ruling class to whose purposes they were almost entirely instrumental.

The groundwork for such an understanding had previously been laid by the Frankfurt School. As is well known, the basic contention of the "critical theory" developed by Max Horkheimer, Theodor Adorno, and Herbert Marcuse is that technological rationality, once a powerful lever for humanity's liberation from want and superstition, has now itself become oppressive. In the "dialectic of the enlightenment," means have usurped ends, the domination of nature has become the domination of man [*sic*], and the forces of production have turned to forces of destruction.[48] Enabled by its technoscientific powers both to generate endless desires and also to fulfill them, capital exercises a control so comprehensive as to produce Marcuse's "one-dimensional man"—a subject incapable of thinking, or even perceiving, beyond the limits of the system.[49]

Although the best work of the Frankfurt School and their colleagues predated the enunciation of postindustrial theory, their critique of science and technology not only anticipated the developments Bell and his colleagues so enthusiastically embraced but also colored an entire line of postwar neo-Marxist response to computers and telecommunications. As the information revolution intensified in pace during the 1970s and 1980s, their analysis of technology-as-domination was extended by a variety of theorists, some following in the steps of Marcuse and his colleagues, others tracking back more directly to Marx. This project developed in two streams—one focused on the labor process, the other exploring the mass media.

The seminal statement of the labor process stream is Harry Braverman's study of the "degradation of work"—a direct reply to the postindustrial claims of progress toward a new and technologically improved era of labor relations.[50] Basing himself firmly in Marx's analysis of the labor process, Braverman argues that the "scientific management" initiated by Frederick Taylor at the turn of the twentieth century, with its separation of conception from execution, managerial monopolization of knowledge, and systematic destruction of skills, is a manifestation of the "great truth of capitalism," namely, that "the worker must become the instrument of labor in the hands of the capitalist."[51] However cosmetically disguised, this remains the dominant philosophy of twentieth-century management.

The rise of "white-collar" work cited by Bell as evidence of an enlightened postindustrial society is for Braverman simply a symptom of the enlarging managerial apparatus of administration, supervision, and planning. Similarly, the new "intellectual technology" of computers and communications, which postindustrialists expected to usher in an era of skilled and satisfying mental work, for Braverman signals a contrary tendency. Whether in the movement of a factory worker following the pace of a preprogrammed tool or the monitored keystrokes of an office secretary, the power of the new technologies to record, store, and reproduce activities previously dependent on embodied consciousness yields only another extension of Taylorist authority. In the hands of scientific management, machinery is seized upon as "the prime means whereby production may be controlled not by the direct producer but by the owners and representatives of capital."[52]

This critique of the computerized labor process has subsequently been developed in a number of studies.[53] Perhaps the most influential is David Noble's work on numerically controlled machine tools—technology central to the vision of the "workerless factory."[54] Noble argues that the drive to automate machining cannot be explained solely by the requirements of a purely technical efficiency but is marked by the managerial imperative to gain total control over the shop floor, and in particular to break the power of skilled, unionized machinists. This is demonstrated by the suppression of technological options that would allow workers an element of control over the newly automated processes. Noble shows that even when this participation might have improved the operations of the system—by allowing for revision of programmed instructions according to circumstances—the managerial desire to eliminate the human element prevailed. The whole thrust of capital's use of information technology in the workplace is, Noble argues, fundamentally antihuman, predicated on a model of "progress without people."[55]

The other strand of the technology-as-domination school is that devoted to the media and communication. In an enormously important move beyond the factory focus so apparent in the work of classical Marxists such as Mandel, Adorno and Horkheimer had argued that the subordination of society to capital is largely the work of the "culture industry"—the entertainment and advertisement conglomerates that create artificial needs, distract dissent, and endlessly endorse the existing order.[56] Subsequently, broadly Marxian scholars such as Herbert Schiller, Vincent Mosco, Dallas Smythe, and Nicholas Garnham have deepened this analysis with detailed research into the operations of the capitalist media.[57] In doing so, they have produced an analysis much more fine-grained than the Frankfurt School's, and sometimes considerably more nuanced in its recognition of possibilities for resistance.[58] Nonetheless, the overall emphasis of these writers falls heavily on capitalism's technological power, producing a picture of domination almost as somber as that discovered on the shop floor, but expanded over a vastly greater sphere.

Here the work of Schiller can be taken as exemplary.[59] Explicitly targeting theorists who claim we are witnessing the transcendence of capital in "an individualized, electronic global commune," he has consistently argued that what is occurring is rather a push toward a "corporate-controlled information society."[60] Focusing on the U.S. situation, Schiller shows how in all areas of information technology—hardware, software, and transmission networks—the flux of innovation follows a path of relentless commodification. The new satellites, fiber optics, and computer networks are deployed to create a media explosion whose apparent pluralism is belied by the near total absorption of thousands of newspapers, magazines, radio stations, TV channels, and cable systems into a few giant media combines.

From ownership flows ideological control. Implicitly following the classic Marxist logic by which economic base must determine ideational superstructure, Schiller insists that corporate domination of communications industries yields a prodigious power over the formation of popular consciousness. While information society theorists claim that a proliferation of technologies and channels democratizes and diversifies opinion formation, Schiller argues that the giant media corporations generate, filter, and refine the flows of imagery, news, and entertainment to exclude anything that might subvert the inter-

est of owners or advertisers and systematically to intensify the commodification of social relations.

"The consequence," says Schiller, echoing Marcuse, "is a national discourse
that is increasingly one dimensional."[61] Although he allows for contradictions
produced by conflict within and among media industries or between such
industries and other sectors of capital, the overwhelming weight of his analysis points to the "systematic envelopment of human consciousness by corporate speech."[62] And since information technologies are seen as a central
instrument in this envelopment, the assessment of them is comprehensively
negative: "It is not a question of 'either-or' . . . good technology or bad technology use. It is solely a matter of developing and using the new communication technology for holding on to the economic benefits derived from a
world system of power. . . . insistence on the potential and positive features
of the current communication instrumentation is disingenuous at best."[63]
"Mind management" in the cultural sphere becomes thus the corollary of
deskilling in the workplace.[64]

These two strands of technology-as-domination analysis—one focused on
the labor process and the other on the media—are melded by Frank Webster
and Kevin Robins in their relentlessly bleak account of "cybernetic capitalism."[65] This makes explicit the connection of Taylorism with media management. Taylorism, Webster and Robins point out in another publication,
was in its original formulation not only a doctrine of shop-floor control but
also an overall social philosophy that pursued increasing productivity as "the
key to future prosperity, harmony and progress."[66] The deployment of information technology represents the realization of this second phase of "generalized or social Taylorism," extending capitalist control of knowledge and
information beyond the factory to society as a whole.[67]

Confronting this prospect, Webster and Robins articulate the deep forebod-
ing characteristic of so much contemporary Marxian analysis of technology:

This . . . is what we foresee in the future: a society in which corporate
capital, using the most advanced forms of I.T. [information technology]
that have been designed to suit its requirements and constantly talking
about the imperatives and promises of a technological revolution, extends
and consolidates its hold in society, strengthening its control over employees (and shedding significant numbers) while intruding further into
the everyday lives of consumers both groups of whom it observes, analyzes and schemes about what changes might be to the company's advantage and perceived as inevitable—by those likely to suffer from restructuring—or desirable—by those able to pay the going rate. Behind,
often in front, and almost always in collusion with this centralized corporate capital, is arraigned a disciplinary state, equipped with the latest
surveillance technologies, able to contain dissent from those minorities

unwilling to accede to the market's control or unable, through unemployment and/or poverty, to participate in its technologies of abundance.[68]

The changes presented by information revolutionaries as liberatory thus signify something quite opposite—greater reach for the "visible hand" of managerial control, now exercised through an arsenal of devices for broadcasting, monitoring, and surveillance to allow the observation and shaping of social subjects as both workers and customers.

Although scientific socialists, like Mandel, had always condemned the uses to which capital puts technology, this critique of the technology-as-domination cut much deeper. For scientific socialists, machines are neutral, although capital's deployment of them is objectionable. For technology-as-domination theorists, however, this apparent neutrality is a lie. Technologies embody social choices made by those with power over their construction. Political intentions are present not only at the level of use, but of research and construction—not merely in what is done with machines, but in how they are designed—and, indeed, in whether potential innovations are realized at all or suppressed.

The thought that technology might, in its very core, incarnate the intentions of the capitalists who make them, while certainly present in Marx, was first enlarged on by the Frankfurt School theorists—who nevertheless clung, somewhat self-contradictorily, to the hope that technological rationality might be rescued from capital's grasp. But in the subsequent development of this line of thought, the redemptive hope largely fades. In a flat contradiction of scientific socialism's technological optimism, machines are seen as buttressing rather than overturning established power. Noble says: "Technology . . . is not an irreducible first cause; its social effects follow from social causes that brought it into being; behind the technology that affects social relations lie the very same social relations. Little wonder, then, that the technology usually tends to reinforce rather than subvert those relations."[69] Increasingly, technological development comes to be seen as so deeply tainted by drives toward domination and omnipotence as to constitute a social pathology—a madness to be resisted at all costs.[70]

From such a position, it is natural that many technology-as-domination theorists look for inspiration to the machine wreckers of the first Industrial Revolution—the Luddites. For Noble, Webster, and Robins the pejorative use of this epithet by information revolutionaries slanders the real nature of a movement that represented a coherent protest against destructive industrialization advanced under the banner of technological necessity. And, just as in the first industrial revolution capital accumulated itself through popular immiseration, so the computerized "second industrial revolution" will expand corporate wealth and control by massive dislocation, deskilling, and unemployment. What is required to confront this prospect is a revival of the resistant spirit of General Ludd—a neo-Luddism for the information age.[71]

Thus for Noble "the essence of the technology question today" is that "there is a war on, but only one side is armed."[72] Notions of technological transcendence peddled by information society theorists are no more than legitimations for the corporate assault on workers. Given capital's control of research and innovation, the immediate possibility of shaping and humanizing the approaching wave of technological change is minimal. Rather, leftist energies should be directed toward an immediate effort at halting, or at least drastically slowing, its diffusion. Pointing to the actual incidence of sabotage among people replaced by computers, Noble declares that "if workers have begun to smash the physical machinery of domination [then] responsible intellectuals must begin to deliberately smash the mental machinery of domination."[73]

Of all the positions examined in this chapter, this neo-Luddite stance seems to me the most insightful. It is the one that most fully confronts the ambition of the information society project, not as a foreordained ascent of civilization, but as a strategy of societal power. This theoretical perspective is backed with concrete studies of the shaping of new technologies to capitalist ends, both in the workplace and beyond it. And the consequent call for resistance has an integrity lacking in the obeisances paid by scientific socialists and social democrats alike to capitalist "progress."

However, such analysis also has serious and ultimately self-defeating limitations. At root this is because the technology-as-domination school overestimates capital's capacity to command living labor with dead labor. It restores the human subject whom objective Marxism banishes, but it introduces this subject primarily as victim. In this respect, the reproach often leveled against Braverman's labor process analysis—that it sees workers only as the passive objects of capitalist designs, and ignores the consequences of their counter-strategies and resistances—is justified. So too are the criticisms made of media analysts who acknowledge audiences only as the cultural dupes of advertisers. On both fronts, capitalism's intentions and its capacities are too easily equated—a conflation that Stewart Ewen has rightly criticized for its belief in "the self-generating potency of . . . technology and domination."[74]

The more persuasively such analysis demonstrates the complete instrumentality of technoscience to capital, the harder it becomes to posit credible opposition or alternative. This of course is where the Frankfurt School encountered a fatal self-contradiction. For if technological dominance was as total as Adorno or Horkheimer suggested, it became difficult to explain even the basis for their own critical viewpoint, let alone how it could possibly mobilize political action. Critical theory relentlessly painted itself into a corner, where hope could be sustained only at the price of heroic inconsistency. This dilemma is repeated by many later theorists, in whose portrait of techno-capitalism revolutionary possibility gives way to dystopian nightmares of indoctrination, surveillance, and robotization. The result is a radical pessimism that, while certainly puncturing the euphoria of information society theory, also concedes its hegemony over the future.

The problem is only partially addressed by the neo-Luddite theorists. In reviving the figure of the machine-smasher, their analyses vigorously reassert the active capacity of capital's subjects—but only in a reactive mode. Such defensiveness can end in the romanticization of forms of labor that are *already* manifestly dehumanizing or, alternatively, that represent islands of relative privilege (the tendency of labor process analysts to focus on the predicament of highly skilled male workers is a case in point). Further, it can take little account of the possibility—particularly apparent in the field of media and communication technologies—that capital's laboring subjects may find real use-values, perhaps even subversive ones, for the new technologies.

Ultimately, this position suffers the deficiencies of all oppositional theories that conceive struggle only as resistance, and not as counterinitiative. Most neo-Luddite authors in fact admit the need to develop perspectives not just of resistance, but of reappropriation.[75] But the theoretical optic they have so powerfully developed cannot really register such possibilities. For if capital does possess such entire, unilateral powers to implant its logic into technologies as neo-Luddites assert, then efforts to recapture these systems or turn them to alternative use are doomed.

It should be noted that although such critiques often begin with a rediscovery of Marx, they frequently end with a repudiation of him. For the more strongly Marx's writings on technology-as-domination are emphasized, the greater the inclination to dismiss or regret his equally undeniable assertions about its liberatory potentials. Although Marx was clearly sympathetic to the Luddites, he was also critical of them—remarking that "it took time and experience before the workers learnt to distinguish between machinery and its employment by capital, and transfer their attacks from the material instruments of production to the form of society which utilises these instruments."[76] For many neo-Luddites, such comments show how deeply Marxism was mortgaged to bourgeois ideas of progress, and its inadequacy to the current crisis. However, in their justified attacks on scientific socialism neo-Luddites have in fact discarded something critical in Marx's vision—his understanding of technological development as a contradictory process yielding countervailing possibilities for contending agencies. To affirm and extend this latter strand, we need theory that, without reverting to the automatism of scientific socialism, can find in technological knowledge empowerment not just for capital, but for those who fight against it.

Post-Fordism: New Times?

The discovery of such a perspective has, however, been complicated by the emergence of yet another line of Marxian analysis, one moving in an almost diametrically opposite direction from neo-Luddism. If this line also leads eventually to a departure from Marxism, it takes its exit by an opposite door: one marked not by despair at the oppressive power of capital's new technologies but by enchantment with their liberatory potentials. And if this tendency marks a return to a "positive" Marxian attitude toward technology, it is one very different from the revolutionary teleology of scientific socialism. For what it looks forward to is not the inevitable victory of socialism, but the technological reconciliation of workers with capital.

Much of this analysis has marched under the banner of "post-Fordism." This is a phrase that has entered a diversity of theoretical positions. Not all analysis that uses the term shares the spirit of compromise that I discuss here: for example, the work of David Harvey and several of the radical geographers who have followed in his footsteps is very different in tone.[77] My remarks here should therefore not be taken as a total rejection of the concept. In fact, later in this work I sometimes use "post-Fordism" as a convenient label to designate recent changes in the way capitalism operates. Nevertheless, here I want to focus on the way in which a certain version of "post-Fordism" has become widely associated with a perspective that brings neo-Marxian analysis surprisingly close to that of liberal academics and management consultants, and to the positions of the information society theorists.

To understand this process, it is necessary to look at one of the roots of the post-Fordist idea, in the work of the French "Regulation School" of political economy. In what seemed in origin a classic Marxian project, theorists such as Michel Aglietta and Alain Lipietz set out to investigate the conditions governing the surprisingly successful and ongoing reproduction of contemporary capitalist society. Capitalism, they proposed, is neither a historically invariant formation nor one teleologically destined to collapse. Rather, it repeatedly overcomes internal contradictions by generating successive "regimes of accumulation"—intermeshed orderings of wage relations, consumption norms, and state intervention that synchronize the overall social prerequisites for the extraction and realization of surplus-value.[78] Consolidation of such a regime depends on the successful development of a "mode of regulation" based on "the institutional forms, procedures and habits which either coerce or persuade private agents to conform to its schema,"[79] and also, in some later versions of the theory, on its integration of a viable "technological paradigm."[80]

The principal application of this theoretical perspective has been to develop the category of "Fordism." Fordism of course takes its name from the integration of a Taylorist division of labor with intense mechanization pioneered in the auto-plant assembly lines of Henry Ford. Inspired by Antonio Gramsci's fragmentary but suggestive essay "Americanism and Fordism," Regulation School theorists expanded the meaning of the term to designate the regime of accumulation characteristic of industrial capitalism during the middle period of the twentieth century.[81] Fordism in this sense was a comprehensive system of social organization, coordinating factory-based assembly-line production, mass markets consuming standardized manufactured goods, and Keynesian stabilization of the business cycle. Under Fordism, capital enjoyed its post–World War II "golden age."

But in the late 1960s and early 1970s, the theorists of the Regulation School argue, Fordism encountered a serious crisis. Their accounts of its causes vary—ranging through a saturation of mass markets, shop-floor disaffection, the fiscal costs of the welfare state, and changing conditions of global competition. Often these factors are combined in an impeccably overdetermined account. But in any event, Regulation School theorists agree that, starting about twenty-five years ago, capital's most successful regime of accumulation began to falter; sliding profit rates inaugurated a period of continuing flux and uncertainty, disintegration, and restructuring in the global economy that continues to this day.

If Fordism is breaking up, the obvious issue is: what will succeed it? This is the question theories of a "post-" or "neo-" Fordist regime attempt to answer. While accounts of the emergent regime vary in detail, it is generally agreed that it will centrally involve the introduction of new technologies—a change in "industrial paradigm." Aglietta himself speculated that a "neo-Fordist" regime would replace the "mechanical principle" of the assembly line with computerized systems based on the "informational principle."[82] His view of such developments was far from rosy: while their arrival unleashed "the most shameless propaganda about the liberation of man in work," they actually meant intensified workplace deskilling and, at the level of society as a whole, "a strong totalitarian tendency."[83] Some theorists drawing on his work retain this skeptical orientation. But others have elaborated the idea of post-Fordism far more optimistically.

Here the Regulation School's Marxism intersects in a remarkable way with non-, indeed anti-, Marxist perspectives. One of the most important of these is the work of Michael Piore and Charles Sabel on the "second industrial divide."[84] Piore and Sabel, far from being Marxists, are, if anything, Proudhonist in their orientation—fascinated by the prospects of escaping the alienation of modern capitalism by return to small-scale, cooperative, artisanal production.[85] For these theorists, the disintegration of Fordism amounts to a moment equivalent in importance to the first industrial revolution. On the other side of this divide lie bright prospects. Information technologies possess a reprogrammability that gives them a plasticity unknown to dedicated industrial machinery. This, Piore and Sabel argue, will allow the restoration to the workplace of the judgment, learning, and variety lost to Taylorism.

New computerized systems of "flexible specialization" can both respond to the disaggregation of standardized Fordist mass consumption into more fluid, niched, and customized markets and at the same time supersede the deadening routine of Fordist mass production.[86] The monotony of the industrial assembly line will give way to versatile high-tech craft work that requires the willing engagement of the operator's knowledge and attention and places a premium on cooperation between management and worker. The result, Piore and Sabel claim, will be to dissolve the alienation and antagonism of the capitalist workplace and lay the basis for a new, artisanal, computerized post-Fordist "yeoman democracy."[87]

By the mid-1980s, the production of such optimistic post-Fordist prophecies had become a veritable academic industry. The concept of a new regime of accumulation was variously married with theories of flexible specialization, Japanese management, or Swedish humanized workplaces to generate a series of predictive models of labor/capital cooperation in the new epoch.[88] With their promise of a new era pivotally shaped by computers and telecommunications, these versions of post-Fordism triggered memories among both critics and supporters of postindustrialism and information society theory. For its proponents on the left, one of the attractions of the concept was undoubtedly that it represented a rejoinder to such theories. It seemed to offer a way of talking about computers that did not pretend capitalism had ceased to exist, yet did not box itself into the relentless pessimism of theories of technology-as-domination.[89] Yet in doing so, it often replicated the most problematic aspects of postindustrial theory. For, as Eloina Pelaez and John Holloway point out in their scathing attack on theories of post-Fordism, in many of these accounts the complexity of Aglietta's original analysis of the crisis of Fordism is simplified into a blunt technological determinism whereby it is the sheer force of new technologies that produces the new era.[90]

A more sophisticated version of the argument—strongly advanced by Lipietz, a founder and the foremost popularizer of Regulation School theory—is that the crisis of Fordism opens the way to a variety of alternative accumulation regimes. Some of these would be better for workers than others. One could have either neo-Fordist regimes—in which informatics duplicate and intensify traditional patterns of exploitation—or truly post-Fordist systems, which take advantage of the new technological opportunities for reskilling and responsibility. For Lipietz, the pursuit of this latter path, the search for "a way out of the crisis" based on "responsible involvement," in which workers gain higher security, higher pay, and/or shorter hours in return for their cooperation in post-Taylorist high-technology systems, represents "the dream of a new deal for the 21st century."[91]

However, many critics have suggested that such dreams of a high-tech "new deal" rest on a very uncritical acceptance of management propaganda about new production systems. Post-Fordist analysis, they charge, de-emphasizes the way "flexible specialization" segments the work force between a "core" of permanent skilled workers and a "periphery" of casualized and temporary employees.[92] It also often glosses over how, even within this "core," the new post-Taylorist work organization, with its "autonomous work teams," peer policing, and internalized competition, has been developed as an attack on trade union strength.[93] Moreover, its customary contrasts between dirty assembly lines and clean computers ignores the reality of stress, repetitive strain injuries, eye strain, and electronic sweatshops.

To this I would add that many theorists of post-Fordism are remarkably silent about the way automation and global communication have been deployed to swell the reserve army of the unemployed, in a way that ferociously undercuts the strength of movements struggling for improved conditions of work and life. Even where these negative features of restructuring are recognized in "post-Fordist" analysis, as they are in some of Lipietz's work, they are seen as contingent options, undesirable alternatives within an array of social options. What is not confronted is the possibility identified by neo-Luddite analysts, namely, that these destructive outcomes might not be subsidiary to capital's logic but rather central to it—that post-Fordist restructuring might be a project predicated on discipline through austerity as a prerequisite for future profit. In this view, the weakening of resistance, on the shop floor and in society at large, is a *central* purpose in the corporate deployment of new technologies, and the chances of negotiating a "new deal" around their use are thus probably illusory.

This tendency to downplay the darker side of capitalist restructuring is even more apparent when post-Fordism has entered discussions on media and popular culture. Just as in the labor process debate the post-Fordist cachet often marked a shift away from pessimism about the degradation of work toward post-Taylorist optimism, so in the field of culture it has been associated with a rejection of somber theories of mind-management in favor of an effervescent enthusiasm for "popular culture." A salient example is the concept of "New Times" proposed in the British journal *Marxism Today* by a cluster of authors including Stuart Hall, Dick Hebdige, Robin Murray, and John Urry.[94] In the New Times analysis the switch from standardized mass consumption to flexible specialization is seen as bringing with it an intensified attention to advertising, design, fashion, media, and market information. This generates a postmodern ambience of sliding signifiers, simulacra, and spectacle, a culture whose volatility and recombinancy both reflects and contributes to the fluidity of post-Fordist production systems.[95]

However, in marked contrast to theorists such as Schiller, New Times analysts do not view this explosion of media and imagery with suspicion or alarm. Rather, the new scope of consumer choice—including the proliferation of media channels—and the energetic experimentation of post-Fordist commercial culture, with its gender-bending advertisements, socially conscious products, global eclecticism, and self-conscious embrace of feminism and multiculturalism, are seen as opening an exciting space replete with possibilities for the forms of life championed in various identity politics. Hall speaks of the disintegration of Fordism catalyzing a "revolution of the subject" and creating an "expansion in the positionalities and identities available to ordinary people." Exhorting the Left to adapt to the pluralizing, decentralizing, and variegating aspect of the new cultural regime, he cites Marx's famous lines about the dynamic effects of "the constant revolutionising of production" in which "all fixed, fast frozen relationships . . . are swept away. . . . All that is solid melts away."[96]

In the eyes of critics such as A. Sivanandan, however, what has melted away in the enthusiasm for post-Fordism is the solidity of Marxist commitments.[97] Even more moderate critics voiced concerns that the "designer socialism" of Hall and his colleagues expressed the limited perspectives of a fraction of the left intelligentsia favored by the growth of new cultural industries, and that their enthusiasm for "new times" was achieved only at the expense of forgetting about "old enemies."[98] And, indeed, the New Times celebrations of post-Fordism's cultural vivacity seem remarkably indifferent to the appearance in Thatcherite Britain of new exclusions and stratification at least as pernicious as the massified divisions of Fordism. Eloquent about the improved choices post-Fordism brings to consumers, it was very silent on the street-level bricolage left for those rendered destitute by the degradation of the welfare state. When this is taken together with an evident distaste for the militancies of miners' strikes or anti-poll-tax riots, a politics hovering vaguely on the left of a Labour Party marching rapidly to the right, and a theoretical rapprochement with specifically "post-Marxist" theorists, it is difficult not to think that the New Times analysis made the title of *Marxism Today* into a very postmodern irony.

Distaste for such positions has led many Marxists to reject entirely the categories of Fordism and post-Fordist as a mystification of capital's perennial, and ugly, features. This may be to throw the baby out with the bathwater. The argument that capital entered a phase of drastic restructuring in the early 1970s is a compelling one. In recognizing this shift, theorists who use the category of post-Fordism have often been more alert to important changes in work, culture, and politics than their more orthodox Marxist critics. They could even be said to have rediscovered a sense of the dynamic, tumultuous, and experimental nature of capitalist development that was salient in Marx's own writings but often forgotten by those who insist that capital is always "the same old thing."[99]

However, to agree with the post-Fordists that capitalism is undergoing a period of rapid change is not to assent to their analysis of the cause, course, or consequence of this transformation. As Julie Graham has pointed out, embedded within the theoretical apparatus of the Regulation School is a deep tendency to downplay the conflict at the heart of capitalist society.[100] For its analysis takes as its focus and "point of entry" the requirements for capital's successful organization of society, not the contestation of its rule.[101] Its research agenda is built around capitalist growth, not class struggle. Once such study is divorced from scientific socialists' teleology of inevitable breakdown, it tacitly enters onto the same ground as non- or anti-Marxist theories of economic development, so that "Marxism becomes another theory of capitalist growth, focusing primarily on those social processes that promote capital accumulation and excluding those that do not."[102] The result, as Graham notes, is a vision that is premised on the "vitality and uncontested hegemony" of capital's reproduction, but "obscures the weaknesses and instabilities of that process (and) hides the failures and unevenness that make non-capitalist alternatives an existing and future option."[103]

This emphasis on the historical adaptability of capital, taken in conjunction with the general demoralization of the Left in the 1980s, has led to a very rapid acceptance that what will emerge from the crisis of Fordism can *only* be another capitalist regime of accumulation. The assumption that restructuring will succeed—an inverse reflection of scientific socialism's faith in inevitable collapse—leads, by gradual but inexorable stages, to a circumscription of left action. Even in the work of Lipietz, shrewdest and most persuasive of post-Fordist reformists, it is impossible not to be struck by how emphatically socialism is ruled off the agenda for the foreseeable future, how complete is the acceptance of the hegemony of the market, or how large the concessions to the corporate agenda for the reorganization of work. The only issue becomes what *sort* of capitalist regime will emerge, and how good a "deal" workers and social movements can cut within it.

This effects what Les Levidow has termed a "foreclosure of the future."[104] By implicitly accepting the success of capital's restructuring, it directs attention away from forms of action that might challenge that completion. It shuts the door on strategies where workers' knowledge of new production systems yield, not partnership with management, but new ways to challenge managerial command, and new ways in which emergent media networks are made to circulate struggles rather than commodities. In doing so, it represses radical potentialities in favor of reformist hopes.[105] This is done in the name of realism. But given the enormous offensive capacity the new technologies allow global business, the expectation that capital will negotiate any reformist compromise unless faced with a serious challenge to its overall control of society is itself utterly utopian. For Lipietz, the task is "to find a way out of the crisis." But the Marxist project has never been to help capitalism find a way out of crisis. It has been to find a way out of capitalism. This is precisely the possibility that much post-Fordist writing abdicates.

Condition Terminal?

In this chapter we have seen how various schools of Marxism have responded in radically contrasting ways to the "information revolution." This diversity of response arises from the complexity of Marx's own writings on technology. The amplification and extension of different aspects of these texts has given rise to very different perspectives on the relation of machines to social change. Scientific socialism has conceived of a teleological interaction of forces and relations of production, leading to the eventual collapse of capital; technology-as-domination theorists, on the other hand, see machinery as consolidating and deepening capitalist power; and post-Fordists have often found in new technologies the promise of a humanization of work that would transcend the traditional patterns of exploitation.

All these accounts suffer major defects as a reply to the anti-Marxist challenge of the information revolutionaries. In a way that uncannily mirrors the logic of their opponents, scientific socialists effectively liquidate human agency and substitute for it an inexorable, and ultimately sinister, technological automatism. Technology-as-domination theorists restore to view the question of the subjectivity constituted by a machine-saturated society—but can conceive of it only as a process of victimized exploitation, to which the best response is a reactive, heroic, but probably hopeless neo-Luddism. Many post-Fordist accounts, on the other hand, have embraced so much of the information revolutionaries' own euphoria about the new subject of technology as to essentially abdicate the negative moment of critique and subscribe to capital's own logic of technological development.

All three perspectives lead, although by different routes, to potential disintegrations of or exits from Marxism: scientific socialism shattered by the confounding of teleological optimism marked by the events of 1989; neo-Luddism descending into a dystopian, radical pessimism; and several versions of post-Fordism converging with a post-Marxist politics that claims to go "beyond" issues of capital and class. Surveying these dead-ends, it would appear that the information age has put Marxism into a terminal condition.

CYCLES

From the great tangle of Marxist theory one thread might guide us through the labyrinth of high-technology capitalism. This red thread, often severed and nearly lost, yet constantly picked up by unlikely hands, goes by a variety of names. Because it traces the conflict between exploiters and exploited, it is often called "class struggle" Marxism; because it contrasts the vitality of living labor with the dead power of capitalist command, it is sometimes known as "subjectivist" Marxism; recently, something close to this tradition has been termed "open" Marxism, because it shows how the insurgencies of the oppressed unseal fixed sociological categories and teleological certainties.[1] But whatever label is attached to it, what defines this strand of Marxism is its emphasis not just on the dominative power of capital, but on people's capacity to contest that power.

The Red Thread

This sort of Marxism is one that, as James O'Connor reminds us, owes at least as much to the passion of Romanticism as to the rationalism of the scientific Enlightenment.[2] It understands capital's crises as arising not from the "internal barriers" to capitalist accumulation, but from an "external barrier"—namely, the working class: its focus is "the condition of availability of disciplined wage labour, or capital's political and ideological capacity to impose wage labour on the working class."[3] This is therefore a Marxism that insists struggle is intrinsic to the capital-relation. In this it contrasts sharply with what Michael Lebowitz terms "one-sided Marxism," which focuses on the activity of capital and neglects the counteractivities of workers.[4] Instead of seeing history as the unfolding of pregiven, inevitable, and objective laws, the class-struggle tradition argues that such "laws" are no more than the outcome of two intersecting vectors—exploitation and its refusal in the constantly recurrent eruptions of fight and flight by which rebellious subjects seek a way beyond work, wage, and profit.

Such a perspective has not been limited to any one group or particular epoch. Rather, it constitutes a heretical strain that time and again has interrupted the hegemony of more mechanistic, objectivist, and authoritarian versions of Marxism and, as often, been savagely extinguished. Such an intermittent and subterranean existence makes construction of a coherent lineage difficult—more a listing of outbreaks than a narrative of continuities. A fragmentary chronology would of course start with passages from the multiplicitous works of Marx and Engels. From the early twentieth century, it would include certain currents within council communism and anarcho-communism, as well as moments in the work of Rosa Luxemburg and the early writings of Georg Lukács, Karl Korsch, and Antonio Gramsci.[5] Later, in the 1930s and 40s, it finds another manifestation in the work of C. L. R. James, Raya Dunayevskaya, Martin Glaberman, George Rawick, and others associated with the Johnson-Forest tendency in the United States.[6] In the wave of activism of the 1960s and 70s, the incidence of this kind of Marxism intensifies, including, in France, the activities of groups such as Socialisme ou Barbarie;[7] in England, the work of E. P. Thompson and other radical historians who investigated the "making" of class through struggle;[8] in Germany, Karl Heinz Roth's analysis of the "other worker movement";[9] and also various groups associated with the Italian ultra-left, to whose contribution I return in a moment. In my view there are broad thematic affinities among these authors and activists—similarities in their emphasis on agency, on struggle, on self-organization and in their repudiation of authoritarian state socialism— that warrant clustering them together.

But is this lineage relevant to the analysis of high-technology capitalism? Many of its makers lived and fought in a world that, though all too familiar with capitalism's command of machinery, is separated from ours by several generations of technological change. They inhabited the world of the assembly line and telegraph, rather than the robot and Internet. Even among those closer to our times, the greatest analytic achievements are often historical and retrospective: Thompson's account of the factory system or James's discussion of the slave plantation, while sharply provocative in their insights about the intertwining of technology, work, and power, do not speak directly to a world saturated with computers, telecommunications, and biotechnologies.[10] Moreover, it might be said, while there are some studies of working-class battles over digital machines and electronic media from a class struggle position, these have usually not offered any theoretical perspectives beyond the neo-Luddism discussed in the previous chapter.[11]

There is, however, a branch of this tradition whose currency and inventiveness on issues of high-technology struggle escapes such objections—the branch often called "autonomist Marxism."[12] As described by its main English-language archivist and chronicler, Harry Cleaver, autonomist Marxism has a genealogy that is deep and wide, stretching out to touch several of the figures I have already mentioned.[13] But of particular centrality is a cluster of theorists associated with the *autonomia* movement of Italian workers, students, and feminists of the 1960s and 70s, including Raniero Panzieri, Mario Tronti, Sergio Bologna, Romano Alquati, Mariarosa Dalla Costa, Franco Berardi, and Antonio Negri.[14] In the late 1970s, *autonomia* was destroyed in one of the most ferocious yet least-known episodes of political repression in the recent history of metropolitan capital. The work of this group of intellectual-activists was violently interrupted by exile and imprisonment. Their brand of Marxism, anathema to neoliberals, Eurocommunists, and social democrats alike, came to constitute a largely clandestine tradition.[15] Yet over the political winter of the 1980s and 90s it continued to develop, undergoing new mutations and making fresh international connections.[16] At a moment when all the accepted verities of the Left are in confusion, heresy can make a regenerative contribution. Transgressing the conventional limits of Marxist thought, but built on the foundations of Marx's work and extending it into the contemporary world, autonomist Marxism proposes not an "ex-Marxism" or a "post-Marxism" but a "Marx beyond Marx."[17]

To pit autonomist Marxism against information revolutionaries is no arbitrary juxtaposition. Groups within the orbit of *autonomia* were among the first to analyze the postindustrial restructuring of capital as a weapon aimed against social dissent. Since that time certain autonomist theorists, most notably Negri, have devoted increasing attention to the vast new informational apparatus of contemporary capitalism. What makes their perspective peculiarly notable is that it grasps the new forms of knowledge and communication not only as instruments of capitalist domination, but also as potential resources of anticapitalist struggle. While autonomists are by no means alone in raising these possibilities, the inventiveness and scope of their analysis has been massively overlooked.

I therefore read autonomist Marxism (and it is worth emphasizing that this is indeed a *reading* of the autonomists' work, just as theirs is an active, inventive reading of Marx) as a subversive counterinterpretation of the information revolution, contributing to the reconstruction of a twenty-first-century communism capable of confronting computerized capitalism with a radically alternative vision of community and communication.

The Perspective of Autonomy

At the heart of autonomist analysis lies Marx's familiar examination of the relation between labor and capital: a relation of exploitation in which workers, separated from the means of production, are compelled to sell the living labor power from which the capitalist extracts surplus value. In elaborating this account, however, most Western Marxisms have emphasized only the dominant and inexorable logic of capital. Its accumulative logic, unfolding according to ineluctable (even if finally self-destructive) laws, figures as the unilateral force shaping the contemporary world. The autonomists' rediscovery—startling enough that Yves Moulier terms it a "Copernican inversion" in postwar Marxism—was that Marx's analysis affirms the power, not of capital, but of the creative human energy Marx called "labor"—"the living, form-giving flame" constitutive of society.[18]

As Tronti put it: "We too have worked with a concept that puts capitalist development first, and workers second. This is a mistake. And now we have to turn the problem on its head, reverse the polarity, and start again from the beginning: and that beginning is the class struggle of the working class."[19] Far from being a passive object of capitalist designs, the worker is in fact the *active* subject of production, the wellspring of the skills, innovation, and cooperation on which capital depends.[20] Capital attempts to incorporate labor as an object, a component in its cycle of value extraction, so much *labor power*. But this inclusion is always partial, never fully achieved. Laboring subjects resist capital's reduction. Labor is for capital always a problematic "other" that must constantly be controlled and subdued, and that, as persistently, circumvents or challenges this command. Rather than being organized by capital, workers struggle against it. It is this struggle that constitutes the *working class.*

This distinction between labor power and working class was originally Marx's.[21] But by reviving it, the autonomists went beyond the sterility of much subsequent Marxist class analysis. For by saying that "the working class is defined by its struggle against capital," they shrugged off elaborate taxonomies circumscribing the "real workers" as some (usually diminishing) fraction of collective labor—manual, industrial, or "blue collar."[22] Rather, they opened a perspective that could see tendencies to incorporation within capital (as labor power) or independence from capital (as working class) as opposite polarities or contending potentialities that permeate the entirety of capital's labor force, understood in its broadest scope. In this view, working-class struggles are the insurgencies of subjects that capital "classes" only as human resources against that categorization—what Cleaver has termed "struggles to cease being defined as either *a* class or as a *working* class."[23]

To analyze such struggles autonomists use the concept of *class composition*.[24] As Cleaver points out, this is a striking instance of their "inversion" of classical Marxist categories.[25] Marx had referred to the way technological change results in a change in the "composition of the collective labourer."[26] But his original account of the "organic composition" of capital focused on the power of capital to direct production through the accumulation of machines. In autonomist theory, however, this emphasis is reversed: the analysis of class composition is aimed at assessing the capacity of living labor to wrest control away from capital.[27] It starts from workers' struggles: how they arise, how they are connected or divided, their relation or lack of relation to "official" workers' organizations, and their capacity to subvert capitalist command.[28] It measures the "level of needs and desires"—expressed in political, cultural, and social organization—that constitute the working class as what Negri terms a "dynamic subject, an antagonistic force tending toward its own independent identity."[29]

Class composition is in constant change. If workers resisting capital *compose* themselves as a collectivity, capital must strive to *decompose* or break up this threatening cohesion. It does this by constant revolutionizing of the means of production—by recurrent *restructurings,* involving organizational changes and technological innovations that divide, deskill, or eliminate dangerous groups of workers. But since capital is a system that depends on its power to organize labor through the wage, it cannot entirely destroy its antagonist. Each capitalist restructuring must recruit new and different types of labor, and thus yield the possibility of working-class *recomposition* involving different strata of workers with fresh capacities of resistance and counterinitiative.

The process of composition/decomposition/recomposition constitutes a *cycle of struggle.*[30] This concept is important because it permits recognition that from one cycle to another the leading role of certain sectors of labor (say, the industrial proletariat), of particular organizational strategies (say, the vanguard party), or specific cultural forms (say, singing the "Internationale") may decline, become archaic, and be surpassed, without equating such changes, as is so fashionable today, with the disappearance of class conflict. Rather than being made once over, the working class is, as Negri puts it, perpetually "remaking" itself again and again in a movement of constant transformation.[31]

In a crucial autonomist formulation, Tronti suggested that it is actually *workers' struggles that provide the dynamic of capitalist development.* In *Capital* Marx had observed that the initial impetus for capital's intensifying use of industrial machinery came from proletarian movements' demanding the shortening of the working day. Building on this, the autonomists argued that capital does not unfold according to a self-contained logic, spinning new technologies and organizations out of its own body. Rather, it is driven by the need to forestall, coopt, and defeat the "other" that is simultaneously indispensable and inimical to its existence, fleeing forward into the future in what Tronti termed "successive attempts of the capitalist class to emancipate itself from the working class."[32]

In this process capital is driven to successively wider and deeper dimensions of control—toward the creation of a *social factory*. Marx had written of capital's tendency to "subsume" not only the workplace but also society as a whole into its processes.[33] Extending this analysis Tronti, writing in the 1960s, argued that capital's growing resort to state intervention and technocratic control had created a situation where "the entire society now functions as a moment of production."[34] To understand these conditions required moving away from the traditional Marxist focus on the immediate point of production (usually the factory) toward the wider perspective suggested by Marx when he wrote of capital as a circuit comprising the moment not only of production but also of distribution and consumption.

This concept was then elaborated by the feminist wing of autonomist Marxism. Mariarosa Dalla Costa and Selma James, anticipating themes now popular in feminist political economy, argued that within the social factory, the *reproduction* of labor power occupied a crucial but unacknowledged role.[35] Without the—to male theorists—invisible labor process of child-bearing, child-raising, cooking, shopping, education, cleaning, caring for the sick, emotional sustenance, in short, "housework," labor power would not be ready for work each morning. This vital reproductive labor, traditionally female and "unwaged," was subordinated to the traditionally male breadwinner.[36] Thus the wage, mediated by patriarchal authority, commanded and disguised unpaid labor time not only in the workplace but also outside it. Other autonomist theorists applied broadly analogous analysis to the situation of other unwaged groups—that is, students, or, in an international context, peasants—within the social factory.

In developing this analysis, Dalla Costa, James, and other autonomists emphasized that the potential unification of workers produced by the universalizing logic of capital has to be understood as cross-cut by a contrary tendency, which Marx recognized, but did not analyze so deeply—namely, capital's drive to divide workers along lines of nationality, gender, and race. As James puts it, "In capital's hands, *the division of labour is first and foremost the division of labourers, on an international scale.*"[37] This systemic organization of "difference as division" was imperative for capital, precisely in order to forestall the unified class movement Marx predicted.[38] Therefore anticapitalist movements, rather than simply mobilizing a unity already supplied by the structure of production, faced the far more complex task of organizing across difference in order to challenge a capitalist totality founded on fragmentation and division.

By extending the analysis of class composition to include reproductive as well as productive labor, and unwaged as well as waged work, autonomists opened up Marxism to radically new theoretical and organizational horizons. For, unlike the Frankfurt School theorists, they did not find the scope of the social factory grounds for despair. If capitalist production now requires an entire network of social relations, these constitute so many more points where its operations can be ruptured. However, autonomists recognized that all of these involved different subjects (factory workers, students, housewives) with specific demands and organizational forms. No longer was the undermining of capitalism the operation of Marx's singular "mole"—the industrial proletariat—but rather of what Sergio Bologna termed a "tribe of moles."[39] The "autonomy" of autonomist Marxism thus came to affirm labor's fundamental otherness from capital and *also* the recognition of variety within labor. This in turn leads away from vanguardist, centralized organization, directed from above, toward a lateral, polycentric concept of anticapitalist alliances-in-diversity, connecting a plurality of agencies in a *circulation of struggles.*

Autonomist Marxism thus sees class conflict moving in what Tronti termed a spiraling "double helix."[40] Working-class composition and capitalist restructuring chase each other over ever widening and more complex expanses of social territory. As long as capital retains the initiative, it can actually harness the momentum of struggle as a motor of development, using workers' revolts to propel its growth and drive it to successively more sophisticated technical and organizational levels. The revolutionary counterproject, however, is to rupture this recuperative movement, unspring the dialectical spiral, and speed the circulation of struggles until they attain an escape velocity in which labor tears itself away from incorporation within capital—in a process that autonomists refer to as *autovalorization* or *self-valorization.*[41] For behind the perennially renewed conflict of capital and labor lies an asymmetry of enormous consequence. Capital, a relation of general commodification predicated on the wage relation, needs labor. But labor does not need capital. Labor can dispense with the wage, and with capitalism, and find different ways to organize its own creative energies: it is potentially *autonomous.*

The autonomist tradition has more often been stigmatized and ignored than given rigorous theoretical examination. But some significant criticisms have been made. Werner Bonefeld, while praising autonomists for breaking with the rigid stasis of structuralist Marxism, suggests that their emphasis on the potential independence of labor from capital can result in a tendency to present workers' movements as entirely external to capital—a sort of pure, uncontaminated revolutionary force.[42] Although this is not the case with the best of autonomist analysis, which clearly depicts such struggles as occurring both *in* and *against* capital, it undoubtedly can manifest in a certain romanticism that underestimates the depths and pervasiveness of hierarchical divisions and ideological assimilation within the working class, and sees every rebellious swallow as a spring of revolution.

Other critics have suggested that the autonomists' focus on the capital/labor contradiction ignores the competitive conflicts and fractures within capital itself.[43] Within autonomist writing one certainly finds relatively little discussion of the rivalries between different sectors of the ruling class, or of the divergence in immediate aims that can occur between sectors such as, say, financial and industrial capital. Moreover, some autonomist analysis seems to suggest that corporate power operates with a single, consciously masterminded battle plan. High levels of planning by transnational organizations such as the International Monetary Fund (IMF) and G7 can make it appropriate to speak of such a capitalist "strategy." But often the anonymous and aggregated nature of the world market's operations make preferable a more impersonal and less intentional term, such as "the logic of capital," used by Michael Lebowitz.[44]

The autonomist emphasis on capital as a totality with certain over-riding systemic imperatives is, however, consonant with the approach of Marx himself, who always emphasized the importance of understanding "capital as a whole" before analyzing the activity of "individual capitals." And this is the only way to perceive what is really at stake in the war against class: people's attempt to free themselves from a structure of alienated and ultimately inhuman power, a process-without-a-subject-but-with-a-purpose, to whose relentless accumulative drive individual capitalists, with all their smart maneuvers and internecine squabbles, are merely petty functionaries.

Interweaving Technology and Power

Autonomist analysis understands capitalism as a collision between two opposing vectors—capital's exploitation of labor and workers' resistance to that exploitation. Its perspective on technology, correspondingly, has two aspects. The first is an analysis of technoscience as an instrument of capitalist domination—a rereading aimed at shattering scientific socialism's myth of automatic scientific progress. The second, however, looks at the situation from the other side and analyzes the ways in which struggles against class can overcome capital's technological control.

In an early essay that established the direction for later autonomist critique, Panzieri broke decisively with left views of technoscientific development as "progress."[45] Rather, returning to the pages in *Capital* on the early introduction of machinery, he reposed that capitalism resorts to incessant technological renovation as a "weapon" against the working class: its tendency to increase the proportion of dead or "constant" capital as against living or "variable" capital involved in the production process arises from the fact that the latter is a potentially insurgent element with which management is locked in battle and which must at every turn be controlled, fragmented, reduced, or ultimately eliminated.[46]

Faced with "capital's interweaving of technology and power," simply to ratify technological rationalization as a linear, universal advance—as the dominant forms of official, Soviet-influenced Marxism did—was to ignore that what it consolidated was a specifically *capitalist* rationality aiming at the domination of labor.[47] It was illusory to believe that the relations of production (property relations) were simply a "sheathing" that would fall away once the forces of production had been sufficiently expanded.[48] There could, Panzieri concluded, be no question of assuming that socialism would arrive as a byproduct of scientific advance: emancipatory uses of machines were possible, but only to the degree that working-class revolt assumed a "wholly subversive character."[49]

Panzieri's perspective was formed in the industrial factory, witnessing the way the Taylorist division of labor and Fordist automation were used to break down worker solidarity. But his analysis of technology as capitalist weaponry has subsequently been applied to situations not only of waged but unwaged labor. Thus, for example, Harry Cleaver has analyzed the so-called Green Revolution as capitalist counterrevolutionary strategy.[50] In the context of widespread communist insurgency in Asia, Cleaver argues, the sponsorship by U.S. development agencies of new plant stocks and agricultural techniques was aimed primarily at breaking down the traditional village structures. This had a twofold aim—to eliminate the communities within which guerrillas moved like fish in the sea, and to allow the creation of an industrial proletariat, fed off the countryside, a prerequisite for capitalist modernization. Agricultural technology served as the civil side to counterinsurgency warfare.

However, autonomists also emphasize that waged and unwaged workers are not just passive victims of technological change, but active agents who persistently contest capital's attempts at control. This contestation can take two forms.[51] The first is sheer refusal. This is the theme of the most famous, and most reviled, of autonomist texts, Negri's *Domination and Sabotage.*[52] Writing in the context of the Italian industrial struggles of the 1970s in the giant Fiat plants and elsewhere, Negri proposes that, confronting the introduction of huge systems of semiautomated technological control, there could be no question of accepting the necessity of modernization, as official trade unions insisted. Instead, workers should stop the innovations used against them—if necessary, by sabotage.[53] This emphasis on the possibilities of sabotage is an important part of the autonomist tradition and puts them close to the neo-Luddite authors discussed in the last chapter, some of whom draw on their work.[54]

However, there is another side to the autonomist analysis that gives it a greater dynamism than that possessed by outright neo-Luddism. This aspect (which Negri develops in his later work) affirms the possibility for workers to use their "invention power"—the creative capacity on which capital depends for its incessant innovation—in order to reappropriate technology. This possibility arises because, in its attempt to technologically control labor, capital cannot avoid creating new types of technologically capable, scientifically literate workers. As Cleaver observes, "The struggles of these workers vis-à-vis their own working conditions as well as vis-à-vis larger social issues can . . . constitute a serious obstacle to successful capitalist planning."[55]

An early instance of this line of thought can be found in the work of Franco Berardi—an activist in the network of politicized "pirate" radio stations that played a crucial role in the Italian *autonomia* movement.[56] Berardi argued that in the course of developing the "technoscientific intelligence" it needed for the control of living labor, capital was unavoidably creating an increasingly "intellectual" work force.[57] With the appearance of this new, scientific form of labor power also emerged the possibility of a "worker's use of science" that would transform machinery from an "instrument of control and intensification of exploitation into an instrument of liberation from work."[58] This manifested in two ways: in workers' insistence on claiming as their own the surplus time created by automation, and in the increasing popular capacity to reappropriate communication technologies, "subverting the instruments of information" and "reversing the cycle of information into a collective organisation of knowledge and language."[59]

Resistance and reappropriation, sabotage and invention power, are, in autonomist analysis, parts of the repertoire of struggle—although different authors, at different times and contexts, may put more emphasis on one than another. Unlike scientific socialists, autonomists find no inherently progressive logic in technological development. But unlike neo-Luddites they do not perceive only a monolithic capitalist control over scientific innovation. Rather, their insistence on the perpetually contested nature of the labor-capital relation and the basic independence of human creativity tends away from attribution of fixed political valencies to machinery and toward a focus on possibilities for counterappropriation, refunctioning, and *"détournement."*[60] If machinery is a "weapon" then it can, as Cleaver says, be stolen or captured, "used against us or by us." Or—to use Panzieri's perhaps richer and less instrumental metaphor—if capital "interweaves" technology and power, then this weaving can be undone, and the threads used to make a different pattern.[61]

This need not imply a crude "use and abuse" concept of technology of the sort that neo-Luddites have rightly criticized. We can accept that machines are stamped with social purposes without accepting the idea that all of them are so deeply implanted with the dominative logic of capital as to be rejected. For if the capital relation is to its very core one of conflict and contradiction, with managerial control constantly being challenged by countermovements to which it must respond, then this conflictual logic may enter into the very creation of technologies.

Thus, for example, automating machinery can be understood as imprinted with the capitalist's drive to deskill and control workers, and *also* with labor's desire for freedom from work—to which capital must respond by technological advance. Similarly, communication technologies have often—as in the case of radio and computer networks—evolved in the course of very complex interaction between business's drive to extend commodification and democratic aspirations for free and universal communication. Along the way communication technologies have been shaped by both forces. This is not to say that technologies are neutral, but rather that they are often constituted by contending pressures that implant in them contradictory potentialities: which of these are realized is something that will be determined only in further struggle and conflict.[62]

In the very course of class conflict, workers will not only, repeatedly, halt and sabotage machines but also challenge capital's unilateral ability to implant *its* logic in technology—and instead bend, twist, and even detach part of the process of technological development to move it in quite different directions. Instead of understanding Marx's "negative" and "positive" visions of machine use in a linear, before-and-after progression—with the same machines that were repressive before communism becoming magically emancipatory afterward—autonomist analysis allows us to reconceive the process of deconstructing and reconstructing technologies as itself part of the movement of the struggle against capital.

From the Professional Worker to the Crisis of the Social Factory

To understand these ideas more concretely, however, we need to look at the three major cycles of struggle that autonomists identify in the twentieth century: those of the *professional worker*, the *mass worker*, and—at least by some accounts—the *socialized worker*. Such a sweeping account will necessarily be highly schematic. As Moulier has emphasized, sensitive use of the cycles-of-struggle concept demands allowance for unevenness, overlap, regional and national variation, and so on.[63] Nonetheless, the very broad-brush version offered here does provide the framework for an analysis of the information revolution that situates it not as the product of ineluctable scientific progress, but of social conflict. In order to clarify this overall dynamic I will proceed through all three of the cycles, moving swiftly at first but then deepening the analysis as we approach the more recent periods.

The era of the professional worker—or what might more generally be recognized as the craft worker—is regarded by autonomists as running from the mid-nineteenth century to World War I. It is so termed because of the strategic position occupied by skilled workers, now absorbed within a mechanized factory system but still in possession of craft knowledge and technical competencies. Such workers are the main protagonists in struggles focused on control of the production process and the preservation of the dignity and value of work. Outside of the factory, capital's subsumption of society remains relatively rudimentary. The state's activity, other than in projects of imperial expansion, is generally limited to policing the operation of the free market, which is characterized by disastrous economic cycles of boom and bust arising from the difficulties of coordinating production and consumption.

Socialist programs in this period are built around the concept of workers' management of industrial production. The role of productive factory labor as the agent of emancipation is unquestioned. Left parties tend to reflect the technical composition of the professional worker insofar as they have a mass membership but an avant-garde leadership—trained cadres of political "experts." Revolutionary organizations constructed on this basis include not only the Leninist parties but also council communist movements based largely among skilled technical workers—such as those of the German metal industries.[64] In the first quarter of the twentieth century such organizations present a mounting threat to capital. With the victory in 1917 of the Bolshevik vanguard party, this threat seems about to attain catastrophic dimensions.

To save itself, capital undertakes a drastic organizational and technological restructuring. This is aimed at decomposing working-class power by destroying the technical base of the professional workers' power and cutting them off from the growing mass of industrial labor. On the shop floor the chronometer and the clipboard of Taylorist scientific management are deployed to break craft workers' control of production. This deskilling, at first attempted primarily through organizational innovation, is subsequently mechanically embedded in the Fordist assembly line. At the same time, in the face of the socialist threat, the first tentative steps are taken toward a more interventionist role for government in social and economic affairs, aimed at stabilizing business cycles and pacifying unrest.

However, this restructuring unintentionally forges the matrix for the emergence of a new working-class subject—the mass worker. The Fordist factory—typified by the huge auto plants that come to form the hub of the advanced economies—spatially concentrates huge bodies of dequalified labor subjected to the brutality of continuous automated machine pacing. In doing so, it creates the conditions for an unprecedented form of class solidarity. With craft skills eroded by Taylorism, the mass worker fights not to uphold the dignity of a trade, but to make capital pay for lives vanishing meaninglessly down the assembly line. No longer able to control production, the worker can still stop it. The vulnerability of the assembly line to interruption and sabotage, and the cost to management of idling the increasingly expensive accumulation of fixed capital, provide the points of attack. In a cycle of struggle that finds its paradigmatic North American moments in the 1937 Flint sit-down strikes, the mass worker finds increasingly effective ways of converting the mechanized factory into a bastion of resistance.

To contain this new working-class strength, capital is forced to further innovation. Here the productivity deal, in which management maintains shop floor control by negotiating with trade unions regular pay raises tied to increases in output, becomes a crucial factor. Although initially only grudgingly conceded, this arrangement was eventually assimilated by business as a way of harnessing working-class strength to accumulation. The link between productivity and pay served to both propel technological innovation and pacify worker resistance. Alongside this institutionalization of "industrial relations" emerge ever more comprehensive plans of social management. Again as a result of working-class struggle, the *factory wage* is increasingly supplemented by a *social wage* of state-controlled payments and amenities—welfare, unemployment, pensions, health insurance, and medical, educational, and recreational facilities. And again capital recuperates these concessions within a new structure of accumulation, as a means to forestall social discontent and guarantee the markets for the volume of commodities pouring off the mechanized lines.[65] Out of this complex interaction of opposition and incorporation there gradually comes into being what the autonomists know as the "Planner State," in which government supports capitalist activity through Keynesian economics and welfare programs.[66]

As John Merrington has noted, autonomists never understood the era of the mass worker as simply a "factory" phenomenon.[67] Rather, they saw it as the moment of emergence of the social factory. Capitalist organization now requires the synchronization of the factory, where surplus value is pumped out on the assembly line, with the household, where the punishing force of such work is repaired, displaced, and hidden, and the pay packet translated into purchases of standardized domestic goods. The gendered division of labor and the pairing of the male mass worker—whose life is to be slowly obliterated on the assembly line—with the housewife, a woman whose lot is to tend the wounds, take the abuse, do the shopping, and raise the next generation of labor power in the isolation of the home—becomes a conscious concern of capital's social managers.[68] The labor of the housewife, whose "consumerist" schedule is organized largely through new organs of mass communication, such as radio and television, starts to become as much the object of corporate planning as the productivity of her male partner on the shop floor—for it is through her activity that the pay increases won by the mass worker are translated into the consumption necessary for a virtuous cycle of continual capitalist growth and stability.

At the end of the Second World War, it seems as if capital in North America and Europe has successfully stabilized itself. The threatening presence of the mass worker is contained in management-union deals, subjected to an increasing weight of mechanical control, and kept ready for work by female reproductive labor in the home. Ethnic minorities and immigrants provide a reserve army available for jobs outside the large-scale industry or in its most antiquated, dangerous sectors. Young people are processed through an expanding educational system that sorts and trains personnel for the increasingly elaborate techno-administrative apparatus required by the Planner State and ever more mechanized production. The threat of the Soviet Union, now turned under Stalin into a ghastly caricature of revolution, is cordoned off with nuclear weapons and a perpetual state of war-readiness. On the basis of this carefully segmented but societywide mobilization, capital secures its golden age of uninterrupted growth.

But then things start to come apart. In the inhuman conditions of the assembly-line factory, the productivity deal always rested on a delicate balancing of capitalist profits and worker anger. In the mid-sixties the tightrope trembles. Mass workers increasingly refuse to restrain wage demands within limits functional to capitalist growth or to tolerate conditions accepted by their unions. Management responds to wage pressures with attempts to intensify the pace and intensity of work, thereby precipitating further resistance. A wave of wildcat strikes, slowdowns, sabotage, and absenteeism—which the autonomists christen "the refusal of work"—sweeps across Europe and North America, concentrated initially in the crucial automobile plants, but spreading to other sectors, rendering factories from Detroit to Turin to Dagenham virtually unmanageable.[69]

Even more alarming for capital, these industrial conflicts start to reverberate with problems elsewhere in the social factory. Students who have flooded the universities to escape a destiny as line workers or housewives refuse to confine their intellectual activities within the limits of the "knowledge factory" and burst into campus revolt. Black and immigrant communities explode against their situation as ghettoized reservoirs of cheap labor. Women, who had in increasing numbers already been abandoning their designated household role to seek paid work, begin a new wave of feminist rebellion against domestic subordination. All these outbreaks are in turn colored by the unexpected challenges in Vietnam and Cuba to advanced capital's global dominance, challenges that generate powerful antiwar and international solidarity movements.

Understood in the light of autonomist analysis, these diverse eruptions, while distinct, are not disconnected. Rather, they appear as a broad revolt by different sectors of labor against their allotted place in the social factory. The new social movements of the era can be understood not as a negation of working-class struggle, but as its blossoming: an enormous exfoliation, diversification, and multiplication of demands, created by the revolt of previously subordinated and superexploited sectors of labor. The swirling social ferment that results certainly involves struggles within and among labor, as those sectors at the bottom of the wage hierarchy—unpaid women, unemployed minorities—assert their equality with those above them—usually white, male, unionized labor. But they also involve a destabilization of the entire capitalist organization of society as a mechanism of surplus extraction.

Complex ricochet effects come into play as demands for improvement in the social wage threaten corporations with higher tax levels and diminished profits, thereby intensifying conflicts over the factory wage. Even more alarming for capital, the multiple outbreaks of dissent begin to be consciously linked with or inspired by one another—as in the interaction of students and workers that occurs briefly in Paris in 1968 and over a longer period of time in Italy; the meeting of labor and antiracist struggles in Detroit and elsewhere; or the rekindling of feminism out of the civil rights and student movements. The result is a circulation of struggles that starts, at multiple points, to threaten the whole intricate equilibrium of the social factory.

Imposing Cybernetic Command

The response can only be counterattack. In a shift that is usually identified with Reaganism and Thatcherism but whose origins the autonomists date to the early 1970s, capital begins another drastic restructuring.[70] In the realm of government, the Planner State is replaced by the "Crisis State"—a regime of control by trauma in which "it is the state that plans the crisis."[71] Keynesian guarantees are dismantled in favor of discipline by restraint; unions hamstrung by changes in labor law; monetary policies exercised to drive real wages down and unemployment up; and welfare programs brought under attack. At the same time, corporate managers take aim at the industrial centers of turbulence, decimating the factory base of the mass worker by the automation and globalization of manufacturing. Dismantling the Fordist organization of the social factory, capital launches into its post-Fordist phase—a project that, however, must be understood as a technological and political offensive aimed at decomposing social insubordination.

It is in the context of this offensive restructuring that the work of the "information revolutionaries" can be situated. As we saw in chapter 2, the first formulations of postindustrial theory by Bell, Drucker, Brzezinski, and Kahn—intellectuals closely affiliated to the nexus of state and corporate power in the most powerful capitalist centers—corresponds precisely to this moment. At that time, George Caffentzis, writing of the apocalyptic calls for a "complete change in the mode of production" issuing from such theorists, observed: "They are 'revolutionaries' because they fear something in the present mode that disintegrates capital's touch: a demand, an activity and a refusal that has not been encompassed."[72] The postindustrialists' futurological reports thus fall into place alongside the infamous report by Samuel Huntington and others on the "excess of democracy" as part of capital's assessment of what is required to reassert command of a deteriorating situation.[73] In the name of irresistible progress and objective prediction, the information theorists propose a program and a legitimization for a great technological deployment whose glittering sheen disguises old, cold objectives: annihilation of working-class power, reduction of wages and social wages, restoration of social discipline.

For Collettivo Strategie, a group within the orbit of *autonomia*, what the new informational doctrines demonstrated was "a militant and revolutionary behaviour on the part of capitalism."[74] Analyzing the projection by Zbgniew Brzezinski—President Carter's national security adviser and a founding member of the Trilateral Commission—of an imminent "technetronic revolution" based on "new technologies, new sciences, microelectronic computers and new means of communication," it noted: "This process is nothing other than a confirmation of the power of capital, as Marx asserted, to impose itself as a force which changes technology or which strikes it down and destroys it violently, thus revealing itself as the least conservative force possible."[75] In fact, Collettivo suggested, the emergence of eminent state officials such as Brzezinski from the culture of think tanks and futurological research institutes indicated that capital had gone "Leninist."[76] Just as the socialist vanguard party was the "organised and theoretical form for seizing power" so, "in the same way capital tries to organise its vanguards into institutions which take the form of a party oriented not toward the destruction but rather the maintenance of power."[77] The project of these informational "vanguards" of capital was a reorganization of production based on "new models of universal communication," launching a new phase of development characterized by the "creation of *uomini mèrce* (humans who have become commodities)" subject to manipulation through "control over the flows of information"—a project Collettivo referred to as the imposition of "cybernetic command."[78]

The military metaphor should not be taken lightly. For what occurs from the mid-1970s onward is that computer and telecommunications devices, developed since the end of World War II primarily as military instruments for the containment of international communism, are transferred for internal application as the "command, control, communications and intelligence" system for the reestablishment of capitalist discipline and productivity. In a classic instance of what Paul Virilio terms "endocolonization," the security apparatus, nominally facing outward to defeat external foes, is turned against the "enemy within."[79] In the United States, a boosting in Pentagon funding, which eventually culminates in the gargantuan Star Wars project, is central to speeding the rate of informatic research and development, and, in some cases, to highly specific injections of new technology into the war against labor. The U.S. Air Force, for example, plays a central role in fostering the computerized automation systems aimed at achieving a workerless factory.[80]

Electronic networking, originally developed as part of preparations to fight a nuclear war, receives its first large-scale civilian application in the emergency management systems used by the Nixon administration to monitor its wage-price freeze and picket line violence in a truckers strike.[81] More generally, both the corporate sector and the apparatus of government increasingly adopt technologies previously nurtured by the military in its quest for battlefield control—microelectronics, computer-mediated communications, video recording, expert systems, artificial intelligence, robotics—now adapted and diffused to provide a similar scope of overview and precision intervention in the workplace and civil society.[82]

Thus the neoliberal transition from "welfare state to warfare state"[83] is supported by a whole new level of intensity and sophistication in the governmental use of information technologies. Mass media and new communications techniques are deployed in depth to measure, massage, poll, and propagandize public opinion preparatory to policy change. Computerization automates and disperses state sector jobs, providing crucial leverage in attacks on public service unions—such as the Reaganite assault on U.S. air-traffic controllers—and creating "lean" institutions attractive to privatization. The same technologies are applied to streamline social programs shaved to levels that monitor, rather than support, and to scapegoat perpetrators of welfare fraud. Last, but by no means least, informatics equips paramilitary security forces with a full arsenal of surveillance devices, electronic intrusion measures, cross-referenced data banks, and field communications for a series of domestic "wars"—on terrorism, on crime, on drugs—that beat down on civil disorders.

The aggressive use of informatics is even more pronounced in the corporate restructuring of work. If the chronometer and the assembly line were the weapons of managerial assault on the professional worker, the robot and the computer network play an equivalent role in the attack on the mass worker. In manufacturing plants, factorywide systems of computerized flow control—Flexible Manufacturing Cells (FMC), Flexible Manufacturing System (FMS), Management Resource Planning (MRP), Computer Aided Process Planning (CAPP), and Just-in-Time (JIT) systems—permit management to sever the solidarity of the assembly line by cutting it into competing "work teams" supplied by robot servers, shrinking the labor force, and in some cases approaching the "lights out" scenario of fully automated factory production. The strategic advantage afforded capital by this disaggregation and downsizing is then reinforced by telecommunications systems that permit the centralized coordination of dispersed operations, making feasible the transfer of work from hot-spots of instability either to domestic "greenfield" sites uncontaminated by militancy or to offshore locations—the first steps toward what would soon be known as "globalization."

On all these fronts the deployments of new information technologies and the restructuring of capital converge so that neither is practically distinguishable from the other.[84] The effects on class composition are devastating. In a series of critical industrial confrontations, informational innovations give capital a winning card, as Italian car workers find their industrial strength destroyed by the total-automation systems of Robogate and Digitron, British miners are undercut by the Minos robot drill, remnants of craft-work strength in London's printers unions are annihilated by computerized typesetting, and the striking clerical workers in the U.S. health insurance industry find their pickets lines overleaped by telematics.[85] Such defeats set the scene for an overall neoliberal attack not only on the wage but also on the social wage, realized through the dismantling of the welfare state.

In the face of this attack, the other movements that had shaken the social factory in the sixties and seventies are themselves increasingly thrown onto the defensive. The most militant—like the Panthers in the United States or *autonomia* in Italy—are destroyed by assassination, imprisonment, and direct repression. But others—such as the student movement—are sapped by insecurity and a lack of resources and time; they are confronted at every turn by the ideological claims of restraint, globalization, and deficit reduction. In the face of cybernetic command, the incipient circulation of struggles disintegrates into a series of atomized rearguard actions.

The effects of this convulsion on Marxist thinking have been devastating. As Caffentzis remarks, "The very *image* of the worker seems to disintegrate before this recomposition of capital."[86] As Fergus Murray argues, in an analysis drawing on autonomist categories, extensive computerization in the factory seems to mark a decisive "decline in the mass collective worker."[87] By permitting centrally controlled, comprehensive factory automation and the splitting up of the production cycle, management can now reduce and disperse workers once concentrated together so they are "scattered territorially, socially and culturally, in different conditions of work and often invisible from one another."[88] In such a situation, Murray observes, "the problem of uniting a single work force, let alone the class, is daunting."[89] There is now widespread acceptance even on the left that aspirations for proletarian autonomy have met a technological nemesis—that capital may have succeeded in achieving its age-old goal of emancipation from the working class.

Socialized Worker?

To stop here, however, would be to omit the most provocative proposal in autonomist thought. Some of its theorists suggest that out of capital's informational restructuring is emerging the subject of a new cycle of revolutionary struggles: the "socialized worker." This term, first used by Romano Alquati in his analysis of student revolt in the 1970s, has been primarily associated with the work of Negri, who describes it as "an innovation in the vocabulary of class concepts" attempting to express the transition from "that working class massified in direct production in the factory, to the social labour-power, representing the potentiality of a new working class, now extended through the entire span of production and reproduction—a conception more adequate to the wider and more searching dimensions of capitalist control over society and social labour as a whole."[90] Negri has progressively deepened and amplified this idea over two and a half decades, from the time of his involvement in the Italian struggles to his exile in France.[91]

The socialized worker is, according to Negri, the subject of a productive process that has become coextensive with society itself. In the era of the professional worker, capital concentrates itself in the factory. In the era of the mass worker, the factory is made the center around which society revolves. But in the epoch of the socialized worker, the factory is, with the indispensable aid of information technologies, disseminated into society, deterritorializing, dispersing, and decentralizing its operations to constitute what some autonomists term the "diffuse factory" or the "factory without walls."[92] "Work," says Negri, "abandons the factory in order to find in the social a place adequate to the functions of concentrating productive activity and transforming it into value."[93]

This diffusion of work unfolds through what he terms "flexibilisation, tertiarisation and socialisation."[94] As the traditional centers of production are automated, enterprises reorganize around flexible models based upon a small core of permanent employees surrounded by a periphery of contingent workers: part-time, temporary, and casual work, dependent subcontracting operations, "black" work, informal work, outwork, and teleworking proliferate. Wage labor is deconcentrated, spatially and temporally dispersed throughout society, and interleaved with unpaid time in new and irregular rhythms.[95]

Simultaneously, as capital reduces its industrial work force, it seeks out new sources of labor in the so-called service or tertiary sector. This process embraces the large-scale conversion of female domestic labor into fast-food, homemaking, day-care, health care, and surrogate motherhood businesses; an extraordinary diversification of cultural industries, turning knowledge, aesthetics, and communications into materials for an explosion of media, music, entertainment, advertising, and fashion industries; and an array of other experiments from massage parlors to management consultancies. This expansion of waged work marks a new order of magnitude in the commodification of human activity.

However, the most radical aspect of this socialization of labor is the blurring of waged and nonwaged time. The activities of people not just as workers but as students, consumers, shoppers, and television viewers are now directly integrated into the production process. During the era of the mass worker, the consumption of commodities and the reproduction of labor had been organized as spheres of activity adjunct to, yet distinct from, production. Now these borders fray. In education, schooling is explicitly reconstituted as job training, lifelong learning as requalification for technological change, and universities as corporate research facilities. In consumption, the integration of advertising, market research, point-of-sale devices, and just-in-time inventory control makes the monitoring of the consumer as integral to the production cycle as that of the worker. Work, school, and domesticity are re-formed into a single, integrated constellation.

The world of the socialized worker is thus one where capital suffuses the entire form of life. To be socialized is to be made productive, and to become a subject is to be made subject to value—not only as an employee but as a parent, shopper, and student, as a flexibilized home worker, as an audience in communicative networks, indeed even as a transmitter of genetic information. The demarcation between the production, circulation, and reproduction of capital is dissolved in a "network of various, highly differentiated, yet confluent mechanisms" that "mixes, in new and indefinite labour, all that is potentially productive" so that "the whole of society is placed at the disposal of profit."[96] "Productive labour," says Negri, "is now that which produces society."[97]

In this situation, where the spatial location of exploitation is no longer the factory but the network and its temporal measure not the working day but the *life-span,* Negri observes that we have "gone beyond Marx."[98] Marx's original concept of "real subsumption," the swallowing of society by capital, has been realized and exceeded. Indeed, says Negri, it is this apparent coextensivity of capital with the social that obscures the "contours of the totality," allowing business to "disguise its hegemony . . . and its interest in exploitation, and thus pass its conquest off as being in the general interest."[99] Facing such an expansion of capital's calculus beyond the point of production we might, he says, now choose to speak of socialized labor power not as a *worker* but as an *operator* or *agent.* Yet, by retaining the traditional Marxist epithet, he emphasizes "an antagonism which has never ceased to exist"—a conflict between the imperatives of capital and the needs and desires of the subjects on whose activity it depends.[100]

For, Negri argues, this intensifying fusion of capital and society has unexpected consequences. Capital "socializes" itself in order to escape the factory-centered conflicts with the mass worker. But the exploitative relation from which that conflict arose—the extraction of unpaid activity from labor—persists. Now, however, it radiates out to inform the extended networks of social activity. Capital persists in paying only for a tiny segment of the life activity it expropriates. But this logic manifests not only in rollbacks and speedups on the shop floor, but in cutbacks to the social wage, the erosion of the welfare state, and the off-loading of the costs of environmental damage. These practices are of course not new. But the intensified integration of capital's circuit sharply highlights the inadequacy of the wage to acknowledge the web of relationships that sustain social production.

The result, Negri says, is that class struggle, transmuted but not eliminated, reappears, refracted into a multiplicity of points of conflict. In a world where capital has insinuated itself everywhere, there is now no central front of struggle. Instead, contestation snakes through homes, schools, universities, hospitals, and media; it takes the form not only of workplace strikes and confrontations but also of resistance to the dismantling of the welfare state, demands over pay equity, child care, parenting, and health care benefits, and opposition to ecological despoliation. In the newly socialized space of capital, a fractal logic obtains, such that each apparently independent location replicates the fundamental antagonism that informs the entire structure— capital's insistence that life-time be subordinated to profit.

The crucial issue therefore becomes whether the scope of socialized labor will manifest as division or alliance, segmentation or linkage. Negri observes that struggles by multifarious subjects at the many sites of the factory without walls—factory workers, welfare mothers, students—manifest their own specificity, their own "concrete autonomy."[101] Yet all encounter a barrier in capitalism's subordination of every use value to the universal logic of the market. Consequently, Negri observes:

> It's either/or: either we accentuate the antagonisms and competitions in the concrete cases or we construct a political and subjective totality dialectical of these segmentations. . . . All this finds its material base if, escaping the myth of factory production, you enter the truth of the process of social production and reproduction, where the functions, the consumption, the elements, the differentiation of the process are fundamental for its own operation, that is, for the operation of producing and circulating wealth.[102]

For Negri, the experimentation with coalitions, "coordinations," "rainbows," "rhizomes," "networks," "hammocks," and "webs" that has been a salient feature of anticapitalist movements in the last decade denotes the search for a politics adequate to "the specific form of existence of the socialised worker," which "is not something unitary, but something manifold, not solitary, but polyvalent" and where "the productive nucleus of the antagonism consists in multiplicity."[103]

The concept of the socialized worker is a conjugation or synthesis of "old" working-class theory and analysis of "new" social movements.[104] Negri argues that the new subject arises at the intersection of "two fundamental axes."[105] One of these runs "from society toward the world of labor" and transmits into the workplace the concerns "of feminism, of ecology, of young people, of antiracist struggle, of social activism, and, in general, a radical cultural modification and a perspective of irreducible grassroots autonomy."[106] The other runs "from the world of work to society" and carries with it not only a critique of capitalist restructuring, of "exploitation aggravated and distributed throughout the most diverse strata of society," but also a demand for increased power in the shaping of the economic order.[107] Out of the fusion of these currents appears the possibility of a "reunification of the traditional components of the class struggle against exploitation with the new liberation movements."[108]

Negri argues that from the 1980s there have appeared the first signs of a *new* cycle of struggles. Focusing mainly on the European context, he and his colleagues look at a series of movements—among nurses, media workers, students—that have challenged neoliberal restructuring. In particular, these theorists have been inspired by the successive waves of social revolt that have shaken French society, from the student protests of 1986 to the interlinked revolts of students, workers, and immigrants in 1994 against proposals to cut the minimum wage to young job entrants, to the massive three-week strike wave of 1996 against the neoliberal Juppe plan. These movements of the socialized worker, Negri says, take forms completely different from the factory struggles of the mass worker, and although historically linked to the first appearance of the new social movements in the 1960s, they are now entering an entirely new phase. This is characterized by "the radically democratic form of organization, the transformed relation with the trade unions (which become more and more just transmission lines for impulses arising from below), the social dimension of objectives, the rediscovery of a social perspective by the old sectors of the class struggle, the emergence of the feminist component, of workers from the tertiary sector and of 'intellectual' labor (above all labor power in training)."[109] Such movements "break with the purely defensive attitude to restructuring."[110] They challenge the Crisis State's managerial control of society, are informed by an ethic that "emphasises the connections of social labour and highlights the importance of social cooperation," and express, in a diffuse but unmistakable form, an aspiration that "cooperative production can be led from the base, the globality of the postindustrial economy can be assumed by social subjects."[111]

Communication against Information

Of particular interest to this study of high-technology struggle is Negri's analysis of the role of communication and information. For he emphasizes that the "factory without walls" is also the "information factory," a system whose operation depends on "the growing identity between productive processes and forms of communication."[112] The conflicts of the Fordist era drove capital to interlink computers, telecommunications, and media in ever more extensive networks the more effectively to subordinate society. While the mass worker labored on a factory assembly line, the socialized worker's productivity emerges at the terminal of fiber-optic lines, as the activity of a nurse monitoring cardiograms, a bank clerk handling on-line transactions, a teacher in a computer lab, a programmer or a video technician, or, indeed, as the audience of interactive television channel or the respondent to a telemarketing survey. The worker's productivity depends on an elaborated network of informatic systems.

However, this technological envelopment does not, Negri claims, necessarily result in a subjugation of social labor. As the system of machines becomes all encompassing and familiar, he argues, the socialized worker enjoys an increasingly "organic" relation to technoscience.[113] Although initiated by capital for purposes of control and command, as the system grows it becomes for the socialized worker something else entirely, an "ecology of machines."[114] The "system of social machines" increasingly constitutes an everyday ambience of potentials to be tapped and explored.[115] The elaboration and alteration of this techno-habitat become so pervasively socialized that they can no longer be exclusively dictated by capital.

In the era of the mass worker, Negri says, the conditions of mechanized labor, concentrated in the factory under the hand of management, led many militants to a "rejection of science." In the age of the socialized worker, however, this situation is "surpassed," as capital is obliged both to devolve and diffuse technological knowledge among its work force. The increasingly social nature of the technological apparatus now makes the tactic of sabotage, crucial to the professional and mass worker, which Negri himself espoused in the 1970s, less central. Rather, expanded possibilities for refunctioning and recuperation appear. Technoscience becomes a site—perhaps, Negri suggests, the principle site—for the reappropriation of power.[116]

This might seem reminiscent of Serge Mallet's earlier concept of a "new working class" based in the skilled cadres of advanced industry.[117] But Negri's theory differs in positing the emergence not of a select intelligentsia of technical workers but of a generalized form of labor power needed by a system now suffused with technoscience. He claims that the new communicative capacities and technological competencies manifesting in the contemporary work force, while most explicit among qualified workers, are *not* the exclusive attributes of this group, but rather exist in "virtual" form among the contingent and unemployed labor force.[118] They are not so much the products of a particular training or specific work environment but rather the premises and prerequisites of everyday life in a highly integrated technoscientific system permeated by machines and media.

Negri suggests that the complexity and scope of the factory without walls creates for capital "a specific social constitution—that of co-operation, or, rather, of *intellectual co-operation* i.e., communication—a basis without which society is no longer conceivable."[119]

Advanced capitalism directly expropriates labouring co-operation. Capital has penetrated the entire society by means of technological and political instruments (the weapons of its daily pillage of value) in order, not only to follow and to be kept informed about, but to anticipate, organise and subsume each of the forms of labouring co-operation which are established in society in order to generate a higher level of productivity. Capital has insinuated itself everywhere, and everywhere attempts to acquire the power to co-ordinate, commandeer and recuperate value. But the raw material on which the very high level of productivity is based—the only raw material we know of which is suitable for an intellectual and inventive labour force—is *science, communication and the communication of knowledge.*[120]

To secure this cooperation, capital must appropriate the communicative capacity of the labor force, making it flow within the stipulated technological and administrative channels: "Capital must . . . appropriate communication. It must expropriate the community and superimpose itself on the autonomous capability of manufacturing knowledge, reducing such knowledge to a mere means of every undertaking of the socialised worker. This is *the form which expropriation takes in advanced capitalism*—or rather, in the world economy of the socialised worker."[121] However, to accomplish this expropriation, capital has to surround the socialized worker with a dense web of communicative channels and devices.

In a rich, if cryptic, passage Negri claims that "communication is to the socialised worker what the wage relationship was to the mass worker."[122] This does not mean that TV programs replace pay. Rather, Negri is suggesting that communicational resources now constitute part of the bundle of goods and services capital must deliver to workers to ensure its own continuing development. Just as in the era of the mass worker Keynesian capital institutionalized wage increases as the motor of economic growth and generalized the norms of mass consumption, so today, post-Keynesian capital institutionalizes the information infrastructure by which it hopes to rejuvenate itself, "plugging in" its socialized work force, multiplying points of contact with the networks, furnishing and familiarizing labor with a "wired" habitat through which instructions can be streamed and feedback channeled.

But the analogy suggests more. In the Keynesian era, attempts to domesticate pay demands as part of capitalist growth plans ultimately failed and became a focus for struggle. Similarly, Negri sees the control of communication resources as an emergent arena of tension. By informating production, capital seems to augment its powers of control. But it simultaneously stimulates capacities that threaten to escape its command and overspill into rivulets irrelevant to, or even subversive of, profit. Insofar as the increasingly "communicative" texture of the modern economy discloses and intensifies the fundamentally "socialized," cooperative nature of labor, it comes into friction with capital's hegemony.

This antagonism can be schematically represented as a conflict between *communication* and *information*—an opposition roughly analogous to Marx's distinction between living and dead labor: communicative activity is "current," information its "imprisonment . . . within inert mechanisms of the reproduction of reality once communication has been expropriated from its protagonists."[123] Information is centralized, vertical, hierarchic; communication is distributed, transverse, dialogic. Capital tries to capture the communicative capacity of the labor force in its technological and organizational forms "like a flat, glass screen on which is projected, fixed in black and white, the mystified co-operative potentialities of social labour—deprived of life, just like in a replay of *Metropolis*," while the direct current of communication takes transverse "polychromatic forms."[124] Or, in a different formulation, "conflict, struggle and diversity are focussed on communication, with capital, by means of communication, trying to preconstitute the determinants of life," while, on the other hand, "the socialised worker has come to develop the *critique of exploitation* by means of the *critique of communication*."[125]

Negri's analysis of this conflict remains characteristically abstract. But one example undoubtedly in his mind is the use of the Minitel computer system by French student protesters. Minitel was originally designed as a one-way videotext service transmitting government and corporate messages—phone directories, advertisements, banking information, timetables—to French citizens.[126] It was changed when hackers converted a small in-house mail system into an open, generalized exchange, an initiative that proved so popular that it was incorporated into the official system—thereby laying the basis for an email system perhaps most famous for its erotic "*messagerie rose.*"

In 1986, however, Minitel attained more political dimensions when students erupted in protest against neoliberal university "reforms" and were met with a police violence that resulted in at least one death. Frustrated by the mainstream media's hostility or indifference to their cause, the Student Co-ordinating Committee, through the daily newspaper *Liberation,* mounted a Minitel service for the revolt. This included disseminating information about the spreading university and school closures, demonstrations, reasons to oppose the proposed legislation changes, and a game service satirizing the government updated news bulletins' appeals.[127] Interactive "enter your reactions" sections received three thousand calls from across France, including questions about reasons for action, the level of student support, the difficulties of government/student negotiations, and the on-line fees charged by the telephone company. For Negri, the significance of the student revolt is that it represents the capacity of labor in training—the emergence of a type of worker who embodies "intellectual cooperation" and technoscientific literacy, and the capacity to use this knowledge in oppositional form.

In the next chapter I give more concrete examples to support Negri's analysis. For the moment, suffice it to say that the struggles between information and communication that he has in mind would embrace the conflicts over the collective organization of work—"team concept," "quality circles," "TQM"—in production; the expansion of alternative media activism contesting the corporate control of news and imagery; struggles in schools and universities between capital's demand for a functionally educated work force and people's insistence in learning for their own purposes; the imposition and transgression of proprietary control over vital medical and ecological knowledge; and the struggle in cyberspace between activists who have diverted global computer networking into an unprecedented form of collective intellect, and capital's attempt to colonize it for commercial purposes.

While the tentative nature of these oppositional projects is evident, Negri would maintain that they constitute the prefigurations of an insubordinate anticapitalist subject whose identity is rooted in the communicative interconnections of socialized production. While neoliberalism has launched a restructuring that has fatally decomposed the traditional bastions of working-class strength and imposed a historic reverse on the left, "nothing," says Negri, "tells us that the journey can be concluded according to the direction established by capital." On the contrary, restructuring has also released "uncontrollable effects . . . perverse from the capitalist point of view, but virtuous from the opposing point of view," creating the conditions for an emergence of new subjects who "even if they escape the historical continuity of the workers' movement, are nevertheless not easily reconciled with capitalist plans for the market."[128]

Fragmented Worker?

It is nothing if not audacious to discern such a recompositional process amidst decades most on the left reckon catastrophic. Many consider it a theoretical whistling in the dark. Alain Lipietz—voicing what is probably a fairly widespread opinion—has accused Negri of a "headlong voluntarist flight into the future."[129] Even many of Negri's political allies dissent from his analysis, suspecting that enchantment with the "cycle of struggles" leads him to find evidence of resurgence where little exists.[130] Several autonomists have been struck not so much by the unification and empowerment of labor in the information economy as by an intensified fragmentation and hierarchization. They have suggested that Negri's work suffers from the defect of so many attempts to periodize class struggle—namely, that an orientation toward what is perceived as the leading edge of struggle leads to a neglect of capital's tendency to pull together into a unified production system very different kinds of labor, in other words to overlook its dependency on what Trotsky referred to as "uneven and combined development."[131]

Thus in an analysis that extends the work of James and Dalla Costa, George Caffentzis argues that capital's decomposition of the mass worker in the mid-

1970s has been accompanied by a redistribution of work in two directions. One is the growth of a high-technology sector focused on the "energy/information" field of oil, electricity, nuclear power, and microelectronics.[132] The other is the emergence of a low-technology "service" sector, built around an influx of women into the work force, and partially transforming traditional, unwaged reproductive labor in the home into a zone for direct exploitation.

The "energy/information" and "service" sectors are functionally complementary for capital, the former providing the cutting edge of profit-taking, the latter the mass employment necessary to stabilize the wage relation. But they differ markedly in conditions of work. While workers in the "high" sector may be technologically skilled and relatively secure, and perhaps may even identify with their work as part of "the brains of the operation," the "low-end" service-sector worker is poorly paid, insecure, untrained, deskilled.[133] Moreover, the sectors are differentiated by the age, race, and especially gender of their labor power—the high sector being predominantly male and white, the low sector disproportionately composed of workers who are young and/or female and/or people of color, often performing a double shift of paid and unpaid reproductive labor at work and in the home. The former gendered division between waged work and unwaged service is now displaced and recapitulated within the wage zone.

Such polarization raises serious questions about Negri's concept of the socialized worker. It obviously affects the "organic" relation to technology he posits for his emergent subject. The grand sweep of the socialized worker thesis often seems to minimize those tendencies that separate strata of relatively well skilled, well paid workers—who may in fact possess strategic technical and communicational capabilities—from the larger mass of a postindustrial service sector—janitors, fast-food operatives, and data-entry clerks—subject to all the most deskilling and isolating effects of technological domination. Since this division of the work force tends to fall along lines of gender and race, to ignore it is to risk universalizing experiences most readily available to labor insofar as it is white and male.

As numerous feminist analyses have made clear, the traditional masculinization of technology—formerly sedimented in the division between house and work—is to a considerable degree perpetuated within the new informational economy. While it is not unusual for women to have positions working with technology, men more often secure jobs in which they control technology, rather than being controlled by it—while female workers experience classic deskilling effects.[134] This can be the case even in situations where workers of different genders use the "same" technology: telework, which can for some—predominantly male—professionals offer significant convenience and control, reveals a very different face in regard to the usually female data processor—poorly paid, outside legislative protection, closely monitored, isolated, and unorganized within an "electronic ghetto." Such patterns of segregation tend to be redoubled where the exclusions of race are compounded with those of gender.

If this is the case, the opportunities for technological reappropriation that Negri identifies may exist primarily for those who are most privileged—and therefore least likely to use them subversively. In not explicitly addressing this issue, the socialized-worker theory invites the accusations—which other autonomists have leveled against Negri's work—of generalizing the experiences of relatively privileged workers in contact with the most advanced sectors of capital and ignoring other strata.[135] Moreover, in his eagerness to identify the leading edge of working-class development, Negri also sometimes seems to dismiss the continued resilience of some "old" struggles—one thinks, for example, of the persistent and, from capital's point of view, very untimely militancy of coal miners in Britain, the United States, and Canada. All this suggests that the divisions within the post-Fordist work force are more complex and significant than Negri allows. Although theorists such as Caffentzis undoubtedly share his hope for an eventual recomposition of the working class, it is with far less optimism about its immediate prospects. In the hands of nonautonomist theorists—including various Marxists and ex-Marxists—the segmentation of the informational labor force is widely adduced as evidence of a final end to class politics.[136]

Negri's writings contain an implicit response to this charge, albeit one that deserves amplification. He emphasizes that in describing the recomposition of socialized labor power he is not talking of "something definitive, concluded," but a "potentiality," a political possibility "which has to be asserted against resistance.[137] Negri's socialized worker is conceived as an agency in process, a subject formed in a struggle that has at stake not only the relation between labor and capital, but also the relation of labor to itself. The countertendency against which this recompositional movement asserts itself is capital's segmentation of the labor market along lines of gender, race, and age, which tends toward a "South Africanization" of society, splitting socialized labor into isolated segments, just as Caffentzis and others have described.[138]

However, Negri believes that this "divide and conquer" strategy for decomposing the socialized worker has some serious limitations. Capital's tendencies to social apartheid, powerful as they are, are contradicted by a simultaneous tendency to subsume labor within a single, unified system dependent on a common infrastructure. Its simultaneous tendencies to "smooth" and "stratify" social space generate paradoxical results, unanticipated interstitialities, upward and downward mobilities and flux. The dissemination of technical knowledge and abilities cannot be limited to safe, reliable strata of employees—who often feel the chill breath of insecurity—but is made catholic by capital's own frenetic processes of circulation. The socialized worker's familiarization with and appropriation of their informational habitat is a process that squirms under and over attempts to strategically contain and stratify it. The system of segmentation leaks.

Although Negri does not elaborate on the point, it is easy to muster examples: the video countersurveillance of police abuses in ghettoized sectors; the development of highly technical modes of politico-cultural expression, such as certain strains of rap music; the importance of community and "guerrilla" radio among subordinated groups; the crucial role of film, video, and media in feminist and antiracist struggle; the increasing use of computer networks—including feminist networks—to publicize otherwise invisible labor struggles; and the remarkable exploration of cyberspace as a medium for the circulation of struggles by some of the most marginalized and dispossessed sectors of the global work force—such as the Zapatistas in Chiapas. It is in the context of such struggles to overcome the segmentation of the work force that the opposition of communication against information to which Negri gives so much emphasis assumes its full importance.

Realistic assessment of the current state of class composition requires taking into account both the recompositional possibilities on which Negri focuses, and the decompositional forces stressed by Caffentzis and other autonomists. Both are present tendencies, and their prominence in any given concrete instance varies. Negri's analysis is clearly rooted in some of the remarkable cross-sectorial linkages made in the French movements—although even there, sectoralism enormously impedes mobilizations against neoliberalism. The more somber perspective of Caffentzis reflects the near-disastrous working-class atomization in the United States. Yet, as we will see, even in the North American context of fragmentation there are important countervailing tendencies. With both these potentialities present, digital capitalism constitutes what Negri calls "an enormous node of strategic contradictions—like a boiling volcano."[139] The next chapter descends deeper into the volcano and more closely observes its eruptions.

5

CIRCUITS

The previous chapter traced the history that led class war onto the terrain of the information revolution. This one makes a map of the contemporary battleground. To do so, it uses one of Marx's central concepts, that of the circuit of capital.[1] Put simply, this shows how capital depends for its operations not just on exploitation in the immediate workplace, but on the continuous integration of a whole series of social sites and activities—sites and activities that, however, may also become scenes of subversion and insurgency. Today, this circuit of accumulation and resistance passes through robotized factories, interactive media, virtual classrooms, biotechnological laboratories, *in vitro* fertilization clinics, hazardous waste sites, and out into the global networks of cyberspace.

The Circuit of Capital

Marx's original account describes only two moments in the circuit of capital. In production, labor power and means of production (machinery and raw materials) are combined to create commodities. In circulation, commodities are bought and sold; capital must both sell the goods it has produced, realizing the surplus value extracted in production, and purchase the labor power and means of production necessary to restart the process over again.

Since Marx proposed this model, however, capital has prodigiously expanded the scope of its social organization. This expansion, and the resistances it has provoked, has made visible aspects of its circuit that he largely overlooked, but which are identified in the autonomist analysis of the social factory.[2] In the 1970s Mariarosa Dalla Costa and Selma James made a crucial revision when they insisted that a vital moment in capital's circuit was the *reproduction of labor power*—that is, the activities in which workers are prepared and repaired for work.[3] These are processes conducted not in the factory but in the community at large, in schools, hospitals, and, above all, in households, where they have traditionally been the task of unwaged female labor.

More recently, another round of struggles has called attention to additional aspects of capital's circuits, previously largely overlooked by Marxists—the reproduction of nature. Capital must not only constantly find the labor power to throw into production but also the raw materials this labor power converts into commodities. As mounting ecological catastrophe catalyzes intensifying protests by green movements and aboriginal peoples, it has become apparent that faith in the limitlessness of such resources is profoundly mistaken. Whether raw materials are in fact available for accumulation depends on the extent of capital's territorial and technological reach, on the degree to which ecosystems have been depleted and defiled, and on the level of resistance this devastation arouses. The *reproduction* (or nonreproduction) *of nature* increasingly becomes a problem for capital and a terrain of conflict for those who oppose it.[4]

Taking account of the insights won not just by workers' struggles but also by feminist and environmental movements, this chapter posits a modified version of Marx's circuit of capital, constituted by four moments—*production, the reproduction of labor power* (which is examined under three headings dealing with welfare, medical services, and schooling, respectively), *the reproduction of nature*, and, finally, *circulation*. At each point we will see how capital uses high technologies to enforce command, by imposing increased levels of workplace exploitation, expanding its subsumption of various social domains, deepening its penetration of the environment, intensifying market relations, and establishing an overarching, panoptic system of measurement, surveillance, and control through digital networks.

However—and this is crucial—the cartography of capital's circuit maps not just its strengths but also its weaknesses. In plotting the nodes and links necessary to capital's flow, it also charts the points where those continuities can be ruptured. At every moment we will see how people oppose capital's technological discipline by refusal or reappropriation; how these struggles multiply throughout capital's orbit; how conflicts at one point precipitate crises in another; and how activists are using the very machines with which capital integrates its operations to connect their diverse rebellions. In particular, I argue that the development of new means of communication vital for the smooth flow of capital's circuit—fax, video, cable television, new broadcast technologies, and especially computer networks—also creates the opportunity for otherwise isolated and dispersed points of insurgency to connect and combine with one another. The circuit of high-technology capital thus also provides the pathways for the *circulation of struggles*. I draw examples primarily from a North American context, perhaps one of the most inauspicious of current contexts for class struggle and, consequently, an acid test for the contention that such conflict has not vanished from the horizons of the information era.

Production: Automatic Systems

Let us start (though not stay) at the traditional heart of Marxist theory, the immediate point of production, the site of work. Here, the information revolution has meant, first and foremost, a leap toward a new, digitized level of automation—an extraordinary intensification of capital's perennial drive to eliminate its dependence on labor by transferring workers' knowledge into machines. Over the last twenty-five years management has invested massively in computerized production technologies—numerically controlled machine tools, robots, automatic delivery devices, and just-in-time inventory systems.

These cybernetic devices first appeared in the workplace shortly after the end of the Second World War, primarily in manufacturing and petrochemical industries.[5] At first, their components were introduced in a piecemeal fashion and only gradually were connected in increasingly self-regulating complexes. This process was, however, accelerated by the industrial revolts of the 1960s and 1970s. Advanced versions of the new systems, aimed at a maximum reduction of the work force and seamless, centralized control from managerially controlled command centers, were brought into the car factories, chemical plants, and steel mills where mass-worker militancy had been strongest. Even where these experimental systems were so expensive as to be, in strictly economic terms, inefficient, their labor-eliminating capacity was frequently critical in crushing the most advanced elements of working-class organization.[6] Today, however, such systems are being experimented with throughout all sectors of work, from nursing to pizza-making to light-house-keeping; while the fully implemented versions are still futuristic islands in a sea of more traditional work methods, their discrete elements are widely disseminated, and the tendency toward integration evident.

The labor-reducing capacities of these "new production systems," in their advanced forms, are truly remarkable. Management in the most sophisticated Japanese automated factories claims to have nearly halved its work force, while simultaneously tripling production; in California, a plant capable of manufacturing a billion dollars worth of computers a year requires only five manual-assembly workers and fewer than one hundred other workers, mostly engineers.[7] Although such levels of automation are only the latest step in capital's protracted substitution of technology for people, it nonetheless seems that computerization does mark a watershed in the relation between worker and machine—a quantum leap in the predominance of fixed over variable capital, dead labor over living. Indeed, with the advent of new production systems, we surely reach that horizon long-ago foreseen by Marx where capital attains its "full development" with the creation of "an *automatic system of machinery* . . . a moving power that moves itself . . . consisting of numerous mechanical and intellectual organs, so that the workers themselves are cast merely as its conscious linkages."[8] When he wrote these lines Marx undoubtedly had in mind the smoky clangor of a nineteenth-century industrial site. Yet they apply with redoubled accuracy to the sterile, silent informational systems with which twenty-first-century capital is now attempting to solve its longstanding "labor problem."

In North America, this solution for many years seemed to be succeeding remarkably well. Throughout the 1980s, capital's massive investments in advanced technology played a vital role in crushing strikes. From airports, where the availability of new levels of automation was critical to the success of the Reagan administration in firing air-traffic controllers, to the meatpacking industry, where extensive technological restructuring reached a climax with the defeat of the two-year strike at Hormel, new production systems repeatedly helped capital prevail in workplace conflict.[9] In other sectors, such as the auto industry, fear of losing jobs to new technology quelled militancy and contributed to a climate of demoralization and defeat in which once-defiant industrial unions acquiesced to concession bargaining and co-operation with management. Capital's technological superiority appeared to be absolute.

Yet although robotized systems have significantly depleted the ranks of the industrial working class, it is clearly false to suggest that cybernetic systems entirely eliminate capital's need for labor. Despite the dreams of wide-eyed digital futurists, the total liquidation of human intelligence from the production process has proven a singularly intractable project. In many manufacturing sectors computerized automation has made production dramatically "leaner." Yet the full "lights out" scenario—in which the final worker replaced by a robot turns out the lights and exits the building, leaving behind a smoothly running automated darkness—remains an unattained goal. And even in the rare plants that approach such scenarios, the operations of such so-called "workerless factories" rest on a surrounding infrastructure of activities—from maintenance to marketing—still dependent on myriad human agents.

If one examines the last quarter century of high-technology innovation, a paradox appears. While in the factory wage-labor has been relatively reduced, in the larger social arena it has, if anything, expanded. Ever-wider areas of human activity—from education to meal-making—are being more widely and intensively subsumed within the capitalist organization of work. This is what is usually described as the rise of the "service sector." As we saw in chapter 2, this phenomenon has long been central to the analysis of information society proposed by theorists such as Daniel Bell and Alvin Toffler. In their hands, however, the process has been so mythologized—as a sublimation of sweaty blue-collared proletarians into suave white-collared professionals— to amount to a near-total mystification of the actual recomposition of the postindustrial work force.[10]

For a more penetrating analysis, it is useful to look back for a moment to Marx. In the *Grundrisse,* while emphasizing capital's relentless drive to replace humans with machines—a trajectory that is of course central to his whole vision of crisis and revolution—Marx nonetheless does not speak of the total elimination of labor by automation. Rather, he refers to its transformation into the "conscious linkage" within a technological system. "Direct production"—the "hands-on" transformation of raw materials into finished products—would be increasingly automated. Living labor would be not so much "included within the production process" but relate to it "more as watchman and regulator"—a description that neatly covers the sort of invigilating and trouble-shooting functions for which human beings are still found indispensable, even in the most sophisticated of new production systems.[11] Moreover, Marx implies, there would remain a field of activities *indirectly* necessary for production, in which human involvement would remain—or become increasingly—crucial. This indirect labor would entail two main types of activity: on the one hand "scientific labor" and on the other "social combination."[12]

Later, in chapters 8 and 9, I will discuss the problems that Marx saw these developments creating for capital. But at the moment I simply want only to suggest that in these cryptic phrases, "scientific labor" and "social combination," he offers some orientation toward analyzing the notoriously amorphous service sector. Applying his lens, we can discern within the category two distinct groups, both of whom are now being systematically assimilated into the capitalist organization of work. On the one hand, there is "scientific labor"—the scientists, programmers, engineers, and designers celebrated in information-society theorists' portrayals of the "knowledge workers" of the future. But on the other, there are the multifarious workers concerned with the tasks of "social combination," involved in facilitating and sustaining the matrix of everyday human intercourse and interaction within which even the most automated production remains obdurately embedded. These tasks of "social combination" comprise some relatively well paid, creative, and prestigious jobs, especially in the media and communications sectors. But they also include the legions of retail clerks, cleaners, janitors, security guards, and fast-food servers who make up the bulk of employment in the information economy.[13] Numerically much more significant than the "scientific labor" they support, but enjoying only a fraction of the rewards, these latter workers constitute the new high-technology proletariat.

Relative to the old industrial working class, concentrated in its factory bastions, these new forms of "social" and "scientific" labor power might appear unlikely contenders in class struggle. They are *disorganized*, insofar as they come into being outside the orbit of the traditional workers' movement, toward whose symbols and institutions they are often indifferent or hostile. They are *dispersed*, across an enormous variety of spatially separated and qualitatively diverse sites. And they are *divided*, in a multitude of ways, but particularly by the lines separating the relatively privileged cadres of "scientific labor" from the superexploited "social" labor that sustains it—a division frequently reinforced by ethnicity and gender. Nevertheless, the presence of these postindustrial laboring subjects, even in the midst of a world of artificial intelligences and information highways, constitutes an ominous spot on management's dream of an immaculate techno-system freed from the insubordinate possibilities of human presence.

In the last few years there have been signs that the postindustrialists' requiem mass for class struggle was premature. Since the early 1990s, strikes and organizing drives in both the United States and Canada have seemed to signal an unexpected revival of labor militancy. In 1996, the number of hours lost to strike action in the United States, after dropping precipitously for decades, began to rise again, although only very slightly.[14] More significant than such quantitative measure, however, were certain qualitative aspects of the new insurgencies. For they were no longer predominantly "mass worker" actions, situated in the classic industrial centers of working-class power, but frequently arose outside the factory, in the diffuse, social labor of the service sector. The continuing militancy of many traditional industrial communities—one thinks of the three-way strike by rubber, sugar, and vehicle-manufacturing workers in Illinois's "class war-zone"—cautions against any quick farewell to traditional terrains of class war.[15] But the wave of labor restiveness also passes through new territories. Often it involves workers at the bottom of the hierarchy of labor power, whose networks of support are founded as much in gender and ethnicity as in the traditions of the labor movement. While established trade unions may provide the organizational form, and sometimes real support and leadership, for these insurgencies, such rebellions constantly bubbled up at a local level below and sometimes in opposition to the upper levels of union bureaucracies, challenging established structures and strategies, and reshaping them from below.[16]

For an example, one need look no farther than Silicon Valley, historic center of the U.S. computer industry.[17] The most well known aspect of the valley's labor history is the emergence of the new strata of highly skilled technical workers—engineers, software designers, and programmers—central to the making of digital technology. Mostly male, mostly white, very highly educated (the valley has the largest concentration of Ph.D.'s and engineers in the world), these are the quintessential "knowledge workers" needed by an industry whose profit depends on a constant stream of innovation. Highly paid, frenetically creative, technologically compulsive, often enjoying substantial entrepreneurial opportunities, this elite work force has been the subject of innumerable adulatory media reports, making their exploits an important part of the information revolution's romantic mythology.

There is, however, another, far less glamorous, face to work in Silicon Valley—that of the janitors, landscapers, cafeteria staff, and microchip assemblers who provide the indispensable support for this technological creativity. Drawn largely from immigrant or ethnic minority communities, these workers—many of them women—are employed at low or minimum pay, outside union organization, without health insurance, maternity benefits, or recourse against sexual harassment. The valley's prestigious high-tech companies, such as Apple, Intel, Hewlett Packard, Oracle, and IBM, could not function without this labor force. But the major corporations try to distance themselves from unsightly superexploitation by a system of contracting-out that allows disavowal of responsibility for working conditions and wages. The workplace segregation between the high-end knowledge workers and low-end service labor is reinforced by residential patterns that divide the valley into ethnically sorted zones. Although Silicon Valley is situated in the most prosperous county in the United States, aggregate wealth on closer examination decomposes into a scene of postindustrial segmentation where "the First World meets the Third in a weird melange of high technology and misery."[18]

For many years, the dispersed nature of the Silicon Valley service work force, its high turnover, and its divided ethnic composition led the U.S. labor movement to deem it unorganizable. In the early 1990s, however, following a wave of worker complaints, Justice for Janitors, an organization of the Services Employees International Union, began a series of campaigns fighting for union recognition, pay raises, and settlement of sexual harassment grievances.[19] These campaigns used a wide variety of tactics—strikes, picket lines, demonstrations, advertisements, leafleting campaigns, hunger strikes—which, although all part of the historic repertoire of the American labor movement, were conducted with an energy and determination that contrasted sharply with the submissive defeatism prevailing in many major trade unions.

Moreover, in some respects the Justice for Janitors campaigns went beyond familiar models of shop-floor activism. They made connections between workplace conditions and issues of race and gender discrimination, and forged alliances with feminist and ethnic community organizations. Because Silicon Valley workers are often directly or indirectly exposed to the highly toxic chemicals used in microchip manufacture, they were also on occasion able to link labor struggles with those of environmental and housing activists challenging the computer industry's poisoning of the local environment through ground, air, and water pollution.[20]

The scope of the Justice for Janitors campaign took employers aback. The turning point in the mobilizing drive at Apple, for example, came when workers threatened to take their campaign into the classrooms of California schools and universities—a major market for Macintosh computers. The result was small but significant victories at a number of high-tech companies— union certifications, pay raises, settlements of harassment cases. Labor councils in Silicon Valley are now speaking about more extensive campaigns that will address not only the terrible conditions of "service" workers, but also some of the grievances of the "scientific" work force, such as frenzied schedules and lack of job security. These new campaigns will, one organizer says, involve "everybody from janitors to technical writers to software gypsies and testers to quality assurance engineers. . . . The janitors were just the first among the contingent workforce. . . . When we talk about doing windows in this valley, we're not just talking about the janitors who clean them, but the software engineers who write them."[21]

The revolt in Silicon Valley—mecca of an industry whose products are specifically intended to free capital from dependence on troublesome humanity—presents an extreme irony. But it is by no means exceptional. During the 1990s, North America's restructured, post-Fordist, informational capitalism has been riddled with unanticipated conflicts. The battle in the computer industry has spread to other areas in the United States and now involves organizations such as the Southwest Network for Environmental Economic Justice, a coalition of over fifty grassroots organizations from Texas, Oklahoma, New Mexico, Colorado, Arizona, Nevada, and California fighting toxic pollution and poor working conditions.[22] In Los Angeles, the same communities that rose up in the 1992 riots generated a surge of labor militancy sweeping the hotel, fast-foods restaurant, and dry-walling sectors.[23] In Las Vegas, janitors and cleaners took on the giant high-technology gambling and entertainment complexes of MGM.[24] Along the United States–Mexico line, women workers fought a mobile garment industry that migrated sweatshop operations across borders.[25] Delivery workers of a partially reformed Teamsters union won a historic victory against United Parcel, at the heart of the increasingly important high-tech communication/transportation industry. In Canada, protests against labor legislation and austerity programs from the Ontario provincial government produced an unprecedented series of rolling one-day general strikes in urban centers, while Quebec unions opened a major drive to organize the youth labor in the McDonald's fast-food chain. Elsewhere, the decade saw major workplace battles waged by airline attendants from Alaska to Miami; newspaper workers in San Francisco and Detroit; teaching assistants at Yale and other universities; and nurses and education workers resisting public-spending cutbacks from New York to Vancouver.[26]

These movements are, in terms of the types of workers involved, extraordinarily diverse—so much so that they at first seem to defy generalization. But this diversity is, in itself, an important defining feature. For these are the revolts of a collective laboring subject that is no longer a homogeneous and concentrated industrial proletariat, but rather heterogeneous and connective, performing the innumerable social activities necessary to maintain the flow of production within capital's increasingly complex and extended techno-systems. And this new positioning of labor gives new organizational form to its uprisings. Situated as the interstitial "conscious linkages" within capital's automated and elaborated chains of production, rebellious workers have been compelled increasingly to seek "conscious linkages" with one another. Recognizing the extreme vulnerability of isolated fights, the new labor movements are frequently to be found expanding the scope of struggle beyond the immediate site of conflict, following the increasingly comprehensive and social scope of capital's own circuits. This tendency takes a variety of forms: increased efforts to organize by sector, rather than in single plants; cross-sectorial connections, such as linkages between striking workers in the telecommunications and garment industries, or the mutual support between airline attendants, construction workers, and bus drivers; and increased resort to consumer boycotts and "corporate campaigns" hitting at every aspect of an employer's investments.[27]

Even more important, workers' organizations have entered into experimental coalitions with other social movements also in collision with corporate order, such as welfare, antipoverty, students, consumer, and environmental groups. The result has been new oppositional combinations. Thus, striking telephone workers join seniors, minorities, and consumer groups to beat back rate hikes, or unionizing drives in the ghettos of the fast-food and clothing industries intertwine with campaigns against racism and the persecution of immigrants.[28] Such alliances are fraught with difficulties and can easily disintegrate. But they expand the boundaries of official "labor" politics, so that the agency of countermobilization against capital begins to become, not so much the trade union, defined as a purely workplace organization, but rather the "labor/community alliance," with a broader, social sphere of demands and interests.[29]

Discussing these developments, Kim Moody (who is connected to the Detroit journal *Labor Notes*, an important node in the U.S. circuits of labor dissidence) suggests that the North American labor movement in the twentieth century has gone through three phases of organization—from "craft" unions, to "industrial" unions, to an emergent "social-movement unionism."[30] For Moody, social movement unionism, the vital current of today's struggles, is an activism whose scope expands beyond the factory gate into a wider arena, overflowing the limits of strictly workplace struggle to include demands for broad social and economic change and alliance with other movements. It is a form of struggle in which "unions provide much of the economic leverage and organisational resources, while social-movement organisations . . . provide greater numbers and a connection to the less well organised or positioned sections of the working class."[31]

This revival of worker militancy in North America coincides with similar, but stronger, tendencies in Europe during the 1990s—the French general strikes of 1995–96, the Italian "Coba" movement, and wave of labor unrest in Britain and Germany.[32] As we have seen, Negri and other autonomist Marxists, writing predominantly in this European context, have also theorized three cycles of class struggle and recomposition: from the "professional" worker of the late nineteenth century, to the "mass" worker of capital's Fordist era, to the emergent "socialized" worker of the current, post-Fordist, informational period.

What Negri and Moody are both suggesting, in different idioms and from different national settings, is that capital's high-technology decimation of the industrial working class does not amount to the end of class struggle. The new production systems have partially chased waged labor out of the factory. In doing so, however, capital has diffused its organization of labor power through society at large. These conditions of dispersal initially appear as the depletion and fragmentation of traditional class solidarities. But they can be reconstituted as conditions of new scope and interconnection. Contrary to postindustrial fantasy, workplace conflicts are not dissolved in the new digital environment; but they *are* decenterd and recomposed with other arenas of activism. However, to understand this dynamic more deeply, we must go beyond the workplace and into the proliferating confrontations between popular movements and the capitalist state.

Reproduction of Labor Power I: The Panoptic State

If labor power is to be available for exploitation it must constantly be reproduced. That is to say, people must be socialized, schooled, trained, prepared, and held in readiness for work, in the quantities and qualities required by capital. Marx noted that "the maintenance and reproduction of the working class remains a necessary condition for the reproduction of capital," but, reflecting both the laissez-faire political economy of his era and the blind spots of his gender, he omitted this process from his detailed analysis of capital's circuits, declaring that "the capitalist may safely leave this to the worker's drives for self-preservation and propagation."[33] Over the course of the twentieth century, however, other Marxists, particularly those within the autonomist tradition, have pointed out that in the course of its development capital has increasingly been unwilling, and unable, to take this reproductive activity for granted. To ensure the proper supply and disciplining of the minds and bodies required for work, it has been compelled to extend systematically its control over society as a whole—a control mediated through the Leviathan-like structures of the state.[34]

Thus the first half of the twentieth century saw all advanced capitalist societies, to varying degrees, respond to the threat of militant working-class movements with a shift from the "Rights State"—where the activity of government was restricted to securing the conditions for the free-market—to the "Planner State," in which the state managed the reproduction of labor power through a vast array of schools, hospitals, welfare offices, and other institutions. Although this transition was set in motion to ward off revolutionary dangers, it also laid the basis for a new stage in capitalist growth. For the schools, health care systems, and various forms of social payments of the Planner State cultivated the increasingly healthy, educated, and peaceful forms of "human capital" necessary for intensive technoscientific development of the Fordist era. The advent of what is generally known as the welfare state represented an ingenious social compromise crafted by reformist business interests, social-democratic politicians, and trade-union leaders, which constituted both a real victory for workers—in terms of a general betterment of living conditions—and a careful containment of that victory within the overall parameters of continuing capitalist accumulation.

In the 1960s and 70s, however, this uneasy settlement began to disintegrate. Movements of workers, the unemployed, welfare recipients, students, and minority groups began to make demands on the vast system of social administration that transgressed the limits set by capitalist logic. They demanded, and sometimes won, increases in social expenditures going beyond those compatible with business's strictly rationed plans for improving its work force. In certain cases, such movements were also able to gain a degree of local control over the administration of social programs so they were, in effect, running the state apparatus from below.[35] These encroachments were intolerable for North American and European capital, whose rate of profit was already being squeezed by shop-floor militancy and international competition. Its response—part of the larger neoliberal restructuring offensive—was to repudiate the postwar social contract and dismantle the Planner State, destroying what it could no longer control.

The new regime of governance, whose full appearance is usually identified with the electoral victories of Reagan and Thatcher, has a double face. On the one hand, privatization, deregulation, and cutbacks systematically subvert the welfare state, slashing the social wage, weeding out enclaves of popular control, and attacking any of labor's protections from the disciplinary force of the market. The costs of reproducing labor power increasingly devolve back onto individuals and households. This shift becomes ever more important to capital as corporate downsizing and automation ejects more and more workers from production, thereby swelling the ranks of the unemployed and impoverished, increasing welfare roles, and diminishing tax revenues. On the

other hand, those aspects of the state necessary to the protection of accumulation—such as the security apparatus or subsidization of high-technology
investment—are strengthened. There thus appears the paradoxical neoliberal
combination of what Andrew Gamble terms "the free market and the strong
state."[36] In what autonomists term the "Crisis State," the governmental apparatus is dissolved in so far as it serves popular purposes, but maintained or
enlarged as the coercive and administrative arm of capital.

Computers, telecommunications, and biotechnologies are embedded at the
very core of the Crisis State, as both means and end. Social programs are cut
to free revenues for assistance to corporations making huge investments in
high technology. Public funds are channeled to private purposes either directly
through subsidization or indirectly through tax breaks. High technology is,
in turn, used to effect cuts to welfare programs that start to be administered
through increasingly precise and omnipresent digitized systems. The delivery of social services is increasingly automated—for example, by computerizing the making of welfare or unemployment insurance claims. This process
not only cuts staff costs, but also reduces payments by imposing daunting
electronic hurdles that have to be surmounted by precisely that sector of the
population least equipped to handle them, and by allowing the digitalized or
biometric monitoring of claimants.

As whole strata of the population are cut off from support, potential social
disorder is kept in check by the technologically intensive policing applied
against the poor, indigent, and ghettoized. Around those convicted of transgression, the web of informational control tightens inexorably. Prisons, as
Foucault so forcefully pointed out, have long been cutting-edge sites for the
development of surveillance techniques. What is not always remembered is
that the original panoptic apparatus that Foucault discussed in a carceral
setting was at first designed for use in a factory setting, as an instrument of
capitalist work discipline.[37] In today's high-technology penitentiaries, however, carceral and the capitalist logic come together. In an increasing number of privatized or semiprivatized U.S. prisons, inmates are put to work for
private corporations, often on electronic data-entry jobs or other forms of
telework, in a process that uses high technology to fuse the Crisis State's drive
to minimize social expenditures with the corporate imperative to cut labor
costs to the bone.[38]

The net tendency is toward a return to the social conditions of the nineteenth century overseen by the technologies of the twenty-first. However, this
regression, bringing with it huge increases in poverty rates, social polarization, and general human suffering, has catalyzed opposition. In North
America, immiseration erupted into rage in the Los Angeles rebellion of 1992,
the most violent urban insurrection in the United States since the mid-nineteenth century. As Mike Davis notes, south-central L.A., a "housing/jobs
ghetto in the early twentieth century industrial city," is now "an electronic

ghetto within the emerging information city"—a "data and media black hole, without local cable programming or links to major data systems."[39] The rioters came from the ranks of the un- and under-employed, in a community whose traditional sources of employment in the aerospace and automobile industries had been gutted through automation and global relocation. This population, dependent on the scanty welfare, casualized service work, or criminal industries that constitute the underside of the information economy, was, on an everyday basis, subject to a regime of draconian police surveillance and brutalization—a regime whose systemic violence, publicly exposed in the videotaped beating of Rodney King, finally triggered a mass explosion.

Its outbreak, in the same city that saw the Watts riot of 1965, was a stunning testimonial to the collapse of a quarter century of capitalist reformism. Framed by the mainstream media simply as an issue of race, the uprising was in fact, as Mike Davis observed, a "multicultural bread riot" involving Latinos, blacks, and whites.[40] Moreover, although the riot was a spontaneous eruption of despair and anger, it was by no means a blind, mindless event, as authorities attempted to represent it. A few days after the uprising, there appeared the "Bloods/Crips Proposal for LA's Face-Lift," a radical, visionary plan for the renewal of the city produced by the infamous street gangs.[41] This document, almost entirely ignored by mainstream media, made extensive proposals for reconstructing the urban environment, and for the introduction of governmentally funded educational, health, employment, and even law enforcement measures to reverse the disintegration of community.

Although the conditions of south-central Los Angeles gave the 1992 rebellion its singularity, it would be wrong to see it simply as a "one-off" event. From the late 1980s to today the intensifying destruction of social safety nets has brought into being a variety of new "poor people's movements," ranging from the squatters of Homes Not Jails, to End Legislated Poverty in Vancouver, to the encampments of homeless in New York.[42] For example, Food Not Bombs is a group whose activities in San Francisco led to over seven hundred arrests from 1988 to 1994. In addition to running the on-street soup kitchens that have aroused the ire of municipal government, it operates its own radio network, based largely on low-watt broadcasting, produces its own audio tapes, and has a World Wide Web site. Through these channels it disseminates information excluded from the mainstream press about the police harassment of its programs and the structural causes of poverty.[43]

In Toronto, a coalition of trade unionists and antipoverty groups has taken aim at a contract between the Ontario government and a private company, Andersen Consulting, to automate the delivery of welfare services. The coalition argues that this contract aims to eliminate social services staff (Andersen gets a "bounty" for each job cut) and simultaneously to make the system increasingly inaccessible to claimants. The coalition has publicized Andersen's record of cost overruns and unfulfilled promises on similar con-

tracts elsewhere in North America, traced its involvement in the privatization
schemes of authoritarian governments from Russia to Nigeria, and exposed
its links to the military industrial complex. In addition to holding marches,
picketing, and civil disobedience actions at the corporation's offices, the
"Andersen Conversion Project" is also bringing forward proposals for the
transformation of the high-tech company to more socially constructive purposes.[44] In such movements, antipoverty groups, trade unionists, and other
social movements take the first steps to turn the technologies developed at
public expense back against the panoptic alliance of state and corporate power.

Reproduction of Labor II: Capital's Biopolitics

The Crisis State's regime of high-technology control is not restricted to the
policing of welfare lines and inner-city streets. It extends further, into homes
and hospitals, where the informational restructuring of capital has been intimately associated with new interventions into the reproduction of labor
power at its most basic levels—motherhood, birth, and, indeed, the basic
biological constitution of human beings.

As discussed in chapter 4, autonomist Marxists have since the 1970s argued that capital benefits from the unpaid reproductive work of women. The
classic nuclear family paired the waged male worker and unwaged housewife
in a relation where the role of the latter was to maintain, repair, and reproduce the labor power of the former. The male worker's wage thus commanded
unrewarded labor time not only in the factory but also in the home. This
conjunction of masculine domination and capitalist exploitation was challenged by the feminist revolt of the 1960s and 70s on a multitude of fronts;
in the exodus of women from unpaid domestic labor in search of waged work,
in demands for "wages for housework," in the rejection of the various medical and psychiatric controls placed over housewives. Among the most important of these struggles was that over abortion rights. Women asserted control over their own fertility and repudiated a "natural" fate as the unwaged
reproductive laborers of the social factory.

The reconsolidating of "family values" and the discrediting of feminism
were thus logical parts in the neoliberal offensive of the 1980s. Limitations
on and recriminalizations of abortion services, legal regulation of the prena-
tal conduct of "unfit" mothers, and experiments in the sterilization of welfare mothers by mandatory Norplant implants were all crucial aspects of
Reaganite and Thatcherite regimes.[45] What is often not fully recognized is that
these apparently "cultural" or "ideological" aspects of the Crisis State were
closely bound up with its economic policies. For as welfare services are de-
graded under the austerity regime of the Crisis State, the resumption of the
traditional female role as a "voluntary" caregiver for the young, sick, and
elderly becomes critical to prevent total social disintegration. Although the

means to this end include both "pro-" and "antinatalist" tendencies, the common theme of these interventions is enhanced state control over maternity—control exercised to ensure the "proper" management of procreation and to reconstruct the household as a costless, reliable site for the reproduction of labor power.

At the same time, however, the most advanced sectors of knowledge-based capital have been experimenting with an alternative system of maternal control—one based on biotechnologies. Already, *in vitro* fertilization, amniocentesis, embryo selection, and artificial insemination are becoming the instruments for an extraordinary experiment—the conversion of motherhood into a domain for the direct extraction of surplus value. As feminists such as Maria Mies and Kathryn Russell have argued, the commercial application of such techniques drives female "labor power"—in the procreative sense—toward the condition of abstraction, divisibility, and alienation traditionally experienced in industrial work.[46] Reproductive engineering applies a technological deskilling strategy, classic in form but unprecedented in intensity, comprehending both conscious knowledge and corporeal capacity, detaching, permuting, and recombining the various moments of pregnancy until the unifying factor governing the conception, gestation, and delivery of a child is no longer maternal but managerial.

This is clearest in the so-called "surrogate mother" business—the ultimate in female service sector labor—in which poor women are, through an entrepreneurial intermediary, paid by rich clients to undergo either artificial insemination or *in vitro* fertilization and carry and bear children.[47] But such obviously exploitative repro-tech arrangements represent the extreme of tendencies evident even in more seemingly benign uses. For example, women who voluntarily attempt *in vitro* fertilization not only pay for the service, but also, in a complex and painful process of self-surveillance and constant testing, often knowingly or unknowingly provide the surplus material—"excess eggs"—required for further commercial experimentation.[48]

Antiabortion crusades and reproductive technology businesses seem antithetical, one resting on a sacralization of procreation, the other on its utilitarian industrialization. And there are real contradictions between these strategies, and between the factions of capital that promote them. But the two strategies of control are also intimately connected. Both counter the reproductive autonomy fought for by women. The "family values" campaign cancels "choice" in an outrightly reactionary manner. But the corporate biotechnologists coopt it as the watchword for the commodification of procreation. Just as in production, capital combines sweated labor and robotics, so "family values" and genetic engineering are poles in a single overarching regime of reproductive control, with biotechnological options commercially available to the rich, and surrogate mothers drawn from the ranks of the poor.

In the very near future, moreover, reproductive technologies promise a spectacular convergence with genetic engineering—the splicing, cutting, and recombination of the genetic code. After a gradual postwar development, founded in North America upon heavy state investment in basic research, these technologies have since the crisis of Fordism in the 1970s undergone an extraordinary acceleration in commercial development as part of capital's overall search for postindustrial sources of investment.[49]

The capacity to rewrite the "code of life" has been applied to agriculture, food production, and plant breeding to produce new strains of plants, new forms of food, and new types of fertilizer.[50] Increasingly, however, genetic engineering has in its sights direct control over human behavior. As Gottweiss argues, the burst of state and corporate interest in biotechnologies during the crisis of the social factory arose because in addition to yielding traditional economic benefits, it was conceptualized as "a potential contribution to a broader social stabilisation, mainly by its expanded capacity to control behavior and bodies."[51]

Today, these ambitions crystallize around the Human Genome Project, the U.S. state-sponsored attempt to map and sequence all the DNA of a "normal" human prototype—a project comparable in cost and scope to the space program of earlier decades.[52] This project is generally promoted as a means of curing hereditary diseases. Eventually, this dream may be realized, and, if it is, the biotechnology industry anticipates lavish profits from the creation of new ways to improve health, longevity, and pleasure for those who can afford them. However, it is important to recognize that currently, genetic engineering's main achievements are neither therapeutic nor even diagnostic but predictive, allowing the probabilistic identification of conditions for which no known remedy presently or foreseeably exists.[53]

Such techniques offer corporate and state managers a way, not of healing, but of targeting subjects with an alleged predisposition to costly disease.[54] The identification of "hypersusceptible" workers with supposed genetic sensitivity toward toxic chemicals or radiation has in the United States already become a significant source both of employment discrimination and of exclusion from health insurance coverage.[55] It also provides an alibi for failure to eliminate such pollutants, which become redefined not as social hazards but as problems of individual predisposition, capable of being handled by genetically "subsensitive" labor. Extensive genetic screening holds out the promise of comprehensive, DNA-level quality control over the reproduction of labor power, control aimed not at the cure of disease but at the discarding of potentially unproductive, oversensitive, or expensive units.[56]

As the Human Genome Project generates the raw data necessary for new "breakthroughs" to enhance the human body, the combination of genetic screening with reproductive technologies offers prospects for the renewal of a eugenic agenda once thought to have been discredited with the fall of fascism. However, the commercial thrust behind the biorevolution means that such a program would probably have a different "feel" from its historical predecessors. As employment possibilities become increasingly dependent on a clean genetic profile, or even on possession of certain bioengineered enhancements, positive and negative selection will be left to the survival instincts and pocket book of individuals. People may biotechnologically reproduce the labor power of themselves and their children in the most saleable form affordable, in the context of an increasingly stratified, privatized, and expensive medical system—a development whose potential is already apparent in the burgeoning market for synthesized human growth hormones, silicon breast implants, cosmetic surgeries, performance-enhancing drugs, and transplantable hearts, livers, kidneys, and corneas.[57] Capital will thus move toward establishing a hierarchy of labor powers in which the various classificatory grades are distinguished not simply by education and training, or according to traditional discriminations of gender and race, but according to fundamental bodily modifications.[58] As Peter Linebaugh has pointed out, in origin, the term "proletarian" designated someone who has no function but to reproduce himself.[59] In Marxist usage, this has conventionally been understood as a person who has nothing to sell but his or her labor power. Soon, however, it may be applied to someone whose only economic assets are gestational capacity and genetic heritage.

However, the emergent neoliberal biopolitics has encountered widespread resistance. In North America, much of this has centered on the revival of the women's reproductive-rights movements. In many cases, its nucleus is the network of abortion clinics and women's health centers, whose defense, both from the harassment, firebombings, and assassinations by the right-to-life movement and from the cutbacks by neoliberal governments, has formed a focus of activism. Women have also attempted to enlarge their own technological control over procreation, through campaigns such as that waged in the United States for access to the abortion drug RU 486. However, largely through the influence of poor women and women of color, the antiabortion movement has undergone a strategic reorientation, sometimes described as a shift "from abortion to reproductive freedom."[60] An earlier emphasis on individual choice has, at least in some sectors of the movement, been gradually replaced by an emphasis on securing the "social conditions necessary for autonomous choice," on the provision of adequate health services, housing, and wages and welfare for women, and on winning control over the research and availability of medical technologies, including opposition to both compulsory fertility and eugenic sterilization.[61]

One aspect of this expanded agenda has been an intensive critique of the repro-tech industry. International feminist alliances such as the Feminist International Network of Resistance to Reproductive and Genetic Engineering (FINRAGE) have exposed the deceptive success rate claimed by the *in vitro* fertilization industry, its exploitation of female labor, the misogyny of sex-selection amniocentesis, and the eugenic potential of the new technologies.[62] They have argued that the "choices" offered by the biotechnologists actually erode female freedom because, as Sue Cox puts it, they "close off women's abilities to refuse various kinds of technological intervention."[63] In Canada, the attempt by the Royal Commission on New Reproductive Technologies to suppress such lines of critique exploded into public scandal.[64] Other points of struggle have involved indigenous peoples, in both North and South America, concerned with the ramifications of the Human Genome Diversity Project (known as the "vampire project"), which has sampled and patented human cell lines from endangered aboriginal communities.[65]

Other groups, with different concerns, have found themselves on a similar collision course with the neoliberal administration of health. In the face of alliance between a state apparatus committed to the reduction and rationing of health care, and a burgeoning, profit-oriented medical-industrial complex formed at the intersection of transnational medical, pharmaceutical, agricultural, insurance, and computer corporations, there have appeared what Patrick Novotny, writing of the environmental justice activism, calls movements of "popular epidemiology."[66] These movements often involve groups marginalized by the industrial-medical complex—people of color, women, gays, and lesbians. They challenge established expertise, demand additional allocations of funding, question the priority of profits over people, reappropriate popular capacities for research, and often seek systemic rather than palliative answers to the causes of ill-health.

A striking example is the extraordinary self-organization of the anti-AIDS movement. In the face of initially inept and callous governmental responses to the HIV epidemic, organizations such as ACT UP and Project Inform attacked the state's underfunding of research and its subordination to commercial purposes. They also reshaped research agendas; amassed and circulated immunological and virological information, both by computer networks and other means; investigated "alternative" treatments; set up guerrilla clinics, smuggling rings, and buyers clubs; clandestinely manufactured commercially patented drugs; and showed enormous sophistication in video-activism and other forms of cultural agitation.[67] Although these movements on occasion cooperated with pharmaceutical companies, they simultaneously criticized these companies unsparingly for either ignoring AIDS research or attempting to extract superprofits from new treatments. These points were underlined by dramatic demonstrations and occupations against companies such as Hoffmann–La Roche, Burroughs Wellcome, Kowa Pharmaceuticals, and Astra. Peter Arno and Karen Felden describe the most famous of such actions, the ACT UP invasion of the New York Stock Exchange protesting AZT price gouging: "Seconds before the 9:30 A.M. opening bell, the activists began to blare portable foghorns. . . . . Fake $100 bills imprinted with the words 'Fuck your profiteering. We die while you play business' were tossed to the traders below."[68]

Over the history of the anti-AIDS movement, these forms of activism, initially concentrated in the white, male, gay community, have become increasingly prominent in movements of people of color and women. In the process, AIDS has been recognized as a disease of poverty, primarily afflicting those whom the disintegration of social infrastructures, community networks, health care, and education render vulnerable. Anti-AIDS struggles have thus been connected to campaigns for improved public health funding, comprehensive medical insurance, and the reallocation of military spending.[69]

As Steven Epstein points out, anti-AIDS activism, which itself draws on the earlier example of the women's health movement, is part of a widening circle of popular mobilizations for the "democratization" of medical technoscience.[70] These movements include those of women seeking to establish causal links between breast cancer and industrial pollution; unions opposed to genetic screening and drug testing in the workplace; and green activists, farmers, and consumer groups concerned about the implications of artificially mutated foodstuffs. Alongside these single-issue movements, and sometimes intertwining with them in complex ways, are broader movements. These aim at preserving the medical services once guaranteed by the welfare state, as in various Canadian coalitions of hospital workers and community groups defending hospitals and clinics against cuts, or at actively extending the socialization of health care, as in the struggle over health insurance in the United States. All of these efforts run athwart the priorities of a state committed primarily to containing social costs, and a corporate logic focused purely on the profitability of life and death.

Reproduction of Labor III:
The Corporate-Academic Complex

At the same time that the Crisis State dismantles the social welfare system, it continues to maintain and enlarge the functions of government as a funding and coordinating agency for capital's technoscientific development. The demands of the information era mean that even as schools, hospitals, and social services deteriorate, business still—indeed more than ever—demands literate workers, carefully socialized technicians, and world-class molecular biologists and software engineers. An integral part of the transition to a post-Fordist model of accumulation has therefore been a major restructuring of public education, a restructuring that has nowhere been more dramatic than in North American universities.

Just as in the workplace, the restructuring of academia has unfolded through a process of revolt and recuperation. Thirty years ago, campuses from California to Paris were in tumult as the postwar generation of students—the first mass draft of the intellectually trained labor-power required by an ever-more socially organized and scientifically oriented capitalism—rose against the rigidities and atrocities of the Fordist regime. After the tear gas, the shootings, and academic purges, the neoliberal response was radical restructuring. Over the late 1970s and 1980s, rates of funding for university education in most capitalist economies were cut. Tuition fees and student debt were sharply raised, measures that, alongside a climbing unemployment rate and general economic austerity, chilled student protest, while academic programs seen as subversive, or simply as inutile to industry, were cut.

With campus unrest apparently quashed, conditions were set for a new, deeper integration of universities and business, one vital to the development of high-technology "knowledge industries."[71] The watchword was "corporate-university partnership." In this new academic order, basic research is sacrificed to applied programs of immediate benefit to the corporate sector. Research parks, private-sector liaisons, consultancies and cross-appointments with industry, and academic-corporate consortiums burgeon. Moneys subtracted from base operating budgets are reinjected back into programs of direct utility to high-technology capital, such as schools of communication, engineering, and business administration, and special institutes for computer, biotechnology, and space research. University administrators move effortlessly between interlocking corporate and academic boards. Enabled by changes in intellectual property laws to exercise ownership rights over patents resulting from government-funded grants, universities become active players in the merchandising of research results. Amidst this intensifying commercial ethos, the internal operations of academia become steadily more corporatized, with management practices mirroring those of the private sector.

This new rapprochement with academia has performed two purposes for capital. First, it has provided business with the facilities to socialize the costs and risks of extraordinarily expensive high-technology research, while privatizing the benefits of the innovations.[72] Second, it has subsidized capital's retraining of its post-Fordist labor force. Rising tuition fees devolve an ever-increasing part of the costs of education onto students and their families, effectively excluding from the universities those sectors of the population whose intellectual advancement is considered irrelevant to accumulation. Those that can pay for entry are trained, sorted, and socialized for the new information economy by increasingly vocational and technically oriented curricula that stress proficiencies in computer literacy at the expense of critical social analysis.

However, the belief that campuses were pacified now appears premature. Rather, the late 1980s and 1990s have seen the emergence of a new cycle of university struggles.[73] As Robert Ovetz notes, this wave of unrest stems from numerous different but interanimating sources.[74] Of central importance is the mounting economic jeopardy in which many students now find themselves. Higher education, rather than guaranteeing personal success, serves to create a standing reserve army of intellectual labor, from whom capital can cull the relatively small number of full-time employees required by the "knowledge economy." With rates of unemployment for college and university graduates high, many find that years of study ensure only lifelong and unpayable debt. These grim prospects have led to a spate of protests against tuition increases, student-aid cuts, and skyrocketing debt loads.

These concerns interweave with a web of other campus protests: against program closures; against commercial development of university lands; against involvement with corporate investment in authoritarian regimes such as those of China or Indonesia. Alongside these run demands by minorities and women for campus centers, day care, and programs of multicultural and feminist studies. The net result has been a slowly mounting campus turbulence, involving picket lines, demonstrations, occupations, national student strikes in Canada, and major confrontations between police and students on several North American campuses. As James Laxer observes, it is likely that in Canada more students were actually "on the streets" in political protest in the mid-1990s than in the 1960s and 70s.[75]

These student protests further overlap with an outburst of campus labor conflicts. Following the overall downsizing logic of post-Fordist capital, academic administrators demand that workers must do more with—and for—less. The one-time ivory tower witnesses an intensification in the rate of exploitation. This logic is usually visited first, and most severely, on the service workers—the clerical, administrative, janitorial, and cafeteria staff—who provide the indispensable infrastructure for the accumulation of intellectual capital. But it eventually arrives at the door of university instructors. Teachers experience increases in the pace and volume of work. A classic strategy of casualization decreases permanent hiring in favor of reliance on pools of sessional instructors and graduate students who form a contingent academic labor force subjected to chronic insecurity and lack of benefits, and required to exercise mind-bending flexibility in pedagogic preparation.

This speedup of academic production has produced a response that, while shocking to academic traditionalists, would come as no surprise at all to workers in, say, the auto industry. On many North American campuses, including some of the most prestigious, regular university faculty are now unionized—something that would have been unthinkable even a decade ago. Strikes by college instructors are no rarity. Graduate students are now an important constituency for labor organizing. Teaching assistants' strikes have spread across North American campuses, involving institutions as famous as Yale and scores of others.[76] The campus activism arising from this combination of factors has a very different flavor from that of the 1960s and 70s—which for most of the participants in today's rebellions belongs to a barely known and faintly mythic past. The revolts of thirty years ago recognized and resisted the movement toward integration of the university "knowledge factory" into advanced capitalism's military-industrial complex. But the fact that this assimilation was only partially completed, together with the relative affluence of the period, gave these uprisings a certain removal from the world of the labor market. Campuses could become temporary red ghettos or autonomous zones, but there was a fundamental divorce between what was experienced in these enclaves and the more general conditions of work and exploitation.

Today, the near-total fusion of academia with business, and the manifest subordination of education to the imperatives of the job-market, removes such relative freedom. But it opens the way for connections between both students and instructors and other waged and unwaged workers, making their conditions far closer to that of the rest of the labor force. The conventional distinction so often made between university and the "real" world, at once self-deprecating and self-protective, becomes less and less relevant. If students and teachers consequently lose some of the latitude of action relative privilege once afforded, they also become potentially participant in and connected to movements outside the university, movements for whom academia can therefore also become a node within the overall circulation of struggles.

Reproduction of Nature: Hazardous Wastes

To grasp the full scope of the opposition running around capital's circuits, however, it is necessary to look beyond struggles over work and wages, or even over welfare, health care, and education. Capital mobilizes technology to control not only labor, nor society as a whole, but also nature itself. It needs not just workers but also raw materials. As it reduces people to labor power, so it reduces nature to a resource: both exist to be used up. And as capital as far as possible avoids paying for the reproduction of labor power it exploits by devolving these costs onto households and communities, so too it minimizes its costs for the repair and restoration of the natural world by assuming that these processes can be left to the regenerative powers of nature. For all that Marx often participated in the scientific triumphalism of his century, he nonetheless clearly recognized the dangers of this trajectory when he spoke of capitalism "simultaneously undermining the original sources of all wealth—the soil and the worker."[77] Today, within a global vista of deforestation, desertification, dying oceans, disappearing ozone, and disintegrating immune systems, the cost of this exhaustive process has become all too apparent, and ecological issues constitute one of the main arenas in which popular movements confront corporate power.

An eruption of such green movements was one aspect of the general crisis of the Fordist social factory in the late 1960s and 1970s. As public awareness of the damage wrought by radioactive emissions, industrial wastes, and pesticide poisoning mounted, capital found its freedom to "externalize" costs by dumping poisons onto the surrounding communities challenged by unfamiliar forms of resistance. At sites from Diablo Canyon to Love Canal, environmental activists stormed fences and blockaded gates, disrupting industrial mega-projects as effectively as labor unrest on the assembly line.[78] In one of the most notable major reverses inflicted on a large-scale capitalist enterprise, development of the North American nuclear power industry was effectively stalled by the ever-rising costs of safety measures demanded by an anxious and angry public.[79] Across many other sectors of Fordist capital both the sheer depletion of easily accessible natural resources and the growing resistance to corporate despoliation began to constitute a serious barrier to accumulation.

The postindustrial leap into the world of computers, telecommunications, and biotechnologies was in part a response to this threat. As the arrival of high-technology on the shop floor was accompanied by promises of liberation from work, so too was it celebrated as the answer to the evils of pollution. Clean information systems would replace industrial smokestacks, recycle wastes, reduce the use of fossil fuels, eliminate paper from offices, replace motorcars with telecommuting, allow for better planning and preservation of natural resources, and dematerialize production into an innocuous flow of bits and bytes. These promises became integral to a succession of strategies—"sustainable development," "Third Wave environmentalism," "ecological modernization."[80] All these announce that technological surveillance, substitution, and surrogacy will deflect ecological apocalypse, enabling capital to manage the continued reproduction of nature by making a move from mining nature to remodeling it—shifting from stripping nature to synthesizing it, recreating a world of artificially generated resources to substitute for the gutted planet left in the aftermath of industrialism.[81]

The problem with such plans, however, is that they do nothing to touch the relentless corporate drive to expand the circle of production and consumption. A system in which the survival of each individual firm depends on its ability to enlarge its market, regardless of collective consequences, capital remains committed to, as Marx put it, "production for production's sake."[82] In practice, therefore, high technology has been used not so much to halt the destruction of nature but to increase the efficiency of the destroying agencies and to circumvent opposition to their activities. Automobile factories, petrochemical plants, and pulp mills have, amidst fanfare about green business, been made more energy-efficient (and hence more profitable)—but have not slackened their search for expanded (and hence more ecologically punishing) global markets. The advanced synthesis of substitutes for scarce natural materials has become a license for the anxiety-free liquidation of animals, minerals, and vegetables. Telecommunications and transport networks have dispersed pollution away from centers of activism and regulation onto the doorstep of those least likely to resist, making the shipment of toxic residues to urban ghettos, native reservations, or the Third World a post-Fordist sunrise industry. Moreover, in many cases, the capitalist development of so-called clean technologies, pursued under the same cost-cutting, profit-maximizing logic that produced enormities of industrial pollution, replicate the very patterns of ecological destruction they purportedly eliminate. The computer industry's use of toxic substances in microchip assembly, for example, has made Silicon Valley home to the highest concentration of hazardous-waste sites in the United States.[83]

Since the new technologies do not, of themselves, halt the devastation of the environment, they also fail to stop green countermovements. While schemes of high-technocratic resource management have played a part in coopting mainstream environmentalism, they have also unintentionally provoked new and radical opposition. Thus in the United States the intensification in the longstanding practice of dumping hazardous wastes—including postindustrial toxins—on the most impoverished and vulnerable sectors of labor has catalyzed the rise of an "environmental justice" movement in communities of color, traditional working-class neighborhoods, Native Indian lands, and regions of the rural poor.[84] Puerto Rican farm workers opposing pesticide poisoning, tenants associations fighting oil and petrochemical industries in Louisiana's "Cancer Alley," mothers battling incinerators in Latino neighborhoods of East Los Angeles, and Latino and African-American students of the Toxic Avengers coalition fighting the transportation of nuclear waste in Brooklyn have brought into being a new round of ecological struggles.[85] Often led by women—whose unwaged reproductive labor deals with the miscarriages, birth defects, and slow deaths created by corporate poisoning—and characterized by strategies that unite class, gender, and race issues, these groups have dramatically challenged the elitism of traditional environmentalism and engaged in a series of head-on confrontations with corporate power.

Generating its own programs of self-education, community research, and communication, the environmental justice movement represents an astounding flowering of popular science among the excluded and dispossessed. In many cases, sectors of the movement pursue objectives going far beyond the established limits of regulation. Their proposals for funds to support workers unemployed by the closing of ecologically destructive enterprises, restrictions on capital flight, elimination of the production of toxic substances, the development of a less-polluting transport system, community economic development, equitable distribution of cleanup costs, and international laws that protect the environment and workers are tantamount to demands for a radically new economic system.[86]

One of the most important aspects of this movement has been its efforts to overcome the rifts between working-class and ecological activism. Since the 1970s, capital, by playing off "jobs versus the environment," has constantly counterposed labor and ecological concerns, often successfully dividing red from green. However, as it becomes clear that high-tech business destroys livelihoods at the same rate as it destroys ecosystems, the falsity of this choice has become increasingly apparent. While the worker-green split remains virulent, in some sectors groupings of industrial and resource workers have developed their own environmental projects and entered into dialogue with ecological activists.[87]

A notable instance involves workers in that most unlikely of industries, automobile manufacturing. Throughout the 1980s an extraordinary coalition of black and Latino trade unionists and community groups in Van Nuys, California, successfully opposed the plan of General Motors to close its local car plant by threatening a boycott in the lucrative Los Angeles auto market.[88] In 1992, the "Save GM Van Nuys" campaign was finally defeated. However, it then underwent a dramatic metamorphosis, providing the nucleus for the WATCHDOG Organizing Committee—a group combating corporate air pollution of working-class neighborhoods, and seeking the conversion of the auto industry to clean, ecologically viable forms of production.[89]

These activists made connections with workers from the Caterpillar vehicle plant in Toronto, who, following an unsuccessful attempt to prevent closure of their plant by occupation, had entered into dialogue with environmental and antipoverty groups to devise a "greenworks" conversion campaign.[90] This alliance has in turn linked with Japanese workers from a joint Toshiba-Amplex high-technology enterprise, where resistance to plant closure led to an eight-year factory occupation.[91] During this time the Toshiba workers not only continued to manufacture and market high-tech media, educational, medical, and industrial operation systems, but ultimately started to redesign these products in order to meet their own criteria of social and ecological environmental responsibility.[92] They were supported in these efforts by the Japanese peace and antinuclear movements, for whom they produced portable loudspeakers for demonstrations, a citizens' Geiger counter, and another special radiation detector, funded by popular contribution, made for the victims of the Chernobyl disaster at half the cost of commercial systems.

Taken in conjunction with the movements against genetic commodification described earlier, such worker-green alliances introduce an extraordinary dimension to struggles against information capital. For what is a stake in such initiatives is nothing less than what Marx termed humanity's "species being"—its capacity to direct its own development as a biological collectivity.[93] The issue today is whether this shaping will be determined by capitalist command and market forces or by broader social logics. In this sense, proletarian struggles, which have, today, become struggles in which people strive to assert collectively a self-determining power over the development of the human species and its natural environment, potentially resume all the universalistic significance that Marx once attributed to them.

Movements fighting at different points on capital's circuit—against workplace exploitation, dissolution of the welfare state, or ecological despoliation—have begun to enter into alliances with each other, creating radical new combinations. The great difficulty facing these struggles, however, remains their fragmentation and separation. Occurring at different points within a vast social factory, and facing different facets of capitalist power, the obstacles confronting the coordination of demands and actions are often prodigious. Moreover, while these movements have a deep-level underlying interest in contesting the corporate subsumption of society, this common ground can easily be obscured by more local, but more apparent, contradictions between them—conflicts between, say, unionists and welfare recipients, or workers and environmentalists. Since capital constantly incorporates these local contradictions into its hierarchical organizations of control, both in and beyond the workplace, its capacity to divide and conquer, isolating points of opposition and turning them one against another, is truly formidable. Paradoxically, however, although informational capital enjoys extraordinary opportunities to overwhelm and disperse its opponents, some of the very technological instruments it deploys to these ends also assist countermovements to overcome this fragmentation. It is to this process that we now turn.

Circulation I: Interactive Media

An explosive proliferation of technologies of communication, from telephone, radio, and broadcast television, through fax, video camera, VCRs, cell phones, cable and satellite television to computer networks, is one of the most prominent features of advanced capitalism today. Some would say it is the most prominent feature. As Fredric Jameson has observed, there is a tendency to identify the benefits of the new media and the virtues of the free market, with each legitimating the other—new communications technologies being praised for accelerating economic growth, and the market exalted for promoting the free flow of information.[94] Yet there is another side to this dynamic. For among the new oppositional movements whose emergence we are charting, alternative uses of all types of advanced communication technologies are becoming a widespread and important element.

To examine this dialectic, it is necessary to move beyond analysis of the production of commodities, the reproduction of labor power, or the destruction of the environment and look at how capital circulates in the marketplace. If it is in the workplace that capital extracts surplus value, it is in the market that this value must be realized through the sale of commodities.[95] Marx repeatedly emphasized that capital had a tendency to integrate these two moments in its circuit, expanding the circle of consumption to match the growing volume of goods its produced, and decreasing the turnover time by accelerating the speed with which goods passed from production to consumption.[96]

In the course of the twentieth century, these requirements have become the basis for a massive project of social engineering—the creation of a consumer society. Capital discovered that, as work requires a laboring subject, so the market requires a consuming subject, a subject that needs what capital produces and believes that these needs can and must be satisfied in commodity form. And as in production it develops automatic machinery to reduce and control subjects in their tasks as workers, so in the market it finds instruments to target and direct subjects in their tasks as consumers—a mission performed by ever more sophisticated waves of media technology.

As so many commentators have pointed out, this commercial development of the means of communication has momentous consequences for public speech.[97] Whether through explicit editorial intervention, journalistic self-censorship, or the demographic imperatives of advertising, market-driven media tend to filter out news and analysis critical of capitalism. This filtration is done with a gross mesh, not a fine one, and is less absolute than the more monolithic models of capital's "media monopoly" sometimes suggest.[98] Competition among various media capitals, or frictions between media empires and other factions of capital, not to mention the occasional refusal of individual journalists or artists to submit to managerial control, mean that *something* usually escapes. Nevertheless, the corporate ownership of the major organs of societal communication tends toward a situation in which, in Marx's classic formulation, "the ruling ideas are nothing more than the ideal expression of the dominant material relationships"—in this case, an air-brushed affirmation of the rightness and normality of omnipresent commodity exchange.[99]

This integration of media into capital's subsumption of society first reached a high level of consolidation in the era of the mass worker. Mass production and mass consumption met in the virtuous circle of Fordism. Broadcast media became indispensable components of this regime, deluging society with the advertising that trained the populace in widespread consumption of standardized commodity goods. In the living rooms of North America the radio and then the television set became the domestic entry point for the same commodifying and conforming capitalist logic that in the factory drove the assembly line and the time-and-motion study.[100]

However, the revolts of the 1960s and 1970s shattered the stability of this arrangement. The rejection of the Fordist factory regime manifested in movements that, as well as demanding better standards of living, asserted diverse needs for self-expression. Social rebellion went hand in hand with experimentation in music, dress, drugs, and art. The cultural tumult of the era exploded the homogeneity of the mass market. When capital reimposed social discipline through austerity, driving down wages and polarizing incomes, not only work but also consumption had to be restructured. One crucial element in this was a major expansion of media industries.

From the late 1970s to the present there have appeared on the market a profusion of new communications devices—cable and satellite TV, VCRs, camcorders, and personal computers. Deployed beneath the mantle of increasingly concentrated, vertically and horizontally integrated media empires, these technologies have been announced as marking a new era of choice, liberation, and personal fulfillment.[101] In practice, they have accomplished two corporate purposes. First, they have provided the channels for an explosive growth of markets for entertainment and information. Here, as on the shop floor, capital has advanced by harnessing the energy unleashed against it. The desire for cultural diversity, subversively expressed in the 1960s, has over the subsequent decades been subjected to an unrelenting commodification, converting rock music, fashion, style, personal growth, and popular culture into highly variegated zones of vertiginous commercial development.[102]

This skyrocketing commodification of culture has been vital as a compensation for a flagging growth in other sectors. In the polarized post-Fordist economy, even those who can no longer buy a house or car can still pay for a CD or cable, while those who already have more residences and vehicles than they need can be persuaded to spend on computers and electronic goods. Moreover, the high rates of obsolescence that obtain in these fields—almost instantaneous in cases of evanescent soft goods, songs, films, and video, scarcely less so in the ever-changing electronic equipment—means that there is little risk of saturating markets.

Second, the new media not only create fresh cultural commodities but also permit extraordinary refinements in marketing other products. Here a central element in the restructuring of capital has been a huge increase in expenditures on advertising, sales promotions, and direct marketing.[103] As the Fordist mass market was fragmented by falling wages and social polarization, corporations sought both to internationalize sales and to segment them, stimulating hyperconsumption among the relatively thin strata of well-paid workers to compensate for the limited consumption capacity of the poor and unemployed. New media systems, such as cable and satellite television channels, are eminently suited to this purpose. They both enlarge audiences (sometimes on a potentially global basis) and make possible this ever more precise targeting of consumers differentiated by taste and income.

This prospect is enhanced by the promise of various kinds of "interactive" media—systems such as computerized video-on-demand or teleshopping, which, unlike unidirectional broadcasting, involve some degree of two-way transaction between receiver and transmitter. One common but underpublicized feature of such systems is their capacity to transmit back to the corporate provider detailed information about consumers' identities, location, consumption habits, and daily schedule.[104] Integrated with other electronic traces left by point-of-sale devices, credit-card scanning, billing and subscription records, and direct polling, this allows the compilation of comprehensive profiles of consumer behavior. Such data then forms the basis for the highly targeted, demo- and psycho-graphic micromarketing required by the increasingly stratified and hierarchical organization of consumption. Furthermore, this data can be fed back into systems of flexibly specialized production and just-in-time inventory control designed for rapid response to shifting market conditions. Interactive media thus hold out the promise of what Kevin Wilson terms "a truly cybernetic cycle of production and consumption."[105]

The implications of this situation were perhaps best recognized two decades ago when Dallas Smythe suggested that the watchers of TV, in "learning to buy," effectively "worked" for advertisers.[106] Electronic capital's expanding media reach meant it exploited not just labor power in the factory but also "audience power" in the home.[107] As the home entertainment center becomes the conduit not only for an incoming flow of corporate propaganda but also for an outgoing stream of information about its viewers, this analysis grows in credibility. The level of surveillance in the home tends toward that already experienced in the workplace, and the activity of the waged "watchman" in the automatic factory, described by Marx, becomes integrally linked with the unpaid "watching time" that he or she passes in front of the television.[108] The rate of surplus-value extraction, dependent on the exploitation of labor power, and the velocity of circulation, dependent on the carefully targeted consumption capacity of the media audience, merely measure different moments in a continuous, overarching, internally differentiated but increasingly unified process of valorization.

However, analyses such as Smythe's often assume that capital's intended exploitation of audience-power is fully successful. From my perspective, the more interesting question is how it fails. If audience power is today analogous to labor power, then it too is a disobedient subjectivity that evades, resists, and reshapes technological controls. There is now extensive evidence that viewers, listeners, and readers do not passively accept hypodermic injection with narcotic messages, but are rather active agents who engage in thousands of little lines of flight and fight—from turning off advertisements to the oppositional reinterpretation of programs and the creation of micronetworks of decommodified cultural activity.[109]

At the very time when innovations in communication are becoming the basis for vast commercial empires, there is apparent an opposite tendency that flouts the logic of the market. People are using the new technologies to get or give out information for free: reproducing, transmitting, sampling, and reconfiguring without respect for commercial property rights. This is known as "piracy." And it is prevalent. As access to the new communication machines becomes more and more thoroughly socialized, we see a wave of photocopying, home taping, bootlegged videos, copied software, zapping, surfing, unscrambling, and culture jamming. Moreover, an increasingly wide variety of groups and movements are using this generalized availability of communication technologies not simply for individual but for collective purposes. This manifests in the development of "alternative" or "autonomous" media.[110] Such experiments first blossomed during the 1960s and 1970s in a wave of radio-activism, guerrilla video, and public-access cable movements.[111] Despite enormous difficulties they have persisted. Radio-activism has continued and spread, reinvigorating itself in North America by the proliferation of inexpensive, low-power, and usually illegal microwatt FM broadcasting by ghetto communities, squatters, and homeless people.[112] Oppositional video making has passed from the avant-garde to common practice among social movements.[113] New areas of activism have opened around television, with the attempts in the United States and Canada to create and sustain public access cable—a medium whose political potential has been developed by the Paper Tiger Television collective and its satellite broadcasting Deep Dish project.[114] Lack of resources means that in most cases the reach of such experiments is limited and their aspirations only very partially realized. But, however raggedly, alternative media do posit something different from, and opposed to, capital's mobilization of "audience power."

Corporate interactivity is ratificatory: it posits dialogue only within the preset limits of profitability. Autonomous media, on the other hand, are, as Rafael Roncaglio puts it, "alterative"—probing the limits of established order.[115] Their practice often includes projects of self-representation, involving subjects in the definition and documentation of their own social experience. They attempt to overcome the restrictions of technical expertise characteristic of capital's division of labor. They experiment with forms of collective ownership. Above all, alternative media often give a voice to those who are excluded or silenced by the commercial logic of market-driven information industries—either because they are not demographically desirable or because they are politically suspect.

Thus, looking back for a moment at the Los Angeles riots of 1992, one remarkable aspect of the uprising was the degree to which the insurrectionaries were able to turn some elements of capital's high-technology surveillance and media apparatus to their own advantage.[116] The uprising was, of course, ignited by a classic instance of countersurveillance—George Halliday's videotaping of Rodney King's beating, and the recording of incriminating police radio conversations. But even before the rebellion, its idiom of anger had already been disseminated by the high-tech cultural inventions of the ghettoized community—hip hop and rap, music whose political significance was neatly demonstrated by President Clinton's subsequent public attack on the rap artist Sister Souljah.[117]

During the riot, the omnipresence of the corporate media, covering the most heavily televised urban uprising in history, had an ambiguous effect: although their representations frequently demonized and distorted the motives of the insurrectionaries, they could not entirely avoid giving voice to the people's outrage.[118] Simultaneously, a variety of autonomous media, ranging from microwatt radio stations in ghettoized neighborhoods—such as the famous

Zoom Black Magic Liberation Radio—to computer networks connecting activists in North America to others in Europe, spread a wider range of news,
analysis, and debate ignored by mainstream media.[119] All this contributed to
the circulation of supporting riots and demonstrations in Atlanta, Cleveland,
Newark, San Francisco, Seattle, St. Louis, and Toronto, and to the perception
of the riot as an indictment of the social policies of the Bush administration.[120]

Autonomous media have also played a significant part in less-explosive but
more-protracted forms of struggle, such as the new waves of labor activism.
In Los Angeles again, in an episode sometimes referred to as "the riot that
didn't happen," Latino and Chicano janitors and maids fighting for a first contract in the hotel industry won a significant victory by threatening to circulate video evidence of abysmal working conditions to potential convention
guests.[121] In Las Vegas, workers involved in a struggle with the entertainment
giant MGM used similar "guerrilla media" tactics.[122] The use by trade unions
of video and film for activist training, worker self-education, and public cam-
paigning has become commonplace. In various U.S. and Canadian cities, this
media activism has led to the establishment of regular labor programming
on community cable and radio stations.[123] This sort of activity is systematically fostered by organizations such as the Labor Video Project, which also
works to connect North American efforts in this field to similar initiatives
globally.[124]

These examples are only a part of a much wider circle of oppositional media activities. Other instances that could be cited, some of which will be examined in later chapters of this book, include the efforts of alternative media
during the Persian Gulf War; the mobilization of support for political-activist
prisoner Mumia Abu Jamal, accomplished almost entirely through alternative
radio, press, video, and computer links; the Vancouver-based "Adbusters"
attempt to infiltrate commercial channels with "subvertisments"; and the international computer networking associated with the transcontinental opposition to the North American Free Trade Agreement (NAFTA), the Zapatista
revolution, and the campaign against the Multilateral Agreement on Investment.[125]

Surveying the scope of this dissident media activity, it appears that capital, in developing its media apparatus, has let the genie out of the bottle. Just
as, by computerizing the factory, capital has not so much destroyed labor as
dispersed it into the wider social sphere, so by wiring the household it has
not necessarily consolidated control over audiences. Rather, in its drive to
extend the scope of the market, it has so thoroughly disseminated and made
familiar the technical means of communication as to open the door to a series of individual and collective reappropriations. This means that on occasion corporate control can be interrupted and spaces opened within which a
multiplicity of social movements, all in different ways contesting the dominance of the market, can be connected and made visible to each other. New

ation technologies therefore appear not just as instruments for the
ation of commodities, but simultaneously as channels for the circulation of struggles.

Circulation II: Struggles in Cyberspace

Today, some of the most dramatic manifestations of this contradiction appear in cyberspace, that notional dimension constituted by flows of electronic
data within computer networks. In post-Fordist capital, these digital flows
are used by "virtual corporations" to link automated machines to just-in-time
inventory systems, connect dispersed production sites, accumulate and mine
data about consumer tastes and habits, and forge new marketing opportunities, coordinating these activities on a global scale and as swiftly dispersing
them.[126] It is in cyberspace that capital is now attempting to acquire the comprehensive command, control, and communications capacity that will finally
allow it to, as Marx put it, "along with labour . . . also appropriate its network
of social relations."[127] And yet at the same time it is also in this virtual realm
that some of the most remarkable experiments in communicational counterpower are being conducted.

Computer-mediated communications, created by the linking of computers and telecommunications, were originally designed under military aus-
pices, initially as part of the U.S. nuclear war-fighting preparations and later
to connect the supercomputing centers vital to Pentagon research. These
origins have led many on the left to see the development of such networks
simply as a quintessential expression of capital's technological domination.
However, there is another side to this process. In an entirely unforeseen development, the technoscientific labor employed in the sites of the military-
academic-industrial complex—faculty, systems managers, and especially
graduate students—extended the network far beyond its original scope, using it for nonmilitary research, designing successive layers of alternative
systems that connected into the main backbone. This accretion of self-organized services proceeded, with the complicity of systems managers enchanted
by the technological "sweetness" of the results, until, as Peter Childers and
Paul Delany put it, "the parasites had all but taken over the host."[128]

Strangely, in the era that supposedly marked the triumph of the free market, the most technologically advanced medium for planetwide communication was created on the basis of state support, open usage, and cooperative
self-organization. A proliferation of autonomous activity transformed a military-industrial network into a system that in many ways realizes radical
dreams of a democratic communication system: omnipurpose, multicentered,
with participants transmitting as well as receiving, operating in near real-
time, having a highly devolved management structure and—since universities and other big institutions have so far paid a flat rate for connection—of-

fering relatively large numbers of people access for little or no cost. On this basis there emerged the unplanned explosion of popular interest in computer networking that by the late 1980s had catapulted the Internet on a trajectory of exponential growth totally unforeseen by corporate planners.

Capital is now of course attempting to contain this outbreak of unanticipated popular inventiveness—most significantly through the U.S. government's National Information Infrastructure initiative. Such a system would rationalize the extant but tangled web of fiber optic, cooper wires, cable radio waves, and satellites that provide the basis for telecommunications, cellular technologies, and cable television into a comprehensive, integrated network. Many companies are interested in this highway for internal purposes: to connect customers with suppliers, improve monitoring of employees, eliminate jobs, cut travel costs, and gather competitive data. The giants of the information and entertainment sector, however, see unprecedented market opportunities. Telephone, cable, video, and software companies are preparing to colonize cyberspace with their "killer" applications—video-on-demand, telegambling, pay-per-computer games, and infomercials. To many, the so-called highway running across the electronic frontier seems closer to the late-nineteenth-century U.S. railway development, complete with informational "robber barons."

However, cyberspace remains an arena of contradictions, in which capital's development is both opposed and spurred by alternative initiatives. To create and operate computer systems, commerce has had to summon up whole new strata of labor power, ranging from computer scientists and software engineers, through programmers and technicians, to computer-literate line and office workers, and ultimately to whole populations relegated to tedious, mundane jobs yet required to be sufficiently computer literate to function in a system of on-line services and electronic goods. As this virtual proletariat emerges, there also appears a tension between the potential interest and abundance it sees in its technological environment, and the actual banality of cybernetic control and commodification.

As so often before, new forms of conflict appear first under the guise of criminality and delinquency—in this case, as "hacking." If, following Andrew Ross, we define hacking simply as the "unauthorized use of computers," then the term embraces computerized sabotage; the reappropriation of work time to play games or write novels or exchange unauthorized email; so-called crimes of data copying, electronic trespass, and information dissemination; and unofficial experimentation with and alteration of systems up to and including the invention of new machines and of alternative electronic institutions.[129] These activities are now giving capital's managers multiple headaches over loss of productivity, theft of trade secrets, cybernetic revenge by terminated workers, and violations of intellectual property laws.[130]

Moreover, the networks are now the site for an array of "virtual communities."[131] These experiments in on-line social relations vary enormously; staggering diversity is perhaps their preeminent feature. However, in many cases participants see such communities as offering escape from the everyday logic of capital. In some cases, they are consciously conceived as constituting a new, electronic form of civil society in which many-to-many cyber-communications undermines the control of established societal gatekeepers—including the giant media corporations—over flows of information. Among libertarian technophiles these prospects sometimes inspire a populist version of technologically determinist information-revolution theory, with computer networks being seen as the solvent that will spontaneously melt the hierarchies of capital into participatory democracy.[132]

Faced with such propositions, many on the left have responded with buckets of cold water. Marxian critics not only stress the Internet's military-industrial roots (the sure mark of original sin) but point to the real demographic limitations on access to personal computers, modems, and technical expertise that sharply segregate computer access, partly by gender, race, and age, but most sharply by income.[133] Feminists, noting the obstacles of time, money, socialization, education, and harassment that discourage the involvement of women with the Internet, have also often been skeptical about its emancipatory potential.[134] Noting that the most likely owner of a personal computer and modem is male, white, middle-aged, and affluent, such critics characterize "virtual communities" as little more than elitist playgrounds of the privileged—the cyberspatial equivalent of walled suburban communities. Observing the corporate drive to market on-line, pay-per services, these commentators anticipate the overrunning of free cyberspaces by commercial development, increasing stratification of information rich and information poor, and relentless state and corporate surveillance. Confronting these prospects, they write off the alleged radical potentials of virtuality as rampant cyber-idealism.[135]

The actual dynamics of cyberspace are, however, more complex than either the virtual communitarians or their critics allow. The relatively privileged status of most (though by no means all) regular inhabitants of cyberspace undoubtedly limits the likelihood of mass subversive uses. There are, however, countervailing factors. Capital's omnipresent deployment of computers as work tools and consumer goods, and the extraordinary pace of planned obsolescence in this field, is making some of the basic equipment for networking readily available. Significant numbers of people still have free or cheap access via universities, schools, and businesses. Moreover, in a political context organizational access—the ability of a movement or group to receive and send networked information, which can then be further distributed via more traditional methods—may be a more critical factor than individual ownership of computers.

Even a rapid survey of the Internet reveals that today it *is* used by a remarkably wide variety of oppositional groups to bypass the filters of the information industries, speed internal communication, send out "action alerts," and connect with potential allies. Looking for the moment just at North America, we see diverse forms of network activism: mailing lists such as ACTIV-L, LEFT-L, PEN-L (the Progressive Economists Network), news groups such as P-NEWS, and World Wide Web sites for a wide variety of social movements. This cyber-organizing has included the construction of independent networks that interface with the Internet but are entirely devoted to social activism, like the Association for Progressive Communications (APC), which arose in the mid-1980s from the coalition of Peace-Net, Eco-Net, and Conflict-Net and now constitutes a global computer system dedicated to peace, human rights, labor, and environmental issues.

Such networks mark the latest phase in the emergence of the autonomous media described earlier. Some social movements have been far swifter to establish a presence in cyberspace than others. Environmental groups, some of which contain many relatively affluent professionals familiar with computers, and student campaigns, which often benefit from their members' free access to the Internet, have been early and frequent users. Women remain significantly underrepresented, but there are nevertheless numerous feminist lists and newsletters. Even if on the left the networks remain to some degree a boy-toy, they nevertheless frequently carry messages mobilizing support for the protection of abortion clinics, the defense of lesbian activists threatened by right-wing violence, the prevention of domestic violence, and the struggles of women workers. Organizations such as the APC have launched projects specifically aimed at supporting the use of computer networking by women from the popular sectors.[136]

It is impossible here to survey the entire range of this cyberspatial activism. But we can get some sense of its growing importance by briefly looking at trade unions' developing involvement in this sphere. So-called "organized" labor has been relatively slow to enter cyberspace, perhaps because of an abiding view of technology as a managerial domain. Nonetheless, as Eric Lee has made clear in his study The Labour Movement and the Internet, this picture is changing rapidly.[137] The early 1980s saw the establishment of the first local North American "labornet," in Canada, by the British Columbia Teachers' Federation.[138] The subsequent years have seen major "Labortech" conferences; the initiation of lists such as LABOR-L and networks such as LaborNet; and a burgeoning of North American union-affiliated bulletin boards, run by teachers, firefighters, plumbers, communications and public service workers, musicians, and journalists.[139] Some, such as the Canadian Union of Public Employees' Solinet, are now well established. Several have connection to similar networks outside North America—Glasnet in Russia, WorkNet in South Africa, Geonet in Germany, and Poptel in the United Kingdom.[140]

The relation of these networks to the internal organization of trade unions varies. In many cases, computer communications are used simply to speed and make more efficient traditional trade union industrial relations practices. Sometimes, access to networked information has clearly been structured to reinforce internal bureaucracy and hierarchies. But on occasion, the networks have become forums for unexpected debate, dissent, or rank-and-file initiatives.[141]

Moreover, in some recent struggles, net-workers have taken the offensive on-line in highly original ways. For example, in the Justice for Janitors campaign in Silicon Valley, strikers attempted to build links across the divide separating the "service" and "scientific" strata of the valley, using the very means of communication produced by the companies with whom they were locked in struggle. With the help of a small number of sympathizers among the core professional staff, they found the email addresses of employees at Oracle corporation, posted information about the exploitation of contract workers, and encouraged their readers to protest the issue to management. They also disseminated news of the strike through the Internet, inviting "netizens" to complain to senior company officers and providing email addresses. This was severely embarrassing to firms such as Apple whose profitability substantially depends on maintaining a benign public image among computer users. One participant in this email campaign describes it as follows:

They had no idea how many people we were sending to. . . . They started answering, "We are not beating our janitors" and it turned out they were beating them, really. Once they started with those answers then people started to ask questions and it created a climate of heightened awareness of what the janitors were doing, though it was not easily visible. . . . The costs of bad publicity, of morale being influenced by email are major. . . . for people like us who live in that world to sense the communications opportunity that exists right now—that email can be used to penetrate barriers that exist for more conventional communications—was rather exciting. Maybe after a while they'll set up filters and they'll get to keep all of our messages out, but we'll be engaged in a lot of measures and counter measures to keep communicating in that fashion. . . . I think it's a creative way to use the technology of the industry to undermine the social relationships that have been built into it.[142]

A similar incident involves the newspaper industry, a business that has felt the full weight of capital's drive to deskill, automate, and shed labor. In 1994, some twenty-six hundred workers from eight unions struck San Francisco's two daily papers. During the strike they produced their own paper—the *San Francisco Free Press*. This was not only distributed within the city but was also made electronically accessible via the World Wide Web, thus making it probably the most widely circulated strike bulletin in the history of civilization. At the same time, the strikers initiated a boycott of companies that continued advertising with newspapers behind picket lines. A computer list, LEFT-L, posted daily lists of "scab advertisers" and encouraged subscribers to call these corporations' 1-800 phone numbers with complaints. This boycott call appears to have been successful, with many companies discontinuing advertising and others having their advertisements run for free as the newspaper proprietors desperately attempted to save face. The eventual settlement was widely seen as a victory for the strikers—an unusual moment in recent U.S. labor history.[143]

Subsequently other labor struggles have pursued similar tactics.[144] However, perhaps of even greater importance than this use of the nets as a weapon against employers is the potential they open for connection and dialogue among movements. Sharing a common cyberspace—such as the widely used Canadian Solinet network—enables participants from different sectors of the labor movement to familiarize themselves with each other's concerns. But this process extends beyond the scope of the labor movement. Lists like ACTIV-L, the major North American activist forum, carry messages from labor, environmental, feminist, and indigenous groups. Sharing such an electronic forum implicitly asserts these movements' interconnections even while participants may still be searching for the explicit formulation of such links. One of the main trade union networks, LaborNet, is housed by the same organization, the Institute for Global Communication, that supports major environmental, human rights, and peace networks, a situation that encourages shared initiatives and informational crossovers.

Just as by creating a common medium for capitalist transactions, digitalization drives toward the merging of once distinct industries, so it creates a momentum for what Jim Davis terms a "popular digital convergence" among different sectors of social labor.[145] Organizations that fought separately for community access to cable television or the employment conditions of phone workers or the artistic rights of musicians and writers now find in their common concern around the "highway" a "new, practical basis for working together."[146] In combination with other autonomous media, such networks provide a channel within which a multiplicity of oppositional forces, diverse in goals, varied in constituency, specific in organization, can, through dialogue, criticism, and debate, discover a new language of autonomy and alliance. In this sense, cyberspace is a potentially recompositional space in which the atomization that information capital inflicts on socialized labor can be counteracted.

Of course, capital is now trying to reabsorb the unruliness of the networks through the corporate information highway drive, an electronic law-and-order crackdown, and the stimulation of a vast moral panic over pornography, terrorism, and other evils on the Net.[147] The brief blossoming of the Internet may well, as Herbert Schiller prophecies, be swiftly "paved over," like the populist initiatives that marked the early days of radio, cable television, and earlier generations of communication technologies.[148] However, it is also possible that this familiar pattern of capitalist recuperation will encounter unexpected problems in the case of computer communication. There are real questions as to whether there is actually sufficient popular demand for commercial projects such as video-on-demand or teleshopping to warrant the enormous investments that the highway demands. All indications are that what people want from the on-line environment is global, communal conversation rather than digital consumer services—in Negri's terms, "communication" rather than "information."[149] To the degree that capital stifles or excludes this possibility, it risks killing the digital goose whose golden eggs it is already counting.

The most adventurous groups in information-age business—such as the cyberlibertarians of the Electronic Frontier Foundation—gamble that they can avoid this impasse by entering into a symbiotic relation with Internet culture, benefiting from the experiments of virtual community constructors, the challenges of hackers, and the widespread interest in two-way communication to spur technological development and perfect a new round of digitally based accumulation. However, such a strategy requires corporate capital to preserve a degree of openness within the networks and to allow at least some continued spaces for alternate digital institutions and experiments. In this case, the networks will continue to serve as a medium not simply for the circulation of commodities, but also for the circulation of struggles.

Cyberspace is important as a political arena, not, as some postmodern theorists suggest, because it is a sphere where virtual conflicts replace struggles "on the ground," but because it is a medium within which terrestrial struggles can be made visible to and linked with one another.[150] Of course, this process is fraught with pitfalls. The European Counter Network, an autonomist network circulating news of struggles by workers, refugees, and antifascists within the EEC, notes the potential hazards of such computer activism: technical fetishism, new hierarchies of expertise, health risks, and the "ultimate nightmare," "a simulated international radical network in which all communication is mediated by modems and in which information circulates endlessly between computers without being put back into a human context."[151] As Dorothy Kidd and I have written, "Attempts to use computers . . . in the struggle require constant, collective reevaluation, to determine which strategies are effective, and which dangerously compromised."[152] Given such ongoing reassessment, however, there is the plausible chance that computer networking can help constitute new forms of anticapitalist combination that do not rest on the directives of a vanguard party, but rather arise out of the transverse, transnational connections of oppositional groupings.[153]

Virtual Commune

It is widely known that in the aftermath of the 1848 proletarian uprisings in Paris, Napoleon III ordered Baron Haussmann to redesign the city, and that a centerpiece of this urban reconstruction was the widening of streets to allow the passage of artillery for the suppression of any future insurrections. What is less well known is that workers employed on this highway development project, impoverished masons and builders housed in squalid Parisian slums, were leading participants in the next revolutionary outbreak—the 1871 Paris Commune that seized the city in its entirety, rocking the stability of capitalist Europe and giving Marx a blazing, prefigurative glimpse of communist society.[154]

Today, in the era of the information highway, capital is constructing its cyberspatial thoroughfares to circumvent or overwhelm the industrial conflicts that once brought it to crisis. Proceeding through its circuit we have seen it deploying high technologies to crush all traces of opposition—enforcing availability for work, commodifying ever-larger areas of experience, deepening social controls, and intensifying the depletion of ecosystems.

Capital has not, however, succeeded in technologically terminating the cycle of struggles. Our travels along capital's data highways have discovered rebellions at every point: people fighting for freedom from dependence on the wage, creating a "communication commons," experimenting with new forms of self-organization and new relations to the natural world.[155] Such movements are incipient and embattled, yet undeniable. Indeed, without in any way diminishing the magnitude of the defeats and disarrays suffered by countermovements over the last twenty years, I suggest that there are now visible across the siliconized, bioengineered, post-Fordist landscape the signs of a strange new class recomposition. This is proceeding on a much wider basis than that traditionally conceived by Marxism. In virtual capitalism, the immediate point of production cannot be considered the "privileged" site of struggle. Rather, the whole of society becomes a wired workplace—but also a potential site for the interruption of capital's integrated circuit.

There is no need to emphasize the present fragility and uncertainty of the various reappropriations, counterplans, and alternative logics whose sinuous course we have traced. In their isolation, each provides only a minor problem to corporate power. But in their proliferation and interconnection they constitute a challenge to its dominion. It is the breadth and variety of such subversions that makes the fields of information and communication so crucial today. For it is by a process of mutual discovery, recognition, and reinforcement—by an accelerating circulation of struggles—that such insurgencies could attain a strength capable of prising apart the coils with which capital now encircles society. However, an assessment of such possibilities cannot limit itself to the most technologically advanced sectors of development; rather, it must take a perspective embracing the truly global scope of information capital—a window that is opened in the next chapter.

PLANETS

Whoever today says "capital" says "globalization." For nothing has been more central to the current restructuring of corporate power than nomadic range of maneuver, deterritorialization from old centers, systematic subversion of national sovereignty, and planetary political planning. And whoever today says "globalization" says also "communication," for the emergence of this new world order would be unthinkable without the telecommunications and computer networks that now form the electronic pathways for the circulation of money, commodities, and power.

The Net of the World Market

Despite the breathlessness of so much contemporary commentary, recent developments represent the culmination of an old logic. Marx in his time saw clearly how capitalism's compulsion to expand production and circulation drove it to successive geographic enlargements of a circuit that would eventually encircle the whole earth, leading to "the entanglement of all peoples in the net of the world market."[1] Observing the telegraphs, railways, and steamships of his age, he observed how capital "by its nature drives beyond every spatial barrier" so that "the creation of the physical conditions of exchange—of the means of communication and transport—the annihilation of space by time—becomes an extraordinary necessity for it."[2]

Marx believed this creation of "intercourse in every direction, universal inter-dependence of all nations," had both exploitative and emancipatory aspects.[3] On the one hand, the "immensely facilitated means of communication" provided the means by which capital imposed its logic worldwide. It "compels all nations, on pain of extinction, to adopt the bourgeois mode of production; compels them to introduce what it calls civilisation into their midst, i.e., to become bourgeois themselves."[4] But simultaneously, this process created the conditions for the success of revolutionary proletarian movements, movements that depended on the "ever-expanding union of the workers": "This union is helped by the improved means of communication that are created by modern industry and that place workers of different localities in contact with one another. It was just this contact that was needed to centralise the numerous local struggles, all of the same character, into one national struggle between classes."[5] Today the net of the world market is made of fiber-optic cables and satellite links. Yet few see in its weaving the dialectical possibilities Marx perceived. Mainstream theorists of "globalization" of course simply celebrate the market-driven march of what they call "civilization" across the face of the planet. But while there have recently been several important critical analyses from a broadly Marxian perspective, nearly all see the recent intensifications in the transnational organization of production, exchange, and finance, and the accompanying developments in new media and communications technologies, only as massively enhancing the power of transnational corporations.[6]

This chapter takes a different tack. It proposes that globalization, rather than simply representing an inexorable deepening of capitalist control, constitutes a defensive corporate response to series of interweaving challenges that in the 1960s and 70s plunged the international structure of accumulation into crisis. Moreover, while the immediate impact of this riposte was profoundly to disarray oppositional forces, it has also opened unforeseen opportunities for their new cooperation and alliance. Not the least of these is the use of global capital's own means of communication and transport to connect a proliferating array of countermovements whose own world-encircling activities of resistance and reconstruction I term "the other globalization."[7]

Three Worlds into One

Capitalism's global expansion has constantly been spurred on by the rebellions of its laboring subjects. Historically, these struggles have spiraled across a succession of expanding territorial spaces: the *national* space, where capital was first challenged by emergent proletarian movements; the *imperial* space, where these challenges were partly defused by capital's capacity to raise domestic living standards on the basis of colonial superexploitation; and the *socialist* space, where Bolshevism, in the midst of interimperial war, seized a terrain within which it was then fatally contained. By the midpoint of the twentieth century, however, the catastrophes of world war, the threat of state socialism, and the mounting pressures of anticolonial liberation propelled capital toward its first exercises in truly international planning.

These took shape at the end of World War II. Under the leadership of a newly preeminent U.S. industry, whose most advanced corporations were rapidly transcending the limits of the domestic market to acquire multinational form, the management of the capitalist world economy began for the first time to be directed and orchestrated by an array of consciously global institutions. Trading arrangements were codified in the Bretton Woods treaties; significant financial controls delegated to the World Bank and IMF; monetary stability assured by the dollar's role in regulating exchange rates; and the whole system held in place militarily by the Pentagon's nuclear might, relayed through various local authorities and regional alliances. The Fordist golden age of capital thus rested not only on the domestic planning of national economies, but also on an unprecedented level of international organization.

The famous tripartite division of First, Second, and Third Worlds describes the success of this international order in segregating the global proletariat into zones of differential control. For the inhabitants of the First World, there was a historic experiment in welfare-state reformism. For the populace of the socialist bloc, the Second World, there was cold war encirclement and forced industrialization. And for the Third World, there was a transition from colonial subordination to European capital to neocolonial penetration by U.S.-based multinationals, with modernization programs courtesy of the Rockefeller Foundation, counterinsurgency from the Central Intelligence Agency, and ongoing mass immiseration. The workers of the world were in effect segregated and exiled to three separate planets with drastically different levels of development and radically incommensurable experiences of work, exploitation, and struggle.

Over the next twenty-five years, however, the stability of this international order was shaken by rebellions converging from different directions. In the Third World, the arrangement was in trouble from the start, as successive revolutionary movements—in China, Algeria, Cuba, Vietnam—fought and won against the sentence of dependency. Ironically, at the same time as Third World movements were establishing state socialist regimes, in the Soviet-bloc Second World the initial rapid growth produced by forced labor stagnated, leading to bread riots and rebellion against police control. Finally, in the First World metropolis, the Keynesian deal started to come apart in the 1960s and 70s. Supposedly affluent workers, instead of being pacified by higher living standards, used these as resources to pursue new levels of struggle, that rolled from inner-city ghettos to industrial shop floors to university campuses, setting off a sequence of mutually reinforcing reverberations.[8]

As Harry Cleaver points out, struggles arising in the different zones of the world system started to circulate.[9] Metropolitan capital had relied on cheap resources from colonial and neocolonial dominions to finance its deals with the mass worker. As anticolonial wars ruptured this control, this domestic latitude of maneuver was diminished. The inflationary effects of the Vietnam conflict, in particular, set in motion a whole series of wage and social wage struggles. Moreover, struggles across the planet began to support one another. Third World revolutions inspired social activists in the metropolis and were supported in turn by international solidarity movements.[10] In the United States, and to a lesser degree in Europe, opposition to the Vietnam war brought on massive internal turmoil.

By the early 1970s, it became clear that, from capital's point of view, the old "triplanetary" division of the world wasn't working.[11] With profit rates in the old centers of accumulation tumbling, the search for a reorganization of capital's global circuits that would allow it to escape worldwide pressures of social unrest was on, both in the probes and experiments of individual corporations and banks, and in the consultations of high-level capitalist planning agencies such as the Trilateral Commission. The U.S. government's abrogation of the Bretton Woods currency agreement in 1971 was a first signal of the abandonment of the postwar international settlement, a departure deepened a few years later with the dramatic redirection of finances and investment occasioned by the first oil shock.[12]

In 1975, Mario Montano argued that what was taking shape was a restructuring that would render previous theories of "development" and "underdevelopment" obsolete. As general capitalist strategies, both underdevelopment and development had failed. For multinational capital, the question now was "how to directly oppose development and underdevelopment against each other, how to make underdevelopment work completely inside development."[13] What was unfolding, Montano suggested, was an undoing of the traditional demarcations by "two opposing dynamics": on the one hand, the "underdevelopment of development"—with the "Latin Americanization" of the United States and Europe—and, on the other a "development of underdevelopment," with the industrialization of portions of the former Third World.[14] The aim of this restructuring was to pit "the starvation of underdevelopment . . . against the living standards of the working class of the metropolis."[15]

While Montano's analysis was necessarily preliminary, it accurately defines the main thrust of the process that is today known as "globalization." To destroy the multiplying threats to its international command, capital has broken out from its old entrenchments, overrun the previous divisions of its world system, and, empowered by its new digital technologies, opened up the whole planet as a field for maneuver. In doing so, it has imploded the Three Worlds into one another. Corporate flight from the demands of the mass worker in Europe and North America has led to the partial Third-Worlding of the First World—deindustrializing manufacturing centers, canceling the Keynesian deal, inaugurating mass unemployment, lowering wages, intensifying work. This has introduced into the metropolis levels of insecurity and destitution previously relegated to the peripheries of capitalist world economy.

The other side of this coin, the selective First-Worlding of the Third World, has equally taken its impetus from the urgent need—mediated through a variety of authoritarian local regimes—to modernize out of existence the threat of revolutionary insurgency. Thus the turbulent energies of immiserated labor of the periphery have been harnessed to the creation of various growth sites—the newly industrializing countries and other development zones—whose appearance controverts cruder models of perpetual dependency. The drive to eliminate the twin nemeses of the industrial wildcatter and the peasant guerrilla links the deindustrialized rustbelts of the North and the new shantytowns of the South in a complementary logic.

At the same time, the one supposed alternative to capitalist development and underdevelopment—the Second World of state socialism—has blown apart and its residues have been allocated between the two poles. Retrospectively, it is clear that the capitalist restructuring of the 1970s sounded a death knell for the command economies of the Soviet bloc. The rigidities of their internal controls proved altogether unable to adapt to the flexibilities requisite for microelectronic, post-Fordist production. When these converging pressures exploded in a series of popular uprisings across Eastern Europe and the USSR in 1989, neoliberalism's market managers rode the wave, channeling movements characterized by an immense diversity of aspirations into marketization and economic shock therapy. Where state socialist regimes have survived, as in China, it is only by bringing to bloom their already present tendencies to act simply as versions of authoritarian capitalism.

The result is the creation of an increasingly smooth and planar world-space of accumulation. The polarities of "development" and "underdevelopment" of course still exist—indeed are massively intensified. And, it is important to emphasize, their distribution continues to fall preponderantly on either side of a North/South axis. But at the same time these poles increasingly designate possibilities of ascendant affluence or abysmal misery that can be visited on *any* point in the planet according to the movement of corporate investment. Inner-city ghettos in North America attain "Third World" infant mortality rates, while cities such as São Paulo, Seoul, or Taipei begin to burgeon with a cosmopolitan affluence matching the one-time "First World." It is beyond the scope of this book to analyze the full unwinding of this transformation, especially since many of its aspects have been adequately addressed elsewhere. What I want to indicate here is the way in which it has been dependent on a massive expansion of the means of communication, and in particular on the development of computers and telecommunications.

Redividing the International Laborer

At the basis of capitalist globalization is the "new international division of labour."[16] This comes into being as capital both flees from and undermines strongly organized, and consequently costly, strata of working-class power—metropolitan, male, industrial—by gaining access to more vulnerable sectors—peripheral, female, domestic, cheapened by destitution and authoritarian discipline.

United States employers have repeatedly responded to cycles of working-class struggle with waves of investment abroad.[17] From the mid-1960s on, this pattern was repeated, not only by U.S. companies but also by European and Japanese corporations, on an extraordinary scale and across an unprecedented range of industries. From the relocation of car production to sites in Warsaw, Teheran, and Brazil, through the shift of light assembly industries to the free-trade zones of Hong Kong, Singapore, Taiwan, and South Korea, to the emergence of Mexican and Latin American *maquiladoras* and the creation of industrial enclaves in postsocialist China and Eastern Europe, this process has proceeded at an accelerating rate. It involves not only smokestack industrial plants but also sunrise microelectronics enterprises. Already it has rendered traditional divisions between metropolitan industry and peripheral hinterland largely obsolete.[18]

This geographical reorganization of labor power has also involved a radical reworking of the gendered division of labor. For the corporate search for inexpensive and reliable labor has largely entailed a switch from male factory hands to the supposedly docile and disposable female "nimble fingers" employed in, say, the garment industries of the Mexican *maquiladoras* or the microchip assembly of the Malaysian enterprise zones. Much of this female labor is organized so as to labor at home, in isolation, while still performing unpaid domestic work in support of male labor power. The global assembly line of many industries—electronics, textiles, light engineering—is thus to a remarkable extent a homework economy, linking transnational contractors and subcontractors in long, shifting chains whose complexity hides responsibility for the abysmal conditions of the new home/factories.

This global spread of female "shadow work" has even darker aspects. Silvia Federici, Mariarosa Dalla Costa, Maria Mies, and others argue that globalization takes its impetus not only from business's attempts to escape the militancy of the industrial mass workers, but also from its flight from, and capitalization on, the First World feminist revolt against reproductive labor work. The transnational explosion of the sex trade, pornography industry, mail-order-bride business, and baby adoption market all represent "enormous quotas of reproduction work which capital has exported in the same way that there has been a strategy of exporting part of the manufacturing process with the free enterprise zones."[19]

All these strategic relocations depend on highly developed systems of transportation and communication. Electronic information systems in particular allow transnational corporations to decentralize operations while centralizing control; executives in Toronto offices open on-screen windows displaying the performance of machine operators in Seoul factories.[20] Manufacturing strategies for products such as Ford's "world car" rely on telecommunications to coordinate production flows at plants on different continents, as well as to perfect the standardization of modularized parts. They depend on fast, cheap transportation and on computerized automation carried to a point where elementary units and simple routines can be performed by unskilled workers. Global homework industries, such as those of Benetton, network computers to tie suppliers to sellers, match production to inventories, monitor dispersed workers, and check quality and speed of supply through every rung of their hierarchy. The same logic, to a greater or lesser degree, allows Canadian supermarkets to sell fresh-cut flowers from Africa, or travel agencies in Bonn and Tokyo to book sex tours in Thailand and the Philippines. In all areas, even though production remains dependent on the most arduous, protracted physical labor—in Mexican car plants, Kenyan agribusiness plantations, or Bangkok brothels—coordination and control is effected through communication flows moving over distances and at speeds that surpass corporeal horizons.

Global Hearts and Minds

The global restructuring of production is complemented by an equally global reorganization of consumption. As we saw in the last chapter, this is necessitated by capital's success in driving down the wage and social benefits of metropolitan mass workers—for this undermined mass consumption within the domestic markets of the First World. Increasingly, therefore, business has turned its eyes to export markets and also toward the population masses of the South. These are so great in numbers that if even a relatively small proportion—the managerial and professional strata benefiting most directly from industrialization—can be brought into the orbit of luxury consumer capitalism, it will more than counterbalance the eroded spending power of northern workers.[21] And for industries whose goods can be sold cheaply enough to penetrate the youthful economies of South African townships and Latin American barrios—cigarettes, soft drinks, cassettes—the potential profits are prodigious.

However, such a global projection of consumerism into zones previously relegated to economic marginality demands a reconstruction of needs and desires—of cultural traditions, religious prohibitions, dietary habits, sexual mores, traditions of self-sufficiency—similar to that experienced by the Euro-American proletariat in the first part of the twentieth century, but exceeding it in scale. In this process the vanguard organizations are the great media corporations—characterized by concentrated ownership, vertical and horizontal integration, and mastery of world-spanning arrays of convergent technologies. These "lords of the global village" are no longer—as in the classic formulations of the "cultural imperialism" thesis—exclusively North American.[22] Although U.S. entertainment and information corporations still generally enjoy a preeminent position, these industries have themselves, to a degree, become globalized, and also include European, Japanese, and even Latin American interests; newly marketized Moscow's most popular soap opera, "Even the Rich Cry," was made in Brazil.

But, whatever their ownership, these corporations—Adorno and Horkheimer's "culture industry" gone planetary—have become the vital agents for a reconstruction of global subjectivity carried out in the interests of transnational capital. Their products—films, programs, music, videos—are quintessentially global commodities, instantaneously broadcastable, evanescent, and demanding vast, worldwide audiences in order to recoup the costs and risks of production. Globalization means that everywhere, all the time, it is "video night in Kathmandu," as the habits of media spectatorship are stimulated and implanted worldwide.[23]

These media commodities in turn provide the vehicle for the global marketing campaigns. During the 1960s and 1970s, the penetration of television to households all over the world had provided multinational corporations with the necessary communication infrastructure to carry out coordinated advertising in Europe, Canada, Latin America, and Asia.[24] In the 1980s such global marketing strategies, promoted by business management gurus such as Theodore Levitt, became the creed of major advertising agencies.[25] These strategies are supported both by powerful campaigns to compel developing countries to lift restrictions on advertising and by the deployment of technologies that can effectively overleap any such barriers. Carried by satellite beam and VCR to the villages of Indonesia, Zaire, and Colombia, Arnold Schwarzenegger and Pamela Lee perform as the simulacral storm troopers of consumer revolutions dedicated to the attractions of soft drinks, hard bodies, high-tech weapons, and high-cut swim suits.

However, this universalization of advertising also goes hand in hand with intensifying segmentation and stratification of markets. Assuming that consumer elites in New York, Rio de Janeiro, Paris, and Bombay will have more in common with each other than with the homeless who in each city swarm on the adjacent blocks, the agencies deploy ever more sophisticated technological resources for surveillance of the world's consumption zones. They also carefully modulate centrally planned campaigns in the light of detailed anthropological, ethnographic, and market research. Arif Dirlik reports a paean from an advocate of this "guerrilla marketing" who declares that "just as the guerrilla fighter must know the terrain of struggle in order to control it, so it is with the multinational corporation of today. Our terrain is the world." This business-administration "guerrilla" goes on to claim that the "world market is now being computer micromapped" into 304 geographical consumption zones cross-referenced with the "unconscious" needs of 507 microconsumption types: "Through an extension of this mapping, even the most autonomous and unconventional desires may be reconstructed for the benefit of market extension and control. . . . we must win hearts and minds. This task can be accomplished by constructing and reconstructing them all the way down in what can only be viewed as an endless process."[26] While the cutting edge of this "hearts and minds" campaign remains the standardized Hollywood style of infotainment, media conglomerates also collect themes from all over planet—world music, ethnic arts, Third World cinema—for conversion into commodities and marketing instruments. The relentless monoculture of Disney and MTV is leavened with multicultural traces of Taiwanese rice farming chants and Indian bhangra dancing. This eclecticism has led some observers of popular culture to enthusiastic celebrations of the diversity and hybridization of the newly cosmopolitan global bazaar.[27]

It is true that, as Marx observed, the world market brings with it a variegation and elaboration of needs and appetites. In a way it does open up new horizons and subjective possibilities. But what too many contemporary panegyrics to this process overlook is the relentless uniformity of the underlying logic, the enormous systematicity that precedes all the apparent differentiation. The order of this system is unequivocal. Every human aspiration, desire, and creative impulse will find its place within the commodity-form: those that refuse are condemned to oblivion.

Money in Command, Worldwide

With globalization, capital cracks the shell of the nation-state. In the Fordist era, national governments had been indispensable for planning and securing the social conditions of accumulation—by Keynesianism in the First World, by neocolonial modernization in the Third. However, the struggles of the 1960s and 70s threatened these arrangements. To varying degrees, ranging from revolutionary power-seizures in the Third World to First World "fiscal crises," social movements undermined business's control of public spending.[28] Capital's reply was to relocate social control outside the national sphere. Over the last decade, a round of regional and global trade agreements—NAFTA, Maastricht, the General Agreement on Tariffs and Trade (GATT), the establishment of the World Trade Organization—have subordinated national policy to supranational agreements favoring unrestricted mobility of capital, deregulation, privatization, and unfettered markets. But such agreements in many ways merely formalize and consolidate a level of a transnational discipline that capital had already won earlier in the globalization process, through another mechanism—that of the international financial markets.

Since the 1960s, these markets have undergone an explosive expansion. This is the result of a number of factors: the collapse of the Bretton Woods agreement and the establishment of floating exchange; the growing importance of offshore or stateless money, such as Eurodollars; the deregulation and restructuring of banking, stock exchanges, and financial institutions; the invention of ever more arcane methods of speculation; and the increasing powers allocated to world-level financial institutions such as the IMF and the World Bank. As Arthur MacEwan points out, what is new in this situation is not the importance and interdependence of financial flows for capital but their degree of integration, speed of transaction, and capacity to escape state control.[29]

Again, these developments are inseparable from the expansion of information technology, which has "probably changed banking and finance more than any other sector of the capitalist economy."[30] Computers and telecommunications accelerate financial flows phenomenally, permitting round-the-clock planetwide investment activity, reducing the costs of transfers, creating a common digital medium for transactions, and spurring mergers and consolidations among monetary institutions. In a sector where a few seconds' advance knowledge over competitors can translate into billion-dollar profits, information systems are hardly less sophisticated than those governing nuclear weaponry. As one observer of electronic trading says, "It's almost like modern warfare, where people sit in bunkers and . . . push buttons and things happen."[31]

This world of virtual finance has become both increasingly detached from and superordinate over material production. As the struggles of the sixties and seventies unfolded, financial trading became very important to capital as an escape from crisis. Faced with loss of control in the shop floor and the paddy fields, many commercial interests simply evacuated the corporeal world, with its mud, blood, and recalcitrant labor power, taking flight not merely by seeking new sites for production but by dematerializing themselves entirely into speculative activity. This migration of money into cyberspace left behind a mundane spoor of abandoned factories and ruined communities and was a major factor in the mounting toll of job loss that undermined labor's capacity to oppose capital.

In other ways, too, the money markets became crucial in driving down social expenses to a level where investment in tangible production would be profitable again. Previously, the financial levers of domestic economies had rested predominantly in the hands of national governments. But when governments failed to discipline their workers, global money bypassed such arrangements. With the valuation of national currencies, interest rates, and credit worthiness determined by international investors and speculators, economic control became immanent within the entire planetary finance system. Governments—national, regional, and municipal—that had previously squared the demands of business and workers' movements by running up deficits now found the continuation of their credit and the stability of their currencies conditional on the implementation of austerity programs. As Christian Marazzi puts it, "the need to preserve credit ratings and currency stability has narrowed in an unprecedented way the margins of manoeuvre— the 'relative autonomy'—of national states, to the extent of dramatically reducing the area of choice within which national policies have to operate. All governments and their oppositions have in this sense been pulled into the narrow area of choice imposed by the logic of international monetary austerity."[32] More and more vital areas of domestic policy are subjected to what Cleaver terms "international adjustment mechanisms virtually invisible to the average worker."[33]

The full enormity of this monetary discipline is of course seen in the "structural adjustment programs" (SAPs) inflicted by the World Bank and IMF on Third and Second World states unable to pay for the large-scale national development programs of the 1970s.[34] However, the debt crisis is not restricted to economies formally placed under the protectorate of the IMF. What is remarkable about globalization is the way in which the pressure of the money markets has resulted in the spontaneous adoption of SAP-style measures within the very heart of the former First World. From the New York deficit crisis of 1975–76, through the rampage of Reaganite and Thatcherite monetarist policy in the United States and Britain, to the retreat of Mitterand's socialism in France in 1982 in the face of financial pressure, to the gutting of the New Zealand and Canadian welfare states in the name of deficit reduction, the imperatives of world money have dictated policies of deregulation, privatization, wage cutting, and welfare reduction adopted regardless of parliamentary regimes' ostensible political coloration. By lifting financial control out of the hands of domestic governments and diffusing it through the electronic nodes of global exchanges, capital has effectively placed economic power in a stratospheric orbit where it cannot be reached by electoral choices taken within the confines of the nation-state. It thus raises to a new level the negation of democracy inherent in the private expropriation of the means of production.

The Resurrection of War

The ultimate disciplinary instrument of the world market is, as it has always been, force. War is always critical to capitalist control, as a means for extending its circuits over new domains, dividing opposition, and destroying any threat to the operations of the market. It is therefore hard to overestimate the significance of the series of military reverses inflicted on capital by revolutionary movements from 1945 on—in Korea, China, Algeria, Cuba, Mozambique, Rhodesia, Angola, and, most seriously, Vietnam, where the preeminent imperial power went down to defeat, partly due to the disaffection of its own populace.

An important part of capital's restructuring has thus been the resurrection of military power as a viable instrument of global command. Amidst privatization and deregulation, one of the few aspects of the capitalist state generally reinforced is the security apparatus. In the United States, which retains its position as the principle enforcer for the world market, the Pentagon has carefully investigated ways to circumvent the unwillingness of citizens to sacrifice themselves in foreign wars. This experimentation has followed two routes. On the one hand, there has been a development of high-tech weaponry—cruise missiles, Star Wars systems, first-strike nuclear missiles, Stealth bombers—capable of fighting highly automated, remote-control wars. On the other, there is increasing resort to the low-intensity, covert, proxy, or mercenary strategies practiced in Afghanistan, Cambodia, Namibia, El Salvador, Guatemala, and Nicaragua.[35]

In both strategies, command, control, communication, and intelligence capacities are vital. Whether in the "black" satellite systems beaming CIA messages from Virginia to the rain forests of Costa Rica and Cambodia or the artificial intelligences unblinkingly monitoring the earth via the orbital platforms of the Strategic Defense Initiative, the new forms of warfare demand omnipresent surveillance, near-instantaneous transmission, and precision targeting. The search for global battlefield supremacy brings to bloom some of the most exotically deadly technological orchids of the information age. At the same time, control of information has also assumed a new prominence on the domestic front. Haunted by the belief that the Vietnam War was lost to the television viewers of America, the Pentagon and its allies have devoted increasing thought to the control of public opinion in time of crisis. The fruit of these deliberations appeared in the Falklands war, the bombing of Libya, and the invasions of Grenada and Panama. Here it was demonstrated that regimes of commercial-style image-management, military marketing, press-pool control, censorship, blackout, and propaganda, combined with the extreme swiftness of operations permitted by massive technological advantage, could largely stifle domestic dissent.

The real flowering of these developments had, however, to await the Persian Gulf War. Here, as Hamid Mowlana says, "the propaganda and communication strategy surrounding the conduct of war entered a new dimension not seen in previous conflicts."[36] Acting as the mercenary agent of multinational capital, the United States terminated the disturbance its former Iraqi client threatened to inflict on the world oil supply by an overwhelming application of information power. Smart weaponry, superior intelligence gathering, radar jamming, stealth technologies, the infiltration of computer viruses, and the annihilation of enemy radar systems gave the Allied forces total battlefield superiority. Meanwhile, a massive media campaign, including fabricated "incubator babies" stories, round-the-clock press conferences and bulletins, in-field interviews, and orchestrated displays of patriotic fervor aimed to win domestic and international support for the war. This was complemented by the targeting of Iraq's civilian telecommunications and other information utilities, depriving its government of equivalent weapons in the propaganda war. In short, front and home lines were interconnected in a near-seamless regime of information control.

In a sense, the whole war, with its spectacularly excessive violence, can be understood as an act of communication. For the message sent to the world by way of bombsight videos was that any interference in the finely tuned balances and flows of the world market would be crushed with lightning force. It is no coincidence that President Bush's announcement of a "new world order," widely understood as signaling not merely a diplomatic but also a politico-economic dispensation, should follow on the heels of the Persian Gulf War. For the underlying logic of globalization is that of war—a war waged by capital to annihilate all points of opposition and permit the relaunching of a new cycle of accumulation.

Globalization as Class Decomposition

The creed of globalized capital is clearly enunciated by Robert Reich, economic adviser and former Labor Secretary to the Clinton administration, in his *The Work of Nations: Preparing Ourselves for 21st Century Capitalism.*[37] Today, Reich argues, wealth generation is entirely dependent on "nomadic corporations" that, having fully transcended any national base, exist only as "global enterprise webs" held together by the threads of computers, fax, phone, and video networks.[38] Nation-states' capacities to control their own destiny is restricted to the creation of infrastructures—such as information highways—attractive to the investment by these mobile corporations. The result is to replace capital's previous territorial divisions of the work force with an increasingly transnational hierarchy of labor power. Those countries, or regions, or cities that can render themselves hospitable in this way will attract well-paying "symbolic analytic" jobs—the knowledge-based work associated with the design and development of new technologies. Those that don't will become the dumping grounds for the unfortunate industrial and routine service workers destined to be devalued by automation and global cheap labor.

Reich, to his credit, expresses considerable anxieties about how this divisive logic will affect the social fabric of the United States, as privileged "symbolic analysts" retreat into fortified enclaves to escape the deepening misery of unemployed service and industrial classes. The rest of the world, however, hardly figures in his optic. But his overall perspective on globalization is both inevitablistic and optimistic. The accumulation of wealth permitted by enhanced trade and specialization, although unevenly divided, will eventually raise global living standards by an inexorable trickle-down process. Even those areas fated to receive the industrial work cast off by the most advanced sectors of capital will be better off than in their previous agrarian situation. Globalization is, Reich insists, not a zero-sum game but rather an "infinitely expanding terrain of human skills and knowledge."[39]

He is correct about the astounding wealth-generating effects of contemporary technology and trade. But his analysis omits the dimension of power—the strategic logic inherent in capitalist globalization. By expanding the division of labor, the capitalist enlargement of the world market allows huge increases in productivity. But it also expands the division of laborers—the degree to which capital can set workers in competition with each other and thereby seize for itself an ever-increasing proportion of this global wealth. The social surplus grows—but so, and probably to an ever-greater extent, does capital's capacity to expropriate that surplus. The "infinite terrain" over which capitalist globalization expands is thus not one of "human skills and knowledge" but of inhuman profit and exploitation.

By seeking out and putting in competition with one another pools of labor power previously inaccessible or isolated because of geographical distance, state regulation, or communal self-sufficiency, capital can repeat a classic strategy—creation of a reserve army of the unemployed, now realized on a world scale. In this context, the nomadism afforded corporations by the "global webs" means that demands for the maintenance, let alone improvement, of wages and social conditions can be circumvented and outflanked. Workers are faced with the choice of acceding to corporate requirements, or seeing the now lighter-than-air means of production—software programs and communication nodes—relocated elsewhere. There is set in motion what Jeremy Brecher and Tim Costello term the "race to the bottom" whereby workers across the world are compelled to cheapen themselves into a job by competitively lowering their wages and conditions.[40]

Thus, although the new mobility of investment shuffles and reshuffles relative positions in the hierarchy of labor with extraordinary rapidity, its overall drive is toward increased power for capital vis-à-vis the global proletariat as a whole. Reich is right that globalization has given some knowledge workers—largely male, largely white, associated with high tech, finance, communication, and information—an exceptional importance. Concentrated in the hubs of "global webs," these constitute a layer of privileged labor on whose loyalty capital can largely rely. But analysis that sees "symbolic analysts" as the crucial actors in globalization does not grasp the speed with which capital tosses yuppies from the lifeboat when cheaper replacements can be found.[41] Even symbolic analysts feel the blast of globalization, as North

American computer programmers are undercut by Lithuanian or Indian competition, and architects, engineers, and professors discover that those who can telecommute can always be teleterminated by cheaper services uploaded from anywhere on the planet.[42] The ultimate benefactors of globalization are not even the symbolic analysts but the power that Reich hardly mentions—that of transnational capital itself.

Beneath the symbolic analysts are the mass of industrial and service workers exposed to increasing insecurity by a mobility of investment that can send jobs catapulting from Oregon to Lima to Jakarta in a matter of weeks. This logic has, so far, primarily been applied against industrial workers in the North—to the temporary benefit of labor in selected growth areas such as East Asia. However this undercutting is a process that can be repeated universally. Workers in Mexico or South Korea who have unionized find their jobs shifted to Bangladesh or Indonesia or China—and when labor there organizes, the work moves on to Vietnam.[43] Latin Americans see investment prospects vanish toward Eastern Europe. At a certain point the deindustrializing process comes full circle, by creating in the old metropolitan areas zones of immiseration so deep that they then become low-wage areas that lure capital back from its flight to the one-time periphery: Scotland and Ireland are now attracting Japanese and Korean investment with industrial wage levels comparable to those in parts of Asia.

At the bottom of the new global hierarchy, in regions and cultures that do not match capital's requirements in terms of wages, work habits, or possession of desirable natural resources, lie the hapless surplus reservoirs of labor power, labeled "not wanted on voyage" in capital's round-the-world restructuring.[44] These populations, still predominantly but not exclusively in the South, uprooted from the land by agribusiness and IMF agricultural rationalization, but bypassed by the electronic paths through which the world market circulates wealth, survive through the networks of the drug trade, prostitution, body parts sales, exotic animal trade, arms smuggling, and other informal economies. To the degree these desperate measures fail, they fall through the holes of the network into an abyss of impoverishment and debasement that is a breeding ground for ethnic, nationalist, and religious wars, in Somalia, Liberia, Chechnya, Afghanistan, or Bosnia. From these catastrophe zones the victims will be rescued only by international "peace-keeping" interventions when the level of chaos threatens to become uncontainable or to interfere with serious investment opportunities.

Their fate—relayed by real-time satellite broadcasts of famine and street fighting—serve as an admonishment to others in more fortunate places not to demand too much, to buckle down, work harder, be grateful for what they have. Such scenes, whether from Mogadishu, Sarajevo, Kabul, Monrovia, Grozny, or south-central Los Angeles, are not incidental to capitalist globalization. They are essential to it. For they represent the ultimate outcome of a strategy of decomposition that empowers capital by intensifying the world-wide competition between workers—dividing labor from itself in a process whose culmination is an internecine violence.

Globalization as Recomposition

The terrible efficiency of this disintegrative strategy should not be minimized. Yet analysis that understands globalization *only* as capitalist triumph is incomplete. For one of the remarkable features of the last decade is the way in which unexpected currents of opposition have started to emerge from the transformed conditions created by transnationalization. Often these new vortices of subversion have started to spin precisely where the victory of the market forces was thought most complete, as in Mexico, where the Zapatistas' challenge to the showcase of neoliberal development (discussed below) has caught the imagination of the world.

As I discussed in the previous chapter, the scope of contemporary capitalist subsumption means that such movements of opposition will no longer be found concentrated at the immediate point of production but will spill across society as a whole. Battles against corporate globalization involve waged workers, but also unwaged labor: women's organizations resisting the deconstruction of welfare services, students opposing the slashing of public spending, movements of indigenous and peasant people fighting eviction from the land, rural and urban communities refusing the ecological devastation of hazardous waste dumps and hydroelectric development projects. The very diversity of these resistances, and the real nature of the contradictions that often divide them, makes the problems of their cooperation and coordination enormous, even on the scale of a neighborhood, city, or region; when viewed internationally, these obstacles might appear insuperable.

And yet the new countermovements are making trans-sectoral and transnational interconnections. In part, this is happening because capital's very success in creating for itself a worldwide latitude of action is dissolving some of the barriers that previously separated oppositional movements geographically. In collapsing the Three Worlds into a single plane of accumulation, capital has introduced from one to the other forms of work, dispossession, and struggle that were previously segregated. Thus the spread of large-scale manufacturing into Korea, Brazil, or South Africa results in the emergence of mass-worker struggles of a sort that were once distinctively metropolitan while the deindustrialization of the United States and Europe is in turn accompanied by social movements resembling those of the "underdeveloped" world; many authors have noted the similarities between the 1992 Los Angeles riots and Latin American urban insurrections.[45] More generally, the global imposition of neoliberal policies has created commonalities of experience for waged and unwaged labor from Warsaw to Cairo, as the destruction of public services and the subjugation of government to supranational financial flows come to constitute a shared lexicon of proletarian existence.[46]

This exchange of experience is intensified by the vast flows of migrants and refugees set in motion by Third World industrialization, war, and environmental catastrophes. Moving legally and illegally, this huge movement of peoples has converted world cities into crucibles of cosmopolitan experimentation, confounding, confusing, and confronting long-held ethnocentric and colonialist perspectives. Capital everywhere tries to harness these exiles as yet another source of cheapened labor, making them the "new helots" of globalization.[47] Yet migrant workers—Turkish autoworkers in Germany, Filipino nannies and Punjabi farmworkers in Canada, Mexican drywallers and janitors in California—also carry with them traditions of struggle, and they often stand at the heart of new militancies. They challenge the racism of established trade unions and social movements and establish new lines of international connection.[48]

Moreover—and this is the point to which the remainder of my analysis will be devoted—capital's own diffusion of the means of communication has inadvertently assisted this connective process. In creating the pathways for its own transnational circuit, it has unintentionally opened the routes for a global contraflow of news, dialogue, controversy, and support between movements in different parts of the planet. To a degree, the very communication channels that circulate commodities also circulate struggles. Despite all the well-known filtering and censorship mechanisms, corporate and state media do carry abbreviated scenes and news of class conflicts across the world. Sometimes—as in the case of the Israeli invasion of Beirut, or the Indonesian genocide in East Timor—shots of a riot, a bombing, or a massacre have been crucial in mobilizing transnational support for resistances that, in a purely national context, face overwhelming odds. However, to a large extent connections and dialogue between globally distant resistant movements depend on the construction of counternetworks that, while drawing on the technologies and expertise diffused by the world market, reconstruct them into radically new configurations.

Thus, while the effect of globalization has often been to divide workers more strictly within a given city, region, or nation, it has, paradoxically, also created the possibility of building alliances across city, regional, and national boundaries. Writing of the transnational linkages established by the indigenous people's movements, Mariarosa Dalla Costa has spoken of how "workers and non-natives, ecological movement militants, women's groups, and human rights activists have been attracted into complex support actions, helping and monitoring from various parts of the world." In this process, she says, a "hinterland of communication and liaison has been constructed . . . across the Americas and in the world" in which "relations of analysis and information have been more clearly and more strongly interwoven," creating a "tissue of communication between and action by different sectors in the working social body."[49] These new connections run counter to the decompositional logic of capitalist globalization. Somewhere between the ethereal activism of radio and computer networks, and the weary odysseys of proletarians trekking from San Salvador to Vancouver or from Manila to Kuwait City, a new global class composition is being born.

Radiating the Information of Struggle

More than twenty years ago, the autonomist Romano Alquati suggested that the movements of working-class struggles could be analyzed as constituting a network, not just regionally or nationally, but on the international level. This network, he proposed, possessed both vertical and horizontal articulations: "*vertical* according to the organisation of the class at points within and against the capitalist circuit of production and reproduction; *horizontal* according to the geographical-territorial distribution and linkage of these movements within and against capitalist accumulation."[50] In one sense, the structure of the network was given by the capitalist organization of production against which it fought, but "the information passing from the apex to the base of the hierarchy . . . does not correspond to that passing in the opposite direction. In other words, the network of the class struggle, like capital, has its own operational information, its own mechanisms for checking and controlling, but the process based on hierarchical command of capitalist accumulation is turned upside down."[51] This combined vertical-horizontal network of struggles has "nodal points" of interconnection, "points of maximum accumulation of information and greatest direct combination of different moments of the anticapitalist struggle." From these poles "the operational information of struggle is radiated." Such communication about "forms, goals, content, organisation of struggle" was, Alquati said, "an indirect, mediated and complex process, operating through a whole series of mechanisms . . . a form of telecommunication which transcends physical spatial contacts between the nodes that are in communication."[52] Today, of course, these connections are often not just metaphorically but literally telecommunicational. And among the many mechanisms by which it proceeds, a crucial one has been the creation of various networks of autonomous media. I outlined the emergence of these networks in North America in chapter 5. But a crucial ingredient of "the other globalization" has been the eruption of similar experiments across the planet. A feature of contemporary struggle is the degree to which many of the crucial "nodal points" from which the "operational information of struggle is radiated" are to be found in the former Second and Third Worlds.

In the Third World, the creation of autonomous communication networks was, of course, integral to anticolonial struggle; one has only to think of Frantz Fanon's observations on the role of radio in the Algerian civil war.⁵³ The residual impetus of these revolutionary experiments propelled the New World Information and Communications Order, the Third World challenge to U.S. global media dominance. But this complex movement was partly harnessed to the interests of postcolonial state and media elites. Its critique of media imperialism was thus compromised by a certain willingness to overlook internal repressions and exclusions. However, over the last two decades, as national governments submit to privatization and deregulation, the focus of information activism in the South has largely shifted from such a state-centered media base toward a proliferation of independent, grassroots initiatives, arising from sectors in struggle with both local and global rulers: Brazilian street television, video training for Korean trade unionists, township-based South African community radio stations. These media often provide the vital channels for movements opposing capitalist globalization within neighborhood, regional, and national boundaries.

The political potential of these forms of activism was strikingly, though ambiguously, revealed in the fall of state socialism in the former Second World. Here the radio activism of Solidarnosc, the rivulets of *samizdat*, underground music and media permeating Eastern Europe, the role of computer networks and radio stations in relaying news of the Stalinist Moscow coup—all played an important part in undermining the rule of the commissars. A similar but more complex dynamic emerged in the deadly dance of subversion and surveillance surrounding satellite-borne images of events at Tiananmen Square. The relative friendliness of the Western media (not to mention the CIA) toward these revolts makes their success a special case. Yet some would argue that they demonstrate a vernacular familiarity with technology and a popular capacity for the self-organization of communications technology potentially inimical to either state or corporate management, and which can be as effectively turned against neoliberal globalism as against Stalinist isolation.

Others are more skeptical. Against hopeful prognoses about international democratic empowerment through information, they identify the formidable global limitations to and inequalities in access to the means of communication in a world where 40 percent of the population is without electricity and 65 percent have never used a telephone. In an essay tellingly entitled "World Wide Wedge," Peter Golding notes that "the terrain occupied by communication goods and facilities is a hilly one, marked by soaring peaks of advantage and dismal valleys of privation."⁵⁴ He points out that all information industries are very heavily concentrated in the developed world. This holds from newspaper publishing, where half the world's production is in the industrialized West, to telecommunications, with more phone lines in Tokyo or New York than in the whole of sub-Saharan Africa. These inequities are even more marked in computer-mediated communications, which are, as Golding notes, "not so world wide after all," since large portions of Africa, Asia, and Latin America currently lack all but minimal connections to the Internet.[55] These problems are compounded by inequities in the availability of technical training and by problems around translation for digital media, in which English remains the *lingua franca.* The implication of such analysis is that the potential for information activism remains limited to a few relatively privileged zones.

These objections are substantial enough to dampen any naive political optimism. But they are not sufficient to dismiss the possibilities for a significant enlargement in the "network of struggles." Capital is, for its own reasons, diffusing and cheapening access to many information technologies. The inevitably socialized aspects of communications—its broadcast and network aspects, which increase in value according to the number of recipients and participants—means that in many areas business is working very fast to extend the reach of its circulatory apparatus: AT&T aims to complete Africa One, the undersea fiber-optic cable that will create a communications ring around the continent, by the year 2000. Televisions, transistors, and Walkmans are already available in areas without schools, running water, or medical care.[56] More of this distorted universalism can be anticipated.

More important, extreme pessimism about global access to communication resources underestimates the ingenuity of the various communities appropriating these technologies for their own purposes. Movements that would seem at the furthest remove from high tech, such as those of the Mayan peasants in Chiapas, or Indian farmers fighting multinational seed patenting, or the Kayapo and other indigenous peoples in the Amazon opposing World Bank development, are interfacing advanced communication networks and highly traditional forms of mobilization.[57] They are constructing hybrids of pre- and postindustrial communication forms. I have seen film of the village-by-village oral education used by the campaigns of Indian farmers and peasants against GATT. These films were brought to North America by Canadian organic farmers, themselves involved in opposition to agribusiness, who in turn cull and relay information about these companies via the Internet to university-based members of the Indian movement.[58] This sort of interaction constructs patterns of activism that defy prediction.

Further, the transfer of technical expertise and experiences in the establishment of countercommunications is itself becoming a focus of political organizing. The transfer of old computers from North to South, for example, has become a commonplace of international solidarity activities. These global connections, both on North/South and South/South axes are taking organizational form. Alternative radio activists have formed the World Association of Community Broadcasters (AMARC).[59] International associations of video activists and producers such as Video Tierre Monde and Videazimut are experimenting with circulation of videos via broadcast and cable networks, independent distribution circuits, and formal and informal networks.[60] While the largest computer of the Association for Progressive Communications is located in Silicon Valley, it has partner networks in Nicaragua, Brazil, Ecuador, Uruguay, Russia, Australia, the United Kingdom, Canada, Sweden, and Germany. It has affiliates from Vanuata to Zimbabwe and subscribers in ninety-five countries, and it runs projects aimed specifically at facilitating the computer networking of peace, human rights, ecological, and labor organizations in underdeveloped countries.[61]

Out of these activities horizontal linkages between various "nodes of struggle" are now being made on a global basis. They include both the transfer of technical know-how and equipment and the relay of political analysis, discussion, and support. Microwatt broadcasters from California assist Haitian activists in setting up radio stations in Port au Prince; video activists in Vancouver draw on the lessons of popular education from Nicaragua; British motorway protesters at Newbury receive faxes of support from Ogoniland in Nigeria; while environmental activists in Europe deluge Shell offices with email protesting the execution of Ken Saro Wiwa. This is the communicational weave of recomposition. Let us look at some of its patterns.

Modem Solidarity

Through globalization, capital attempts to pit First against Second and Third World workers, undermining the wages and conditions of the former via the immiseration of the latter. However, this paradoxically opens the way to a reverse logic in which workers of the one-time metropolis, losing the position of relative privilege that gave them a partial stake in capitalism's international system, acquire an interest in raising the living standards of those in previously peripheral zones. Northern workers might—and often do—attempt to insulate themselves from globalization by traditional forms of protectionism. But this strategy depends upon business support, at the very time when multinational capital has decisively rejected such an option. An alternative is for (ex-) First World workers to seek alliances with their counterparts elsewhere in the world. Over the period of capitalist global restructuring, the slogan "When they win, we win" has begun to be heard within the most progressive sectors of U.S. and European trade unionism, and there has appeared a tentative web of connections between metropolitan and peripheral labor.[62]

Instances include the ten-year solidarity campaign by U.S. trade union and church groups supporting the occupation of a Coca-Cola plant in Guatemala City; networks of international sugar workers formed around issues of land reform and crop diversification; autoworkers' conferences involving U.S., European, Malaysian, Brazilian, Japanese, and South African delegates; U.S.–Japan worker-to-worker meetings in the computer industry; South African unionists painting murals for striking Minnesota meatpackers fighting an apartheid-implicated employer; West Virginia steelworkers forging links with Swiss, Dutch, and Eastern European unions and green movements to beat their multinational employer; the international connections spun among maritime labor around a Liverpool dockers' strike; and the burgeoning networks of transnational support among female homeworkers, discussed in the next section.[63]

Reviewing such initiatives among U.S. labor in the late 1980s, Kim Moody suggested that their strengths included an activist orientation and the bypassing of bureaucracies in favor of direct communication among militants. Their corresponding weaknesses were lack of resources and frequent suspect status of participants within their own unions.[64] A decade or so later, these obstacles are far from dissolved. But such projects of international solidarity have—largely under the impact of free trade agreements such as NAFTA—become increasingly common. They turn not only on person-to-person contact, but also on communication via film, video, fax, and computer networks.

In a series of important articles, Peter Waterman has analyzed the role of communications in forging the new labor internationalism, focusing particularly on labor's growing involvement in computer-mediated communications.[65] By the mid-1990s this had produced two major networks, the U.S.-based LaborNet and the European Geonet, both devoted to union matters, and a number of activist conferences, in locations from Manchester to Moscow. This labor interest in computerization, Waterman says, arises largely from the obvious need of trade unions facing multinational corporations to possess communication capacities matching those of their managerial antagonist. Increasingly, such unions use large-scale databases to track information on companies' financial status, health and safety regulations, and collective bargaining practices; email for internal communications and solidarity appeals; and bulletin boards for membership orientation and discussion. But while such networks facilitate the conduct of traditional trade union activities on a larger scale, with greater speed, Waterman observes that their operation also often replicates the classic limitations of business unionism: centralized control, a purely corporate basis of organization, and narrow or nonexistent political aims.

However, he goes on to suggest that in Europe, over the 1980s, a new, more expansive approach to modem solidarity began to emerge. This resulted from the interaction of international trade union organizations and a loose ensemble of radical activists, NGOs, communication specialists, and researchers, whose base lay not so much in the European unions as in a variety of Third World collectivities. While the perspective of on-line unionists from the core tended to be pragmatic and utilitarian, those from the periphery were more innovative and experimental, opening up "alternative visions and utopian prospects."[66]

These wider visions included the use of networks in alliance building between unions and other social movements, recognizing differences in needs
and skills among the various potential participants, and emphasizing improved communication as a component of inter- and intra-organizational
democracy. As representative of this broader style of on-line solidarity,
Waterman cites examples such as South Africa's WorkNet, developed by the
alternative press in the antiapartheid struggle and subsequently used by trade
union, church, media, and housing movements; the Asia Labour Monitor
Resource Centre, started by radical church activists in Hong Kong on the basis
of "U.S. computer familiarity and ever cheaper East-Asian computers," to
circulate information about worker struggles in China, Taiwan, South Korea,
and Hong Kong; Mujer a Mujer, a collective of Mexican, U.S., Canadian, and
Caribbean women's groups representing waged and unwaged female labor who
use on-line communication in their transcontinental opposition to neoliberal
restructuring.[67]

One particularly telling example involves Glasnet, the Moscow affiliate of
the Association for Progressive Communication. In the second Moscow coup
of October 1993, where Yeltsinite forces of free-market capitalism re-repressed
democracy, three members of the independent Russian Party of Labor, including author Boris Kargilatsky, were arrested by police, charged with planning
to attack a radio station, systematically beaten, and threatened with death.
A criminal released from jail told the wife of one of the prisoners, who contacted an Australian correspondent for the *Green Left Weekly,* who in turn
reached a Russian union officer with access to Glasnet, who sent an international email alert via a series of computer conferences. "Within hours the
police station was inundated with calls," Kargilatsky writes. "We were watching from the cell. . . . One of the first was from Japan. The police didn't seem
able to believe it. After that the calls seemed to be coming from everywhere—
there were quite a few from the Bay Area in the United States."[68] Email attention was reinforced by the arrival of a Moscow TV crew from the program
"The Individual and the Law." Within hours the detainees were released and
the charges dropped. Waterman comments that "through the concrete and
steel" of state socialism and "out of the shit and blood of an increasingly
globalised information capitalism," there "appears to bloom one flower of
global solidarity—an electronic one."[69]

As Waterman admits, labor's electronic networking is barely nascent, directly involving only a relatively low number of specialists. While it has had
some manifest successes—in pressuring states to free imprisoned militants,
in providing negotiators timely access to strategic information—it is far from
matching, let alone beating, the power that business has discovered in
cyberspace. Nonetheless, he argues that its potential for reorienting workers'
organizations is significant. Drawing on the formulations of media theorist
Fred Stangelaar, he suggests that the realization of these possibilities depends

on labor computer networks becoming a relay in "spiral flows" of alternative communication that both laterally connect a wide range of oppositional groups and vertically heighten their degree of coordination and support.[70] Given this condition, Waterman suggests computer networking could become a vital element in the constitution of what he calls a "fifth international"— a transnational connection of oppositional groupings that does not, like the four previous socialist Internationals, rest on the hierarchical directives of a centralized vanguard party, but rather arises from the transverse communications of multiplicitous movements. Waterman's account corresponds closely to the autonomist concept of the circulation of struggles. Let us examine some further turns of the spiral.

Electronic Boycotts

Movements contesting global capitalization extend beyond the immediate workplace and engage corporate power in the sphere not just of production but of consumption. This manifests in a number of ways—"adbusting," cultural jamming, media piracy—but is perhaps best exemplified in the growing number of transnational boycott campaigns.[71] Groundbreaking instances of this tactic include the boycotts against Nestle's infant formula, South African wines, and Chilean grapes. Recently, these examples have been widely emulated by human rights, feminist, environmental, and labor groups. Targets include clear-cutting by forest companies from British Columbia to Sarawak; Indian carpets made by child labor; U.S. coffee bars supplied by Guatemalan plantation workers; toys produced in the superhazardous factories of China, Hong Kong, and Thailand; and North American clothing chains selling garments manufactured in Taiwanese-owned sweatshops located in El Salvador.

In many ways, capital's own globalizing momentum opens the door to such counterattacks. By making the effects of sweated labor and intensified environmental destruction reverberate worldwide, planetary corporations unintentionally prompt the making of connections between conditions at the point of production and decisions at the point of sale. The heightened combativeness of the international market, the consequent corporate dependence on image and public relations, and, above all, the very communication networks vital to global production and global advertising have made business vulnerable to challenges in the world marketplace. Thus, for example, a campaign waged by labor, religious, and other groups across North America against the exploitation of sweated labor by the sportswear giant Nike could focus on the fact that in 1992 the company paid basketball superstar Michael Jordan more for his promotional efforts than the combined yearly income of thirty thousand young Indonesian women who toiled to piece together the sneakers he advertised. The same campaign also used the Internet to coordinate international global "phone zaps" of Nike headquarters.[72]

An even more striking example is that of the British "McLibel 2." Two British activists were sued by the McDonald's hamburger chain for distributing leaflets denouncing the corporation's low-wage labor practices, child-targeted advertising, involvement in rain-forest destruction, animal welfare record, and promotion of unhealthy diet. By assembling a volunteer defense of international experts that substantiated their accusations they turned the five-year civil trial—the longest in British history—into a public relations fiasco for the company. Worldwide "McLibel" support groups have distributed over 1.5 million copies of the original leaflets, as well as sponsoring numerous demonstrations and disruptions at McDonald's sites across the world.

They have also created McSpotlight, a World Wide Web site combining text, graphics, video, and audio materials in a thoroughgoing critique of the corporation. *The Guardian* newspaper reported that this site "claimed to be the most comprehensive source of information on a multinational corporation ever assembled—and that doesn't sound like an exaggeration."[73] McSpotlight, in addition to documenting the McLibel trial and the claims of the original leaflet, contains news of other anticorporate campaigns and discussions of alternatives to food production by multinational corporations. It reported 190,000 hits in its first week, email responses at a rate of forty a day, and was widely reported by the mainstream press, further discomforting its corporate adversary.

The use of new information channels has also been important in throwing the light of public attention on the shadow work of domestic labor. This has been particularly telling in the highly image-conscious fashion industries, where contracting and subcontracting allow major corporations to distance themselves from slavelike conditions of production. Here feminist organizations have built alliances both among the internationally dispersed home workers and between these laborers and the shoppers—themselves predominantly women—who purchase the products they make.

In doing so they have availed themselves of the most up-to-date means of communication. Thus on the World Wide Web one can find the well-appointed home pages of organizations such as the Clean Clothes Campaign, a movement started in 1990 by Dutch women supporting striking Filipino garment workers in the Bataan Free Trade Zone. Its web site carries information about homeworkers unions and support organizations; coverage of strikes in Lesotho, Nicaragua, Vietnam, Egypt, Jamaica, Sri Lanka, and California; discussions of the strengths and weaknesses of corporate "codes of conduct" and "social clauses" in free trade agreements; news of boycotts and other ways shoppers can support workers, for example, through the use of "clean clothes scorecards."[74]

This is not the place to analyze all the strengths and pitfalls of boycott tactics. Boycotts can, without careful agreement among the different parties involved, lead to disastrous contradictions.[75] But the experiments described here do seem significant. They show electronic communications deployed to link labor, ecological, and feminist perspectives, connecting oppositions to capital across the fields of production, consumption, and reproduction. Aimed at specific products, they nevertheless inevitably prompt questioning of the consumerism that is the complement to capital's doctrine of endless work. And they do so by mobilizing withdrawals of consumption power over the same global terrain on which capital attempts to stimulate it, taking the same technological means corporations deploy to coordinate exploitation and depredation in lonely and underreported places and turning them into instruments of exposure and contestation.

Cross-Border/On-Line

The scope of oppositional networking exceeds resistance to specific corporations. Capitalist globalization entails the subordination of state policy and public spending to international financial flows and treaty obligations. Consequently, opposition to it, whether insurrections against structural adjustment programs or mobilizations against free trade agreements, tend to catalyze the formation of broad movements involving diverse sectors of the working class with interests in resisting privatization, deregulation, and austerity. Further, the transnational logic of capital gives a powerful impetus to the connection of these revolts in regional and multinational organizations. However, such coalitions require the resolution among potential participants of real contradictions and conflicts of interest resulting from capital's international division of labor. They thus depend on communicational channels for information, discussion, and debate.

This was apparent in struggles around the North American Free Trade Agreement (NAFTA). When the final draft of this treaty was announced by the governments of the United States, Mexico, and Canada in August 1992, it became clear that an agreement that gave capital unlimited mobility across borders, pitted labor forces in direct competition with one another, and dismantled a wide range of public services would encounter resistance in all three countries.[76] However, coordination of a trilateral opposition faced serious obstacles. In the United States and Canada the anti-free-trade movement often tended toward a national-chauvinist protectionism. Development of an alternative direction partly depended on contact with and understanding of Mexican social movements. Such efforts would, however, run contrary to both the corporate media's pro-NAFTA predisposition and neglect of issues in the South, and the Salinas regime's state control over Mexican news flows.

In fact the NAFTA debate spawned a wide variety of alternative communications across the Canadian, U.S., and Mexican borders. Visits, personal contacts, conferences, tours, and transborder exchanges, particularly visits to *maquiladoras* by U.S. and Canadian workers, became frequent among activists. While there were important organizational nodes for these transfers, they proceeded from a multiplicity of points in complex and interweaving paths. This circulation of struggles and perspectives was not only carried by on-the-ground contacts but was also made known through newsletters and journals, videos, alternative radio and television broadcasts, and computer networks. These provided the media for the discussion of strategy and tactics, reports on conferences, announcement of cross-border exchanges, organizing efforts, and human rights appeals.

Focusing again on computer networks, John Brenner and Howard Frederick have both made inventories of the anti-NAFTA organizations using on-line communication.[77] These include the North American Worker-to-Worker Network, supporting the connections, within and without official union frameworks, between U.S. and Mexican workers in the automobile, telecommunications, and electronics sectors. They also number feminist organizations, such as El Paso's La Mujer Obrera, fighting to improve the conditions of women workers in the border regions, and Mexico City–based Mujer a Mujer; green organizations, tracking pollutant flows across three borders, or funneling information from North American sources to Mexican opponents of medfly spraying in Chiapas; and a variety of U.S.- and Mexican-based services that specialized in disseminating critical analysis of the free trade negotiations.[78]

The anti-NAFTA coalitions, while mobilizing a depth of opposition entirely unexpected by capital, failed in their immediate objectives. But the transcontinental dialogues that emerged checked—though by no means eliminated—the chauvinist element in North American opposition to free trade. The movement created a powerful pedagogical crucible for cross-sectoral and cross-border organizing. And it opened pathways for future connections, including electronic ones, that were later effectively mobilized by the Zapatista uprising and in continuing initiatives against *maquiladora* exploitation.

While the intensity of transborder networking catalyzed by NAFTA was perhaps exceptional, both because the treaty so sharply posed issues of capital mobility and because of North America's situation as a center of communications technologies, the phenomenon is by no means unique. Thus, if we turn from the Americas to Asia, we can see a similar process unfolding, albeit in a more diffuse way. Over the last five years, India has been systematically opened up to the world market under a New Economic Policy, adopted under pressure from the IMF and the World Bank.[79] In 1992, an estimated fifteen million workers participated in a one-day nationwide industrial strike to protest this process.[80] Resistance has taken a number of forms—some fundamentalist and fascist, such as the Bharatiya Janata Party, others of a broadly left nature.[81] Among these latter there has emerged a variety of transnational alliances, solidarities, and contacts with oppositional movements both in other Asian countries and in the North. These connections flow through multiplicitous channels of oral, written, film, video, and computer communication.

Thus we find world-spanning alliances between Northern environmentalists, Indian "tribals," and urban intellectuals opposing the World Bank's
Narmada dam project—fed by a circulation of films, videos, and email; Indian peasants fighting GATT's intellectual-property clauses visited by Canadian organic farmers, who carry with them books analyzing the activities of
multinational seed corporations, return with films and videos of these same
corporations' offices sacked by million-strong demonstrations, and keep in
touch by email; Internet solidarity appeals from Indian workers occupying
jute factories; Northern NGOs electronically scanning data banks for details
of commercial plans to patent plant and animal species and transmitting the
news back to the resistance organizations of Indian, Thai, and Sri Lankan
farmers; and Indian labor and human rights organizations sending delegates
and films to North American trade unions supporting boycotts of Wal-Mart
megastores selling carpets made by child labor.[82]

These initiatives proceed without central focus. They constitute a diffuse
coalescence of microactivisms contesting the macrologic of capitalist globalization. I would suggest that similar constellations could probably be found
forming at virtually any point on the planet. They exist as a sort of fine mist
of international activism, composed of innumerable droplets of contact and
communication, condensing in greater or lesser densities and accumulations,
dispersing again, swirling into unexpected formations and filaments, blow-
ing over and around the barriers dividing global workers. In the next section
we will consider some of the thunder and lightning that accompanies these
clouds.

Netwars and Antiwars

At its cutting edge, capitalist globalization means war—not only the immediate violence of military attack, whether in the form of imperial invasion
or low-intensity conflicts, but also the sustained social and environmental
violence of starvation, social disintegration, and deprivation that in turn sets
the scene for ethnic rivalries and internal conflict. Consequently, the circulation of struggles between a multiplicity of movements—trade unionists,
feminists, ecologists, indigenous peoples—has increasingly taken the form
of a front arrayed in the name of peace: life against death, a refusal to accept
the sentence that says what is not profitable must be erased. The great international mobilization of the antinuclear movement of the 1980s, which was
colored by the particular interest of the inhabitants of the North in avoiding
the punctual holocaust of nuclear Armageddon, has in a sense broadened and
deepened to become a demand, enunciated from a wide variety of sites, for
the end of the everyday holocausts proliferating all over the planet. This perspective is not strictly pacifist, since it usually entails recognition of the right
of resistance against exploitation and degradation. But it seeks to block the

infinitely greater exterminatory violence brought to bear on such revolts, in order to defend a space for the creation of alternative social options. The new counternetworks transmit an old slogan: "Bread and Peace."

Communication is, again, vital—for exposing the actual or potential atrocities that capital prefers to have executed in secret. The most striking example is of course the uprising of the Zapatista Army of National Liberation against the Mexican government in 1994—a revolt that specifically denounced capitalist globalization as the culmination of a centuries-long dispossession of the people of Chiapas. In an important analysis, Harry Cleaver has suggested that the success of the EZLN in avoiding the normal fate of peasant revolts in Mexico—outright massacre—was partly due to their weaving of an "electronic fabric of struggle."[83] Despite the Zedillo regime's control of Mexico's mainstream media, the EZLN was able to disseminate rapidly its own communiqués not only within Mexico but globally. This was accomplished largely through the network of electronic contacts established via the Internet during the NAFTA campaigns. EZLN documents and news reports flashed into conferences and lists on networks such as Peacenet and Usenet. They were then rediffused, accompanied by additional information, analysis, and discussion from those familiar with the situation in Chiapas, into other parts of the Internet, and from thence into left-wing newspapers, magazines, and radio stations, and, eventually, into the mainstream press.

"Communicative action" then passed into "physical action," not only in a worldwide series of protests at Mexican embassies and government offices, but in an influx of Zapatista supporters—journalists, human rights observers, delegations—into Chiapas.[84] This occurred in a context where NAFTA had made Mexico an exemplar of capitalist development and an object of intense scrutiny by international investors. Cleaver suggests that, together with the many protests within Mexico, it was this focusing of global attention that made it impossible for the Zedillo government to impose a purely military solution, and compelled it to switch to cease-fire and mediated negotiations.

After the initial moments of the revolt, the "electronic fabric of struggle" was strengthened with new threads. Videos made in Chiapas have gone north: microwatt broadcasting has gone south, as radio-activists from Free Radio Berkeley assist Zapatistas and local autonomists in setting up their own microwatt transmitters. The translation of entire books of EZLN documents has been coordinated in cyberspace, and the Zapatistas have established their own "Ya Basta" World Wide Web site.[85] And these electronic flows have in turn attracted interest in the *encuentros* organized by the EZLN with the explicit aim of stimulating global opposition to neoliberalism—international meetings whose discussions have in turn been relayed across airwaves and networks.

For capital and its advisers, such activity is a threat. This was acknowledged by some of its own analysts. In the aftermath of the Gulf War slaughter, two

RAND corporation analysts, John Arquilla and David Ronfeldt, had written a paper suggesting that the conflicts of the future would take the form of "cyberwars" and "netwars."[86] Cyberwar, waged at a purely military level, might "be to the twenty-first century what blitzkrieg was to the twentieth." It would be a type of conflict "in which neither mass nor mobility will decide outcomes; instead, the side that knows more, that can disperse the fog of war yet enshroud an adversary in it, will enjoy decisive advantages." Netwar is a broader concept of "societal-level ideational conflicts waged in part through internetted modes of communication" and entails "trying to disrupt, damage, or modify what a target population knows or thinks it knows about itself and the world around it": "it may focus on public or elite opinion, or both[,] . . . involve public diplomacy measures, propaganda and psychological campaigns, political and cultural subversion, deception of or interference with local media, infiltration of computer networks and databases, and efforts to promote dissident or opposition movements across computer networks." Cyberwar and netwar are "forms of war about 'knowledge,' about who knows what, when, where, and why." Both "revolve around information and communications" and imply that in future conflicts "whoever masters the network form will gain major advantages."

Shortly after the outbreak of the Zapatista revolt, Ronfeldt was interviewed on the situation in Mexico. Although in his earlier writings he had focused on information technologies as instruments of interstate conflict, he had also noted that netwar applies to "low intensity conflict" by "non-state actors, such as terrorists, drug cartels, or black market proliferators of weapons of mass destruction." By "making it possible for diverse, dispersed actors to communicate, consult, co-ordinate, and operate together across greater distances, and on the basis of more and better information than ever," netwar might create a terrain favorable to what would otherwise be small and conventionally weak organizations. Ronfeldt now emphasized that social activists were on the cutting edge of the new "network" system of organizing. Noting that the Zapatistas and other opponents of Zedillo used the Internet, fax, and video, he suggested, "At a time when the political and economic crisis has created widespread disaffection . . . network style organizing will enable the opposition to overcome its traditional factionalism. The greatest threat to the government could be hundreds or thousands of independent groups united in their opposition but accepting of each other's autonomy."[87] Although the "decentralization" of this oppositional force meant it could not "take national power," Ronfeldt suggested its activities could make Mexico "ungovernable": "The risk for Mexico is not an old-fashioned civil war or another social revolution. . . . The risk is social netwar. The country that produced the prototype social revolution of the 20th century may now be giving rise to the prototype social netwar of the 21st century." What Ronfeldt calls "netwars" I would rather call "antiwars"—the mobilization of worldwide communications to hold open spaces within which experiments in autonomy can escape extermination.

Three Examples

Subcomandante Marcos, inputting communiqués to a laptop plugged into the lighter socket of an old pickup truck, has by now become something of a mythical figure both for the Left and its enemies. But the communicational logic demonstrated by the Zapatistas is not limited to Chiapas. I will point briefly to some other examples, from Asia, Africa, Europe, and the South Pacific.

One is East Timor. Here, until a very few years ago, the Indonesian government's invasion and genocide could proceed in quiet obscurity thanks to the huge interests of multinational capital in the development of one of Asia's most populous and resource-rich markets. In the early 1990s this changed, largely due to three events: the filming by British television journalists of the massacre of student demonstrators in Dilli; the circulation of the independently produced film *Manufacturing Consent*, giving central place to Noam Chomsky's analysis of mainstream media silences about Timor; and the establishment of several computer news groups, email lists, and web sites giving information about the situation on the island.[88]

This dissemination of news and analysis has encouraged a proliferation of international actions in support of the Timorese resistance, including civil disobedience and sabotage in England against an aerospace company supplying fighter jets to Indonesia; North American student protests against university cooperation with the Suharto regime; and contacts between Timorese resistance leaders and U.S. workers in Charleston, Illinois, striking against a company with commercial links to the Indonesian government.[89] Furthermore, the illumination of the Timorese situation has spilled over to shed light on other human rights abuses in Indonesia, including the repression of trade unions and students, and the implication of mining corporations such as Freeman McMoRyan in the ravaging of Irian Jaya.[90]

The second case is that of Nigeria. Here again, there is a long history of struggle against the military regime whose self-styled "wasting operations" have swept across pollution-drenched landscapes, protecting the operations of Shell Oil from a population whose living standards have dropped 25 percent in the last twenty-five years.[91] And again, this struggle was shrouded in a handy—from the viewpoint of Shell and General Abacha—oblivion. Until the execution of Ken Saro Wiwa and nine other activists. For Saro Wiwa's role as an author and television playwright placed him at the center of a web of cultural and communication networks. As these transmitted the news of his death, they stimulated an unprecedented volume of mainstream analysis of the Nigerian situation. This provided the opportunity for international solidarity groups to set underway major demonstrations and boycotts against Shell, actions that were publicized and organized through alternative networks of computer, print, and film.[92] This activity in turn built pressure for other campaigns driving for trade sanctions, and all rolled together to create an unprecedented attention to the cost in Nigerian blood of corporate oil.

In Timor and Nigeria, unlike Chiapas, this flow of information has brought no immediate lessening of horrors. But it has resulted in an intensified circulation of struggles. It is, I emphasize, resistance on the ground, in the streets—the willingness of people to fight and die—that lies at the base of these situations. But when the cries of the wounded, the crackle of machine-gun fire, and the pop of tear gas enter into global communication networks, they can create a series of feedback effects and noise very unpleasant to capital. Business went global to find stability and predictability. In search of these goals it will turn a blind eye to, and pay for, unspeakable atrocities. But when such atrocities become visible, capital's very mobility can destabilize its own operations. Facing imponderable risks—the costs of public relations, the consequences of international protests, the rising morale of the local resisters—money sometimes finds it easier to migrate than fight, relocating production elsewhere or evaporating into financial speculation.

This volatility can create difficulties for the local authorities whose task it is to maintain the conditions of accumulation at gunpoint. On some recent occasions, the flight of private funds from "hot spots" has created the need for massive intervention by the highest levels of capitalist organization. In Mexico, partly as a result of the war in Chiapas, and in Russia, partly as a result of war in Chechnya, global financial institutions have had to siphon in billions of dollars to uphold the regimes they are depending on to secure the open market. The funds available for such rescue operations are vast, but not limitless; this is a game that can be repeated once or twice simultaneously, but perhaps not five or ten times.

Let us consider one other example of oppositional networking. In 1995, France's government announced an austerity plan aimed at meeting the Maastricht treaty's requirements for European financial union. The response was a four-week strike-wave that put millions of French workers, students, and citizens into the streets in what has been termed—a tad Eurocentrically—"the first revolt against globalization."[93] These domestic actions coincided with an international outcry against French nuclear testing in the Pacific, which included mass rioting in Tahiti and other islands in the region, worldwide demonstrations outside embassies and airline offices, and a boycott call against French wines. A few months later, the shipment of French nuclear wastes across European borders precipitated three days of pitched battle between German protesters and police.

The link between these apparently disparate events was made in a novel way on the computer list "counter@francenet.fr" that circulated news of the strike. Here an Italian group, Strano Network, proposed a "net strike" ("*grève en réseau*") against French government internet sites, to be conducted by inundating them with hits to the point of paralysis. The proposal reads:

> The French government shows a total contempt for its people, for the international community and for ordinary people who want to see their children grow up in a better world. It carries out nuclear tests in the Pacific. It continues to use "civilian nuclear power." It maintains its projects of "social reconstruction" despite demonstrations of massive opposition. For these reasons we intend to take away (although partially, and for a limited period) from the institutions of the French government the privilege which all the powerful—the lords of war, famine, and social injustice—seek: access to the ever more powerful means of communication and the channels of information, those same privileges which are denied to the vast majority of the global population.[94]

The proposal stirred some on-line debate about its utility or desirability as a tactic, but Strano Network persisted with its initiative, and later issued a report claiming the participation of "several thousands of strikers" and success in shutting down numerous French government sites.[95]

The significance of the strike does not lie so much in its immediate effectiveness—a point on which the critics of Strano Network are probably right—as in the linkages it made, tying together in a worldwide electronic bulletin the austerity inflicted on French workers and the nuclear fallout imposed on Pacific Islanders, pointing to the value placed by neoliberalism on military as against civilian expenditures and to its disregard for popular opinion, global or domestic. Connecting the marchers in the streets of Rouen, the rioters in Papeete, and, prefiguratively, the German antinuclear protesters, it thus created an optic within which the French government's partial retreat from its domestic cutbacks and its abandonment of nuclear testing could be grasped as twin victories against a common enemy. The logic of France's Juppe government and its business and financial backers is that of capitalist globalization. The logic of Strano Network, of the French strikers, and of the German and Pacific Island rioters is that of a transnational counterpower that is mobilizing by means of the very networks capital has created.

The Globalization of Others and the Other Globalization

In an earlier era, prospects of breaking through the net of the world market were often thought to lie in the piecemeal withdrawal or disassociation of liberated zones, which would succeed first in peripheral zones and gradually surround and destabilize the capitalist center. This concept was given classic expression in Samir Amin's theory of "delinking"—often interpreted, and in some cases implemented, as a program of nationalist autarky.[96] In today's situation, where the integration of economic activity has reached an entirely new level and the positions of metropolis and periphery become profoundly intermingled, such concepts become increasingly problematic. At the very least, it is paradoxically apparent that any localized *delinking* can only succeed as a moment in a series of highly *linked*, mutually supportive regional and transnational projects of withdrawal.

In the current context a more promising line of initiative is what Jeremy Brecher and Tim Costello call "globalization from below."[97] This refers to the activities of "peoples transnational coalitions," formed across national boundaries by social movements aiming to fulfill mutually supportive objectives for workers in different parts of the world.[98] Brecher and Costello suggest that such movements will come to oppose the "downward harmonization" of wages, social wages, human rights, and environmental standards effected by free trade agreements and financial discipline with demands for "harmonization upward"; they will have as a priority the democratization of economic institutions and be oriented toward the creation of "a multilevel one-world economy" with regulation above and below the level of the nation state.[99] Such proposals are often presented within a reformist perspective that obscures the depth of confrontation with capital that their realization would require. Nevertheless, "globalization from below" seems to correspond roughly to many of the tendencies in transnational struggles identified in this chapter.

I have suggested that the increasing circulation of struggles during the crisis of the 1960s and 70s compelled capital to a fundamental reorganization, one that broke down the previous "triplanetary" segregation of the globe into First, Second, and Third Worlds. The objective of this maneuver was to unify and integrate the circuits of profit while severing and destroying connections among the working class, decomposing points of opposition and unrest from the industrial factory to the jungle paddy field. This process has, however, unintentionally created the terrain for a new recomposition of oppositional forces—not least by its fabrication of a worldwide net of communications, a net formed to facilitate the operations of the market, but increasingly expropriated by oppositional forces for very different purposes. The result has been to produce not *one* but *two* globalization processes—simultaneous, superimposed, interdependent, and antagonistic.

The first is capitalist globalization. Its tendency is to create incredible wealth and power for the few controlling the flows of international investment and finance; improvements in living conditions within a persisting context of exploitation for some; and, for very many, a chaos of immiseration. Celebrated, with partial truth, as the unification of the planet, this globalization also carries within itself a lethal acceleration of divisions and antagonisms. For its mechanism is an intensification of competition within a planetary market, an intensification of polarities and hierarchies in a "one world" economy, a relentless setting of labor against itself—a globalization of "others."

The alternative, opposing tendency is that of the worldwide countermovements confronting transnational capital. As Waterman points out, these movements appear to have "no international headquarters, no organization, . . . no obvious terrain of battle"; but "alternatively, one could say that they have many headquarters, many organizations—and many terrains, forms and levels of struggle."[100] Appearing first as a series of sporadic and localized neighborhoods of survival and communities of resistance, these struggles are generating a series of connections, contacts, coalitions, and networks of cooperation. They aim at the creation of a world space that, rather than being subject to the monologic of capital, contains within it the conditions for the interaction of diverse ways of living and organizing. This is "the other globalization."

7

POSTMODERNISTS

Notions of information revolution
carry around with them, like a flickering aura, that most shimmering of contemporary concepts—"the postmodern." Theorists of a "postmodern condition," such as Jean François Lyotard, Jean Baudrillard, and Gianni Vattimo,
explicitly or implicitly base their claims about radical changes in today's
society on the analysis of postindustrialism previously posited by Daniel Bell
and other futurologists.[1] So deeply embedded in postmodern theory is the
belief that computers, telecommunication, and other high technologies are
a vital element distinguishing our epoch from the fading modern age that it
can be seen as offering a new inflection of the earlier distinction between
"postindustrial" and "industrial" eras, now reworked to stress the epistemological, philosophical, and aesthetic consequences of this transformation.[2]

Postmodern Postindustrial Proletarians?

It is hardly surprising that Marxists' encounter with postmodern theorists has
largely followed the trajectory of their earlier meeting with the postindustrialists—ending in hostile collision. This is unfortunate. For although postmodern theory often accepts too easily the idea that high technology inaugurates a historically unprecedented era, it does not usually look on this

prospect with naive enthusiasm. Indeed, it includes highly critical perspectives. Moreover, postmodern theory is a beast with several heads, some venomously anti-Marxist, but others much more conversational. Thus while there are very substantial issues at stake in Marxist/postmodernist polemics, such arguments can also sometimes constitute a disabling fracture of intellectual forces antagonistic toward high-technology capitalism.

Recently, certain lines of theory, emerging from both the Marxist and postmodernist camps, seem to reach across this divide toward new dialogues, *rapprochement*, or even synthesis. Such efforts have concentrated on identifying certain aspects of postmodern culture as manifestations of capitalist restructuring. However, they have had relatively little to say about the *sine qua non* of Marxist analysis—the possibility of opposition and subversion. In this chapter I suggest that autonomist Marxism, as developed by theorists such as Antonio Negri, Gilles Deleuze, and Felix Guattari, can supply this deficiency. These theorists offer a sort of recombinant postmodern/Marxism that, without sacrificing the Marxist emphasis on class struggle, admits important postmodern insights into the variegated and technologically mediated aspects such conflict assumes today. In doing so, they open important perspectives on the postmodern proletarian condition—a disturbing and exciting scene of simulacra, cyborgs, net-nazis, and rhizomatic alliances.

Hostilities: Postmodernity versus Marxism?

The hostilities between postmodern theory and Marxism have important historical roots. Many of the Parisian progenitors of postmodern theory—Jean François Lyotard, Michel Foucault, Jean Baudrillard, Julia Kristeva—were onetime Marxists for whom the defeat of the student-worker uprisings of 1968, and particularly the total failure of the French Communist Party to comprehend or respond to these revolts, was a watershed of disillusionment.[3] The theories they subsequently developed can in part be seen as an attempt to understand the nature of conflicts apparently beyond the ken of orthodox Marxism—conflicts in which, for example, the leaders of dissent were not factory workers but university students—and also to comprehend why these new movements failed in their revolutionary aspirations. There was thus implanted at the root of postmodern theory an anti-Marxist tendency, which, although it in many cases turned in outrightly reactionary directions, also contained strong radical impulses.

In their attempt to grasp the problems Marxism had apparently failed to address, the dissident Parisian intellectuals looked, somewhat incongruously, to conservative American sociology, and concepts of postindustrialism. Just as, according to postindustrial theory, contemporary societies are passing beyond industrialism to informationalism, so, according to the prophets of postmodernity, we are now speeding past the limits of modernity, with its

confidence in reason, progress, and universalist political projects, into unknown territory. Among the most important features of this postmodern
world is its communicational texture. Signifiers are supreme over referents,
images more powerful than substance, symbols trump things. The real is
constituted by a play of texts, discourses, language-games, or codes. While
this inseparability of world from word may perhaps always have been the case,
what now intensifies and renders it apparent is the growing prominence of
information technologies that saturate society with messages and images and
break down the solidity of the material world into an immaterial flow of digits
and data subjected to infinite processings and reprocessings.[4]

The result is an ambience mobile, multiplicitous, and elusive in the extreme. The proliferation of media channels throws all stable and authorita-
tive accounts of the world into crisis. This collapse of what Lyotard calls
"metanarratives" may be seen either as potentially liberatory diversification
or as profoundly atomizing and disintegrative cacophony, but it is in either
case inescapable.[5] Its force is such as to explode the possibility of any unitary or totaling perspective on society as a whole, leaving only a contingent
juxtaposition of incommensurable perspectival shards and fragments: "playing with pieces, that is postmodernism."[6] Lamentation for lost unities and
stabilities is beside the point: all that is possible is clear-eyed acceptance of
a transformation that has shattered pretensions to theoretical mastery of
society and, with it, all grand projects of political emancipation.

In such postmodern theorizations, Marxism is depicted as fatally anachronistic—usually elected as the exemplary case of modern thought, only to
be immediately consigned to the dustbin of posthistory. Lyotard catches the
prevalent tone: "The mere recall of the well-known guidelines of Marxist
criticism has something obsolete, even tedious, about it. . . . the ghost has now
vanished, dragging the last grand historical narrative with it off the historical stage."[7] The decisive influence of the "mode of production" is superseded
by that of what Mark Poster terms the "mode of information."[8] Marxism
claims that the economic sphere constitutes a ground-level base of which
other cultural superstructures are mere epiphenomena. These claims expire,
however, as it becomes apparent that the real is made, not in the material
transformation of the world, but in the immaterial play of signification. Consequently, the importance attributed by Marxists to class—that is, location
within relationships of production—is dissolved in favor of concepts of social identity as decentered, transitory, and heterogeneous. Furthermore, in a
world now revealed as containing innumerable and incommensurable accounts of the real, the Marxist ambition to "grasp the totality"—that is, to
gain a comprehensive overview of the societal whole—becomes not merely
unattainable, but intensely suspect. It is seen as a manifestation of a dominative will-to-power deeply related to totalitarian schemes of social control—
a megalomaniac theoretical dream that leads straight to the Gulag.

It is hardly surprising, then, that the first—and often last—response of many Marxists to postmodernism is withering hostility. Postmodernist tendencies have been denounced by Marxian scholars as a "mystique . . . which strives to cultivate ignorance of modern history and culture" and serves to "echo the ruling-class self-delusions that it has conquered the troubles and perils of the past";[9] as a linguistic idealism that has "strafed meaning, over-run truth, outflanked ethics and politics and wiped out meaning";[10] as an irrationalism that "challenges the very notion of emancipation" and "produces an anxiety-ridden sense of chaos and isolation";[11] or as just "the smoked-out butt-end of . . . theory."[12]

Marxists have pointed to the many self-contradictions into which post-modern theory lapses as it dismisses totalizing theories while itself indulging in the most airily grandiose gestures of historical speculation. They have challenged the credibility of the information-society theory whose accuracy so much postmodernist thought simply assumes, with its implausible claims that capitalism has quietly succumbed to ineffable postindustrial evaporation. They have pointed out the lack of self-reflexivity postmodern theorists often display about their own class-situation. Alex Callincos, for example, has tellingly suggested that the popularity of postmodernism owes much to the fact that it elevates to the level of general theory the experiences and habits specific to a particular stratum of intelligentsia immersed in cultural production and anxious to arrive at an accommodation with an apparently triumphant capitalism.[13] Others have effectively demonstrated how destructive is the belief—which some postmodernists certainly flirt with, and which Marxists generally ascribe to all of them—that it is impossible to know anything beyond the images dominating contemporary life.[14]

I tally this passage of critical arms as a bloody draw. Marxists have effectively ridiculed postmodern theory's hyperboles and inconsistencies. This, however, cannot cancel out the fact that such theories identify, often in intentionally ironic and provocative style, aspects of life in an information-intense, technologically enveloped society that have previously escaped Marxist analysis. Foucault's concept of "panopticism," Baudrillard's discussion of "simulation," or Lyotard's account of "immateriality" all speak to phenomena that are neither immediately dismissible nor already defined in the standard dictionaries of historical materialism. At the very least, they touch on crucial aspects of what Raymond Williams called the "structure of feeling" of contemporary life in advanced capitalism.[15]

Moreover, while Marxists are right that the postmodern rejection of "metanarratives" is untenable (so that, as Fredric Jameson notes, the refusal of totalizing theory simply results in its surreptitious and unacknowledged reappearance via the back door), this does not answer the postmodernists' point that something is seriously amiss with the *specific* metanarrative of classical Marxism—namely, that its central protagonist, the industrial proletariat, seems to have gone absent, missing in action in a field of robots, computers, and telecommunications.[16] Postmodern theory's undeniable insights into new mechanisms of power and new social subjectivities have thus been thrown up against Marxism's equally unanswerable arguments about the persistence of capitalism and the implacable consequences of commodification, generating a profound theoretical impasse.

Rapprochements: Beyond the Great Divide?

Certain attempts to surpass this impasse have emerged, proceeding from both sides of the postmodern/Marxist divide. From the Marxist camp, the pioneering example is Fredric Jameson's now-canonical essay "Postmodernism, or the Cultural Logic of Late Capitalism."[17] In this essay, Jameson argues that the emergence of a distinctively postmodern culture, rather than marking a break with capitalism into a new era, corresponds to the "late" or "multinational" stage of capitalism analyzed by Ernest Mandel.[18] In this phase, previously untouched domains of social activity are penetrated by the forces of a technologically integrated world market. One aspect of this process is a surge in the commodification of cultural and communicational forms. Advertising, design, marketing, fashion, and entertainment become a primary focus of commercial activity. Consequently, the distinction—valid for earlier stages of capitalist development—between an economic base and cultural superstructure collapses. Capitalized culture envelops all aspects of the social in an omnipresent wrap of imagery whose multiple surfaces extinguish material reference or sense of history. Subjectivity becomes, as postmodern theory suggests, increasingly decentered and unstable—experiencing a condition not so much of alienation as fragmentation, induced by the fluctuating stimuli of electronic media and the malleable spaces of commercial architecture and urban design.

This analysis—whose boldness is indicated by the fire it drew from partisans of both Marxism and postmodern theory—has subsequently been elaborated in a variety of ways, most notably by connecting postmodernity to the concept of a post-Fordist regime of accumulation.[19] The most impressive of these efforts is that of David Harvey, who relates postmodern culture to post-Fordist "time-space compression."[20] Capitalism, says Harvey, is periodically compelled to flee the risk of overproduction by both expanding the geographical horizons of the market and accelerating the circulation time of commodities. At such moments, society undergoes a massive speedup in the pace of daily life and a dramatic expansion in spatial horizons. Since 1972, the passage from Fordist mass-production to a post-Fordist regime of flexible accumulation has precipitated such a convulsion in North American and European culture, such that "spaces of very different worlds seem to collapse upon each other, . . . and all manner of sub-cultures get juxtaposed."[21] Postmodern culture—with its cosmopolitanism, eclecticism, and volatility—is both reflective and constitutive of this shift: its emphasis upon "ephemerality, collage, fragmentation, and dispersal . . . mimics the conditions of flexible accumulation," and also stimulates the new images, fashions, and styles of thought that are so central to the restructuring of production.[22] Although Harvey is fiercely skeptical toward postmodern theory, which he believes fails to distance itself critically from the transformations it records, he does allow that it recognizes, albeit in mystified form, important alterations in the structuring of subjectivity and perception. Postmodernism registers a "sea-change" in culture caused by a "shift in the way in which capitalism is working these days."[23]

Jacques Derrida, the leading poststructuralist philosopher, has in a way met these lines of Marxist analysis from the other side of the hill. To the infinite dismay of many of his disciples, in 1994 Derrida broke his decades-long silence on the topic of Marxism, not to issue one more declaration of its obsolescence, but, on the contrary, to affirm its unsurpassability of a horizon of contemporary thought. In fact, Derrida suggests, it is precisely the immaterial or "spectral" conditions of contemporary production, on which so many postmodern theorists have dwelt so extensively, that throws into new salience certain features of Marxist analysis.[24] In particular, the internationalization of production through telecommunication has made the issue of the world market, and with it issues of exploitation and inequity in the distribution of global surplus, inescapable. Rather than agreeing with Lyotard that "the [Marxist] ghost has now vanished," Derrida argues that the "spectral" conditions of the new global economy, an economy predicated on media empires and telework, in fact summons up the continuance of Marxism as a "spectral" presence, a certain spirit of resistance against injustice that obdurately refuses to vanish from the world stage.

These various postmodern/Marxist conversations are of considerable importance. Yet they lack a crucial dimension. While all in various ways identify aspects of what might be called "postmodern capitalism," all are virtually silent on the question of opposition to such an order. Derrida calls for a New International, but does not specify how or where this might emerge. As Adrian Wilding points out, Derrida's reasserted Marxism is undermined by his insistence that the specter of revolution can never be conjured in full presence, that communism is an ever-deferred futural project, "urgency, imminence, but, irreducible paradox, a waiting without horizon of expectation."[25] Jameson suggests that postmodern culture has to be seen dialectically as *both* a mystificatory veil over the realities of contemporary exploitation *and* a field of emancipatory potential, but says almost nothing about how this latter potential might manifest.[26] Similarly, Harvey evokes a revival of historical materialism but gives no indication of where this regenerated Marxism might find its protagonist or translate into political practice.[27]

These silences signify a major problem. For, if Marxism cannot under contemporary conditions locate agents of contestation and practices of opposition, its analysis of postmodern capital amounts only to a reiteration (albeit on a more political economic basis) of the chief point of anti-Marxist postmodern theory: that under postmodern conditions, the game is over. The struggle does *not* continue. What is therefore required is analysis not of postmodern capital alone, but also of the subject(s) potentially antagonistic to it: an analysis of the postmodern proletarian condition. For at least some hints in this direction we can look to autonomist Marxism, and in particular to the work of Negri and his collaborators, Gilles Deleuze and Felix Guattari.

Recombinancy: Postmodern Class Struggle?

To situate the autonomists within the Marxist/postmodernist debate, some historical perspective is again useful. As we saw earlier, the *autonomia* movement emerged from the wave of struggles that swept Italy during the 1960s and 1970s, starting in industrial plants but rapidly involving universities, schools, homes, urban squats, radio stations, transportation networks, cultural organizations and every facet of their society—struggles similar to, but more protracted than, the French student-worker revolts that provided the seedbed of postmodern theory. However, unlike both the official French and Italian communist parties, the Marxists of *autonomia* did not reject the widespread uprisings outside the factory as marginal and incorrect, but rather embraced them and tried to adapt their theoretical perspective to encompass these new points of conflict. Many postmodern theorists—such as Michel Foucault, Paul Virilio, and, especially, Felix Guattari, who was actively involved with dissident radio in Italy—had sympathies with *autonomia*. When the movement was repressed and its leaders were put on trial, they joined in the international campaign against persecution. Negri, fleeing Italy, found refuge in France through the assistance of Guattari, with whom he has subsequently worked collaboratively.

Negri has referred to his own work as a theory of "class antagonism in the postmodern world."[28] From what we have already seen of his work, it is perhaps not hard to understand why. For while Negri reaffirms the Marxist analysis of the war between capital and labor, he reinterprets this antagonism within a horizon that emphasizes both the diverse sites over which this conflict is fought and the importance to it of communicational practices.

It will be remembered that Negri, like other autonomists, traces class conflict through a series of cycles of struggle—from the "professional" or craft worker at the end of the nineteenth century to the mid-twentieth-century "mass," industrial factory worker. Each of these cycles of conflict has driven capital to adopt successively more highly organized and technologically intense forms. This trajectory has today led to a situation where "the factory spreads throughout the whole of society. . . . production is social and all activities are productive." However, according to Negri, such a development only inaugurates a new cycle of struggle—that of the "socialized worker."[29]

For, says Negri, capital's self-enlarging subsumption of society also multiplies the potential points of resistance. When the locus of production shifts from the factory to society as a whole, anticapitalist antagonism is no longer concentrated in the mass factory, but radiates out to households, schools, hospitals, universities, media, and so on. Struggles at each site manifest their own specificity, yet all encounter a barrier in capitalism's subordination of every use-value to the universal logic of exchange. Thus, unlike the relatively homogenized, factory-based "mass worker," the "socialized worker" arises from a pluralistic, variegated form of labor power whose ranks include not only diverse forms of wageworker (in the service as well as industrial sector) but also the unwaged workers (homemakers, students) whose activities are indispensable for the operations of the social factory. As Negri puts it, in a formulation that clearly shows his convergence with characteristically postmodern themes of heterogeneity and diversity, "The specific form of existence of the socialised worker is not something unitary, but something manifold. The paradigm is not solitary, but polyvalent. The productive nucleus of the antagonism consists in multiplicity."[30] Moreover, Negri argues, the social expansion of capital gives both its operations and the struggles against them an increasingly communicational nature. Avoiding the base/superstructure metaphor, whose baggage of mechanical materialism has so plagued Marxism, Negri's rests instead on Marx's observations about the importance of "laboring co-operation." For Marx, a central feature of capital's enlarging organization was its attempt to impose despotic managerial control over a work force whose activities depended on "collective unity in co-operation, combination in the division of labour."[31] Developing this theme, Negri says that the advent of the "social factory" produces "a specific social constitution—that of co-operation, or, rather, of *intellectual co-operation* i.e., communication—a basis without which society is no longer conceivable."[32] To coordinate its diffused operation, business must interlink computers, telecommunications, and media in ever-more-convergent systems, automating labor, monitoring production cycles, streamlining turnover times, tracking financial exchanges, scanning and stimulating consumption in the attempt to synchronize and smooth the flow of value through its expanded circuits. It is only through the elaboration of this vast information system that "advanced capitalism directly expropriates labouring co-operation":

> Capital has penetrated the entire society by means of technological and political instruments (the weapons of its daily pillage of value) in order, not only to follow and to be kept informed about, but to anticipate, organise and subsume each of the forms of labouring co-operation which

are established in society in order to generate a higher level of productivity. Capital has insinuated itself everywhere, and everywhere attempts to acquire the power to co-ordinate, commandeer and recuperate value. But the raw material on which the very high level of productivity is based—the only raw material we know of which is suitable for an intellectual and inventive labour force—is *science, communication and the communication of knowledge.*[33]

The preeminence of "communication" as a category in postmodern theory, Negri claims, registers this process. In the *Grundrisse* Marx explains that the discovery of "labor" was an historical event. Although the category "labor in general" represents an "immeasurably ancient relation valid in all forms of society," nevertheless it had to await formulation until capital's forcible "abstraction" of labor power—technologically reducing craft skills, homogenizing the work force, stripping workers of all attributes other than as a factory "hands"—gave it "practical truth."[34] Today, Negri suggests, the incorporation of a variety of informational flows and interaction into production is imposing a similar "abstraction" on the concrete variety of communicative practice. This is perhaps most readily recognized in the creation of a universal digitalized idiom into which all forms of communication can be coded and transcoded as "information"—a flow of bits and bytes that can be measured and monitored as the stuff of workplace productivity and pay-per services.

However, Negri says, this development has a double face, each side of which is recognized by a different branch of postmodern thought. One side is the harnessing of all sorts of communication to ever-expanding commodification, the reduction of social relations to a series of exchange relations, and the consequent hollowing out of meanings and relations: "In the circulation of values, every commodity has become money, every reference appears in a circuit of equivalents . . . every singularity has lost all significance and the sense of being has become pure paranoia."[35] This is caught by what Negri calls the more "banal and pessimistic" version of postmodernism, in which the novel features of the age lie in "the total disintegration of received language, of its meanings and expressions . . . the tectonic slippage of all foundations."[36] This negative moment of postmodernism arises from the sense of immersion in capitalist subsumption—a vast apparatus whose sole purpose is, in Marx's terms, "production for production's sake." This situation, Negri says, produces "a painful . . . perception of the total insignificance of the being in which we are immersed; a being whose framework and directions we no longer perceive."[37]

However, Negri suggests, there is another aspect to capital's extraordinary development of its informational apparatus—namely, that its channels can potentially be used for purposes other than those for which it was intended. It is these creative openings that are glimpsed by what he regards as the more

"sophisticated and positive" versions of postmodernism, attuned to the "plurality of languages, the uncertain role of judgements, and the becoming-ever-
more absolute of the horizon of communication."[38] At its best, Negri says,
such postmodern theory "presupposes not merely an enormous, fluent universe of communication, but throughout every stretch of this mass of communicative threads it identifies contradiction, conflict, and, above all, new
power."[39] In this version, postmodernism constitutes "a primitive but effective allusion to the . . . new subjects which appear in the Marxian phase of
general circulation and communication."[40]

In Negri's view, the negative and positive moments of postmodern theory
between them present a portrait of the contradictions that run through a
capitalism predicated on a vast communicational infrastructure—"simultaneously the ruin and the new potential of all meanings."[41] Both offer important insights, yet each provides only a partial perspective. The first responds
to the deepening reach of computerized commodification but nihilistically
denies the possibility of resistance; the other recognizes the "socialized
worker's" potential for experiments in diversified and democratic communications but occludes issues of exploitation and capitalist control. Only when
the two tendencies are seen counterpoised in ongoing conflict does an adequate perspective emerge. Thus in Negri's view postmodern thought is "ambiguous"; although "eclectic," it does identify "certain conditions on which
it is possible to construct the concept of new subjectivities."[42]

Negri's Marxism thus enters into a tentative rapport with postmodern
theory. Yet his insistence on the universal and progressive goals of struggle
is also reminiscent of the postmodernists' major modernist opponent, Jürgen
Habermas. Negri's contrast between dominative information and insurgent
communication owes an acknowledged debt to Habermas's theory of communicative action, which upholds an "ideal speech situation" of democratic,
symmetrical dialogue unobstructed by inequities of power and skill as a yard-
stick against which to measure emancipatory social change.[43] However, for
Habermas economy and workplace lie outside the orbit of such judgment and
are subject to an instrumental logic that finds inexorable embodiment in
capitalist rationalization. The consequence is a purely defensive social democratic politics that aims to protect select areas of the "life-world" from the
encroachments of the "system," but abandons any fundamental challenge to
capital's dominance of productive activity.[44]

For Negri, in contrast, the advent of the "factory without walls" makes it
impossible to split work from life. The increasing prominence of communicative action is a result of the socialization of production. Conflict between
instrumental and communicative logic crystallizes around the contradiction
between capitalist command and collective labor; and the horizon of the
"ideal speech situation" can only be reached by way of a full-blown revolu-

tionary project whose ultimate objective remains the demise of capital. In the next two sections I elaborate this point by looking briefly at examples of what Negri would consider "negative" and "positive" moments of postmodern analysis, and their relation to autonomist theory.

Simulacra: The Reality Gulf

What Negri terms the "banal and pessimistic" school of postmodernism is best represented by the school of Jean Baudrillard and his followers. Baudrillard, after starting from a brilliant critique of orthodox Marxism's base/superstructure dualism and an incisive analysis of cultural commodification, has since gone on to develop an ever-more nihilatory analysis of the power of media.[45] In an age of advanced information technologies, he claims, signs, which once pointed to reality and then served to mystify it through advertising and propaganda, have now come to substitute entirely for it. We enter a world of simulacra, where models come before originals. In this hyperreality, "The territory no longer precedes the map, nor survives it. Henceforth, it is the map that precedes the territory. . . . The real is produced from miniaturised units, from matrices, memory banks and command models—and with these it can be reproduced an indefinite number of times." Subjectivity is no more than an effect of an omnipresent "code," produced by a shadowy "neocapitalist cybernetic order."[46] A "cyberblitz" of advertisement, propaganda, television shows, and polling techniques produce the very needs, desires, opinions, and identities to which they ostensibly respond. A media apparatus that effortlessly recuperates opposition as spectacle annuls every antagonism. With reality itself constituted by wall-to-wall media images, the epistemological ground for distinction between actuality and imaginary, truth and lies, fabrication and authenticity evaporates. Social existence undergoes an "implosion," becoming a "black hole," a spongy, infinitely absorbent mass that soaks up media images from Bobbits to Bosnia, indifferent to veracity but hungry for ever more intense waves of sensation.[47]

The recent culmination of this line of thought comes in Baudrillard's articles on the Persian Gulf War.[48] Written at the time of the conflict, these focused on the role of the media in a war where "our strategic site is the television screen, from which we are daily bombarded."[49] Baudrillard claims that propaganda and disinformation make it impossible to know what was actually going on in the sands around Kuwait: epistemological certainty, including even the confidence that what was occurring constituted a "war," was swallowed up in an abyssal "reality gulf."[50] While he admits that large numbers of people were killed and cities bombed, the "virtual" nature of the electronically mediated hostilities makes any "practical knowledge of this war . . . out of the question."[51] All that skeptical intelligence can do is "reject the probability of all information, of all images whatever their source."[52] The aim of this is not to "seek to re-establish the truth"—for which, Baudrillard insists, "we do not have the means"—but rather to "avoid being dupes."[53]

Despite the denunciations that these and other of Baudrillard's writings have rightly attracted, they should not be lightly dismissed. His account of the simulacra has, as Negri puts it, "a very high degree of descriptive power."[54] In many ways it more fully acknowledges the enormous challenges facing oppositional movements today than many more conventional Marxist accounts of "ideology." For it registers a situation in which control of the media often (if not as uniformly as Baudrillard suggests) gives established power the capacity not just to promulgate specific beliefs and values, but to set the very parameters of perception. Moreover, Baudrillard's account recognizes— even as it reinforces—one of the most problematic aspects of the postmodern proletarian condition, namely, that awareness of such manipulation may take the form of a deep-seated cynicism and relativism, inimical to activism. Indeed, his account of "social implosion" can be seen as a percipient account of an advanced state of class decomposition in which solidarity and agency have broken down in favor of atomization and spectatorship.[55]

Negri himself uses Baudrillardian language to describe this capitalist "duplication" of reality. Discussing the neoliberal state (which they also term "the postmodern state"), he and Michael Hardt suggest that one of its central roles in capitalist restructuring has been to disintegrate the institutions of civil society (trade unions, political parties) so as effectively to annul political debate. However "this void must be covered over by the construction of an artificial world that substitutes for the dynamics of civil society." Thus "even while the real elements of civil society wither . . . its image is proposed at a higher level."[56] Here, they remark, "The new communicational processes of the so-called information society" play a vital role, with a move "from the democratic representation of the masses to the representative's production of their own voters"; "through the mediatic manipulation of society, conducted through enhanced polling techniques, social mechanisms of surveillance and control, and so forth, power tries to prefigure its social base."[57]

Where autonomist analysis parts company from Baudrillard is, of course, on the possibilities of challenging and subverting the reign of the simulacra.[58] Underlying Baudrillard's fatalistic cynicism is a highly structuralist view of the subject as simply an effect of the dominant "neo-capitalist" cultural code. An autonomist perspective would understand the operations of this dominant code not so much as constructive as reductive—something that selects, limits, and constricts the possibilities of a more expansive field of social practices that always includes at least some elements "other" than capital. If the self is always fabricated, some fabrications promote a subjectivity of passivity, dependency, and indifference, while others foster agency, autonomy, and inquiry.

In a rather cryptic phrase, Negri has suggested that in the face of the "duplicatory" power of capital, the task of opposition is nothing less than "a Socratic task—that of reimposing the principle of reality."[59] This need not imply a naive objectivism. But it is, in the flickering world of postmodernity, an important affirmation of the possibility of distinguishing between truer and falser depictions of reality—in the sense of identifying more or less coherent and comprehensive accounts, and more or less manifestly self-interested narratives.

Even advanced capital does not so completely or efficiently monopolize the channels of communication as to make this activity impossible. As Christopher Norris has argued, even in the midst of the Gulf War propaganda blitz the activities of a few reporters of integrity did occasionally make it possible to discern the discrepancies and omissions of official accounts.[60] And Europe and North America also saw some remarkable uses of video, alternative television, and computer networks to transmit news and analysis marginalized or excluded from mainstream accounts.[61] Although these efforts were, in Robert Hackett's phrase, "engulfed" by the U.S. military-marketing campaign, they nevertheless point to potentialities that in other circumstances could be more effective.[62] Indeed, Negri would argue that one of the characteristics of the socialized worker—or postmodern proletarian—is his/her increasing ability to reappropriate capital's communicational machines in order to contest its simulations. But to consider this possibility we should turn to a more optimistic version of postmodern analysis.

Cyborgs: Living/Dead Labor

For such an example, we can do no better than to look at the notion of the "cyborg" presented by Donna Haraway in her "Manifesto for Cyborgs: Science, Technology, and Socialist Feminism in the 1980s."[63] For Haraway, the figure of the cyborg—a cybernetic organism—provides an "ironic myth" expressing contemporary possibilities for political activism in an era when capitalism operates through a high-technology "informatics of domination."[64] To refer to the inhabitants of this global system as "cyborgs" is to suggest that in a society permeated by media, computers, and genetic engineering, subjectivity has in a profound and irreversible way become technologized—formed at the interface between human and machines. Drawing on postmodern theory, Haraway argues that in such a technological world, identities cannot be predicated on some "essential" nature, but are instead relentlessly artefactual and constructed. However, in a spirit diametrically opposite to the antitechnologism of much left and feminist thought, she does not find this prospect defeating or dispiriting. As she puts it, "cyborgs . . . are the illegitimate offspring of militarism and patriarchal capitalism, not to mention state socialism. But illegitimate offspring are often exceedingly unfaithful to their origins."[65] At its most literal level, cyborg politics means refusing a "demonology of technology" and embracing the possibilities of reappropriating the instruments of information capitalism for alternative purposes, reconstituting the boundaries of daily life by "both building and destroying machines."[66] More broadly, the border-transgressing figure of the cyborg is for Haraway a metaphor for the hybrid identities emerging in a situation where the "elementary units" of "race, gender and class . . . themselves suffer protean transformations" within a global high-technology capitalism.[67] Cyborg politics thus also means discovering new forms of organization adequate for an era when a "new industrial revolution" is "producing a new world-wide working class."[68] This project, Haraway suggests, involves rejecting vanguard parties but fostering affinities and alliances. "Oppositional, utopian and completely without innocence," she writes, cyborgs are "wary of holism, but needy for connection—they seem to have a natural feel for united front politics."[69]

Haraway's concept of the cyborg has a distinct affinity with Negri's theory of the socialized worker. For Negri, the socialized worker is a figure operating at variegated sites throughout the circuits of capital, immersed in a technoscientific environment where computers and communications have become so commonplace as to constitute a second nature. S/he (Negri specifies the feminization of the work force as a feature of the socialized worker) inhabits an "ecology of machines."[70] Computers, videos, faxes, and other media become so quotidian that workers have "organic" familiarity with them.[71] Capital is thus unable to stop socialized workers using these technologies for their own purposes—of which the most politically significant is the establishment of communication across the divisions that segregate sections of the work force. Hardt and Negri specifically declare this parallelism with Haraway's line of thought, saying that the increasing interface of the laboring body with technological appendages means that "the cyborg is now the only model for theorising subjectivity."[72]

Several cases that would serve as examples of such "cyborg" activism have already been discussed in earlier chapters: Subcomandante Marcos plugging in his laptop; French students appropriating Minitel; video countersurveillance in Los Angeles or East Timor; the mobilization of biomedical knowledge in struggles around AIDS, abortion, and environmental health. Andrew Ross, in an article inspired by Haraway's line of thought, cites a case that would also serve well as an instance of the "organic" connection to technoscience that Negri sees in the socialized worker.[73] This involves a group of Michigan autoworkers that had been promised courses in computer programming as part of their on-the-job training by General Motors. When the company abruptly terminated these courses, declaring that such depth of technical knowledge was excess to functional workplace requirements, the workers—who included veterans of the Flint sit-down strikes—launched a lawsuit, hinging on the corporation's use of state-provided public education funds for private purposes. But they also formed their own Usenet news group and email bulletin, "The Amateur Computerist." This bulletin was devoted simultaneously to practical self-instruction in computer lore, criticisms of the corporate use of technology, arguments for the reduction of the work week, support of autoworkers' strikes, and "netizen" arguments for the democratic, rather than commercial, organization of cyberspace. It eventually came to command a relatively wide following—a prime example of cyborg struggle.

Although there are strong similarities between Negri's and Haraway's lines of thought, there is a difference in emphasis. Haraway's work is characteristically postmodern in its refusal to nominate any central axis of conflict along which activism might be arrayed. This refusal results (particularly in elaborations by later authors) in the discovery of cyborg resistance in every aspect of contemporary technoculture, with little attempt to make strategic or tactical differentiations about its political significance. Negri's appropriation of the cyborg concept reinscribes it within a Marxist horizon of capital/labor conflict, but to heretical effect.

Marxists have always emphasized that capital is a system that tends to supplant living labor with dead labor, replacing the variable capital of human workers with the fixed capital of machinery. This tendency now appears to be reaching a culmination in genetic and computerized technologies, where machines are infiltrated deep into organic life itself while artificial intelligences promise to assume many of the attributes of consciousness. One interpretation of this situation is to see in it a necrotic apogee of capitalist control—a near-total subjugation of living to dead labor, the ultimate victory of fixed over variable capital, a nightmare of technological exploitation extended to the point where the very biological integrity of the species is subordinated to the imperatives of accumulation. This is the theme of some Baudrillardian strains of postmodernism, such as the brilliantly graphic, but ultimately voyeuristic, accounts of technocapital's virtual "harvesting" of human flesh offered by Arthur Kroker and his colleagues.[74]

But from Negri's perspective this is only half the story. Against it must be set countervailing tendencies, in which the increasing interface and infiltration of living by dead labor opens toward a quite different outcome: a prosthesis of labor and machine that loosens capital's unilateral control of technology. Expanding his point, I would say that capital, in its drive to automate every function of the workplace—mental as well as manual—has been compelled to develop machines of extraordinary versatility, technologies that in their potential universality emulate the very flexibility and plasticity of living labor itself. In this respect, information society theorists are right to emphasize the difference between mechanized and information systems. However, this protean quality of computers and communication systems—

their reprogrammability, their interactivity—is often taken as simply marking a new, intensified level in capitalist development. What such analysis
omits is the possibility that this flexibility might be used, not to augment
capital, but to subvert it. For the malleability of the new technologies means
that their design and application becomes a site of conflict and holds unprecedented potential for recapture. These are the possibilities recognized by
Haraway and Negri, possibilities of which any Marxism confronting postmodern culture must take account.

Rhizomes and War Machines

Although we can find elements of a postmodern/Marxist recombinancy in
Negri's work, to see a sustained exploration of this possibility we should turn
to the oeuvre of his allies and collaborators Felix Guattari and Gilles Deleuze.
Of course, despite the explicit insistence of these authors that they are indeed Marxists, many would feel that the chaotic playfulness, exotic vocabulary, and celebrations of desire and schizophrenia found in their writings are
far removed from the sober business of historical materialism.[75] On the other
hand, Guattari has specifically divorced his work from postmodernism—
denouncing the ideas of a "postmodern condition" promoted by Lyotard and
Baudrillard as "the very paradigm of every sort of submission, every sort of
compromise with the existing status quo"—yet is regularly included in anthologies of postmodern thought.[76] This confusion if anything confirms the
accuracy of a hybrid designation—"postmodern Marxists."

Deleuze and Guattari's work is now the topic of a growing number of excellent analyses. I will therefore give only a brief overview of their position
before looking more specifically at how it bears on our discussion of information capitalism.[77] In the universe of Deleuze and Guattari, all social reality is constituted by desire. Desire is not good or bad, just productive and
dynamic. It is fair to say that Deleuze and Guattari's desire is the principle
of transformative, constitutive action that Marx called "labor"—prior to its
appropriation within a structure of surplus-value extraction.[78] Desire is heterogeneous and mobile. Social order is built on its homogenization and stabilization—the organization of the small, fluid, multiplicitous "molecular"
forms of desire into big, institutional "molar" macrostructures: "To code
desire is the business of the socius."[79] This binding of desire is a "territorialization"—a fixing in place, setting of boundaries.[80] But desire is "nomadic,"
always seeking lines of fight or flight, pursuing more objects, connections,
and relations than any society can allow.[81] Consequently "there is no social
system that does not leak in all directions."[82]

Capitalism "deterritorializes" more stable archaic social orders based on
landed property or tribal community, but "reterritorializes" everything in
terms of exchange value.[83] Constantly adding or subtracting organizational

"axioms" and altering its combinations of labor process, political organization, and cultural apparatus, it is more flexible than any of the social systems it supplants.[84] Its most recent form is "integrated world capitalism," in which "the single external world market [is] . . . the deciding factor."[85] The global economy emerges as a "universal cosmopolitan energy which overflows every restriction and bond": "Today we can depict an enormous, so-called stateless, monetary mass that circulates through foreign exchange and across borders, eluding control by the states, forming a multinational ecumenical organisation, constituting a de facto supranational power untouched by governmental decisions."[86] Characteristics of "integrated world capitalism" are a reshaping of the international division of labor, with the appearance of areas of underdevelopment appearing within the developed world and limited development within the underdeveloped world; a declining number of jobs; intensified integration of the upper, privileged strata of the working class and the appearance of new strata of great insecurity—"immigrants, hyper-exploited women, casual workers, the unemployed, students without prospects, all those living on social security"—and a "constant reinforcement of control by the mass media."[87]

However, over the same period that capitalism has consolidated this global, "molar" structure, there also appears what Guattari terms a "molecular revolution"—"a proliferation of *fringe groups, minorities and autonomist movements* leading to a flowering of particular desires (individual and/or collective) and the appearance of new forms of social grouping."[88] These are movements appearing beyond the ranks of the industrial working class among the unemployed, women, ecologists, homosexuals, the old, the young. These, Guattari says, "constitute 'fighting fronts' of a quite different sort from those that have always marked the traditional workers' movement."[89] For these movements, it is "not just a matter of struggling against material enslavement and the visible forms of repression, but also, and above all, of creating a whole lot of alternative ways of doing things, of functioning."[90] The undecidable factor today is whether these microrevolutions "remain contained within restricted areas of the socius" or establish "a new inter-connectedness that links one with another" and end by producing "a real revolution . . . capable of taking on board not only specific local problems but the management of the great economic units."[91]

Deleuze and Guattari speak of revolutionary organization as the creation of "machines of struggle."[92] This has to be understood carefully. For Deleuze and Guattari, any assemblage of desire—at a subjective or social level—is a "machine." The term is aimed to break with humanist concepts of natural identities, to emphasize (as Haraway does with her concept of "cyborgs") the constructed, produced, and collectively fabricated nature of psyche and society. Thus when they speak of radical political organization as the creation of nomadic "war machines," while they certainly do not preclude armed struggle, the phrase has a far wider dimension. They are thinking in terms of aggressive, mobile, decentered organizations, capable of being built or dismantled as needed, that can harry and erode the structures of established order—"state machines." At the same time, given their affirmative attitude toward the subversive use of technology, which we will examine in a minute, there is also a certain literal embrace of the machine as an instrument of struggle.

The characteristic form of a contemporary "machine of struggle" is a "rhizome."[93] By this name Guattari and Deleuze designate decentered, divergent, transverse, nonhierarchical, lateral, or transverse modes of organization—contrasted with "arborescent" or rigid, linear, vertical, and hierarchical patterns.[94] Deleuze and Guattari apply the term "rhizomatics" to modes of philosophy and psychoanalysis, but the phrase also has clear political implications. The experimentation with coalitions, rainbows, networks, and webs that has been a salient feature of anticapitalist movements in the last decade are all experiments with rhizomatic forms of organization. Guattari speaks of the needs for the "molecular revolution" to find forms of organization in which "the different components will in no way be required to agree on everything or to speak the same stereotypical language."[95] In doing so he reiterates a persistent theme of autonomist Marxism: Sergio Bologna has similarly spoken of the search for "a set of recompositional mechanisms that start, precisely from a base of dishomogeneity," while Sylvere Lottringer and Christian Marazzi emphasize "multi-centered" forms of struggle that "stress similar attitudes without imposing a 'general line.'"[96]

One characteristic of "rhizomatic" organizations is that the distributed nature of their decisions and actions makes rapid and efficient communication very important. Thus the possibility of using information technologies becomes significant. Guattari, himself involved in politicized pirate radio, was particularly aware of this possibility and he repeatedly emphasizes the liberatory possibilities of new machines. On the one hand, high technology offers "integrated world capitalism" the opportunities of extending "a generalized machinic enslavement" in which humans operate as input-output relays within elaborated information systems dedicated to speeding the circulation of exchange values. However, this situation also abounds in "undecidable propositions."[97] There is a "shared line of flight of the weapon and the tool: a pure possibility, a mutation"; "there arise subterranean, aerial, submarine technicians, who belong more or less to the world order, but who involuntarily invent and amass virtual charges of knowledge and action that are usable by others, minute but easily acquired for new assemblages."[98] Guattari specifically rejects "media fatalism," arguing that as a result of declining costs, continued technological advancement, and continuous labor market retraining there is a growing "potential use of . . . media technology for non-capitalist ends."[99] Media, he says, can be tied to different types of "group formation"—one based on "standard identifications and imitations, the father, the leader, the mass media star," the other more open and creative, leading to dialogues that can break down received stereotypes and encourage diverse collectivities to form their own discourses and self-representations. The first, Guattari claims, is encouraged by the unidirectional broadcast technologies of the "mass media," the latter by the new capacities of a "postmedia age" in which the communication technologies can be "reappropriated by a multitude of subject-groups."[100] Computerization, in particular, has "unleashed the potential for new forms of . . . collective negotiations, whose ultimate product will be more individual, more singular, more dissensual forms of social action."[101]

Harry Cleaver has made an interesting application of the "rhizome" concept to the Zapatista networking discussed in the previous chapter.[102] For another example of the "rhizomatic," "postmedia" movements of the sort Guattari envisages, one might think of the antiroads struggles that have snaked their way through post-Thatcherite Britain across sites like Twyford Down, the M11 Extension, and Newbury.[103] These campaigns, aimed at blocking the new motorways built largely to facilitate integration with the European Economic Community, involve highly diverse groups—Earth Firsters, middle-class conservationists, local property owners, Marxist militants, Greenpeacers, the Donga Tribe, and so on. They also interweave loosely with other movements, such as the very nomadic struggles by Gypsies, "travelers," antihunt saboteurs against the draconian restrictions on civil liberties and personal mobility imposed by the Tory government's Criminal Justice Bill, or various "New Digger" groups such as "The Land Is Ours" attempting to reappropriate the one-time "commons" from corporate ownership.

One feature of these antiroads struggles has been their pervasive use of various forms of high-tech communication: personal computers to coordinate rapidly assembled blockades and demonstrations; video to record and publicize protests and for countersurveillance against police and security guards; the dissemination of such film through alternative television producers, such as the celebrated "Undercurrents" programs; and, more loosely, the construction of a cultural ambience of protest closely associated with various forms of technomusic. One reporter on the "postmodern tendencies" of this "media-friendly, technologically-literate" movement comments: "Anti-roads activists phone up the media to give interviews from the top of cranes while videoing the behavior of police and security guards swarming beneath them. The action footage is replayed at clubs and festivals or broadcast on the Internet across the world. As the electronic icons . . . are appropriated for protest, the information technology revolution is being pressed into service in the name of further widening the scope of political communication and participation."[104] This, I suggest, is exactly what Guattari thought "the molecular revolution" would look like.

Cyber-Nazis and Nizkor Projects

It is important to recognize that the potentialities recognized by Deleuze and Guattari also have a malignant side. One of the salutary aspects of these authors' work is that they take seriously the possibility of a postmodern fascism, in which the very communicational and nomadic capacities so rich in anticapitalist possibilities are recuperated in appallingly destructive form. Guattari and Deleuze have always emphasized that molecular rebellions can turn negative, becoming paranoid or suicidal, and they have taken conventional Marxisms to task for their failure to recognize the unconscious and preconscious paths in which longings for emancipation and freedom become twisted into racist, sexist, and homophobic hatreds and authoritarian dependencies.[105] Like Baudrillard, they speak of "black holes"—in this case, meaning the turning inward of revolutionary aspirations toward internecine hostility.[106] In this perverted form, they become available to capitalism as a weapon against movements of autonomy, providing the basis for fascism— "without doubt capitalism's most fantastic attempt at economic and political reterritorialisation."[107]

Today, it is very evident that desires for autonomy from "integrated world capitalism" can take "right" and well as "left" forms. The proliferation in North America and Europe of neo-Nazis, the Klan, Aryan Nations, Patriot Militias, Holocaust deniers, and fundamentalist churches, mobilized both in official forms, such as the movements headed by J. M. Le Pen in France and Pat Buchanan in the United States, and in clandestine, underground networks of the sort responsible for the Oklahoma City bombing or the burnings of immigrant hostels in Germany, represents a significant popular response to the social and economic costs of neoliberal restructuring. Recruiting their membership from sectors of the working class dramatically devastated by the advent of the information economy—the unskilled, rural white males at the base of the U.S. militias, the masses of European unemployed—these movements present an analysis that often mixes percipient analysis of globalization with extremes of pathological fantasy. Unemployment is attributed to aliens and immigrants; disintegrations in family security and social infrastructures to the activities of feminists and homosexuals; capital's overrunning of national sovereignty is deciphered as the result of Jewish banking cabals; real intensifications in security-state activity appear as fantasies of black helicopters commanding take-overs engineered by the United Nations; and desires for release from deepening immiseration translate into programs of vengeance against every form of social "other."

These movements have proved at least as adept as the Left, probably more so, in availing themselves of the widely socialized capacities of information-age capitalism. "You may ask 'why the computer technology?'" wrote one

Aryan Nations leader as early in 1984: "The answer is simple, because it is our Aryan technology just as the printing press, radio, airplane, auto, etc. etc. We must use our own God-given technology in calling back our race to our Father's Organic Law."[108] Such uses extended from the sophisticated computer network linking the armed cells of various North American white supremacist groups and militias; the Usenet newsgroups such as alt.skinheads, alt.politics.white_power, or alt.politics.nationalism.white; Holocaust-denial World Wide Web sites, such as the trilingual "Stormfront"; the distribution by German and Austrian neo-Nazi groups of children's computer games based on genocidal scenarios; and the extraordinary success of the far right in colonizing talk radio in the United States.[109] The considerable communication power of protofascist groups has meant that combating their high-technology propaganda itself has become an important focus of information activism—one thinks of the Nizkor Project (from the Hebrew word for "we will remember") operated by a fifty-four-year-old Vancouver Island store-clerk and self-described "modem junkie," Ken McVay, who has over the years compiled a vast electronic archive (or what has been described as "the information equivalent of a gigantic weapons dump") devoted to refuting Holocaust revisionism on the Internet.[110]

The relations of far-right groups to the central institutions of capital are complex. On the one hand, the threat to order posed by their armed wings has meant that such movements are targeted by the state security apparatus, which often brings to bear on them the most violent forms of repression (Waco, Ruby Ridge), while at the same time making their activities a pretext for a more generalized repression (censorship of the Internet). At the same time, there are undoubtedly sectors of capital—for example, the corporate backers of the Republican right in the United States—that look to either tolerate or actively harness the energies of such movements to the project of paralyzing and destroying working-class unity. Out of such complicity emerges the real possibility of a fascist "reterritorialization" of capital.

Deleuze and Guattari note that "what makes fascism dangerous is its molecular or micropolitical power, for it is a mass movement: a cancerous body rather than a totalitarian organism."[111] As they observe, the success of Nazism in Germany lay in its creation of microorganizations capable of penetrating every cell of society, organizations that both predated its assumption of state power and, persisting afterward, gave this power an insidious and omnipresent grip on society:

> Fascism is inseparable from a proliferation of molecular focuses in interaction, which skip from point to point, before beginning to resonate together. . . . Rural fascism and city or neighbourhood fascism, youth fascism and war veteran's fascism, fascism of the Left and fascism of the Right, fascism of the couple, family, school and office: every fascism is

defined by a micro-black hole that stands on its own and communicates with the others, before resonating in a great, generalised central black hole.[112]

As Douglas Kellner and Steve Best point out, it is not hard today to perceive the potential for such a North American fascism, which would surely combine racists, "pro-family" groups, fundamentalist Christians, skinheads, antienvironmentalists, MIA groups, and gun lobbies in a deadly resonance.[113]

The condition of the postmodern proletariat thus includes what Negri calls "alternative subjectivities."[114] One powerful tendency is for the destructive effects of capital's offensive to translate into intensified competition between different groups of workers. To the degree that this tendency prevails, the various limbs of the collective laborer will be turned against each other in the mutual dismemberment of neofascist populism, religious fundamentalism, ethnonationalism, gay-bashing, and sexist backlash. In this situation of extreme decomposition, the absorption and appropriation of new technologies could serve only to provide fresh instruments for internecine self-destruction—Nazi hate lines, homophobic computer bulletin boards, fiber-optic evangelism, and right-wing grassroots radio. Above this wreckage of class politics, the multinationals will glide through the global networks, swooping down to gut and abandon successive sites for profitable exploitation. No one witnessing recent events in Europe and North America can doubt the plausibility of this outcome.

The other possibility is for the different segments of social labor to connect and interanimate their struggles against capital. In this context the reappropriation of informational technology has a special significance, not only as an inroad upon capital's control over what is now a vital force of production, but also, simultaneously and inseparably, as a means to open the channels through which the "socialized worker" can overcome segmentation and constitute itself as a subject of radical cooperation. Communication—through contestation and infiltration of established channels, alternative media, autonomous radio, tactical television, culture jamming, and computer counternetworks—spins the life thread of awareness, negotiation, dialogue, criticism, self-criticism, and solidarity by which the variegated agencies of the collective worker develop their basis for alliance, create a recombinant politics, and recognize each other as members of a compound subject capable of reclaiming the direction of society from capital.

Post-Marxists . . . or Communists Like Us?

In 1985 Negri and Guattari coauthored a work published in France as *Les nouveaux espaces de liberté* (New spaces of liberty) and in North America (in 1990) as *Communists Like Us*.[115] Their declared objective was "to rescue 'communism' from its own disrepute," and to challenge a situation where "the 'ethic' of social revolution has become instead a nightmare of liberation betrayed, and the vision of the future is freighted with a terrible inertia."[116] Against the devastating effects of "integrated world capitalism" they urged "reunification of the traditional components of the class struggle against exploitation with the new liberation movements."[117] Rejecting both Leninism and anarchism, Negri and Guattari propose the creation of multicentric "machines of struggle."[118] This would require discarding the Marxist habit of nominating some agents as central to anticapitalist struggle and others as marginal. Instead, it would involve constructing a system of "multivalent engagement" between movements, "each of which shows itself to be capable of unleashing irreversible molecular revolutions and of linking itself to either limited or unlimited molar struggles."[119] In this process, the development of communicational links among movements, using the advanced technologies that capital is unavoidably disseminating, would be of crucial importance: "All the current catchwords of capitalist production invoke this same strategy: the revolutionary diffusion of information technologies among a new collective subjectivity. This is the new terrain of struggle."[120] Negri and Guattari offer a number of "diagrammatic propositions" about the issues around which the new rhizomes might cohere. These include struggles on the welfare front, for the establishment of a guaranteed equalitarian income, and against poverty in all its forms; shortening and reorganizing the time of the work day; "a permanent struggle against the repressive functions of the State"; campaigns against war, particularly antinuclear movements; and the construction of North-South alliances among movements.[121] These, they say, would all be steps toward the rediscovery of communism not as "a blind, reductionist collectivism dependent on repression" but as a "process of singularization."[122] They write, "Real communism consists in creating the conditions for human renewal: activities in which people can develop themselves as they produce, organisations in which the individual is valuable rather than functional."[123] The struggle for communism could regain the universality Marx attributed to it if "Truth 'with a universal meaning' is constituted by the discovery of the friend in its singularity, of the other in its irreducible heterogeneity."[124]

This postmodern Marxism can usefully be contrasted with the very influential "post-Marxism" advocated by Ernesto Laclau and Chantal Mouffe in their *Hegemony and Socialist Strategy*, published in 1985.[125] In post-Marxism, the importance Marxists traditionally attribute to struggle against capital is dismissed as crudely economistic. Instead, the social is seen as an open, fluid, "unsutured" field, constituted by a plurality of power relations and struggles—over class, gender, race, homophobia, the environment—none of which can be said to have any priority over, or intrinsic connection with each other, although they may be contingently linked together. Socialism is redefined in such a way as to diminish the importance of reorganizing the relations of production, which simply becomes one part of a program of "radical democracy" that seeks to promote equalitarian relations across the whole social spectrum. From this point of view, eliminating capitalism no longer claims any centrality among emancipatory projects.

Laclau and Mouffe believe that in moving the focus of social analysis outside the factory to embrace this wider field of conflicts they have decisively gone beyond Marxism. And indeed, in acknowledging the importance of struggles around issues of gender, race, and a multitude of other oppressions they have transcended the workerist logic of the Second International (their constant, and perhaps slightly outdated, target). In this respect their project does constitute an important break with sclerotic Marxisms.

However, to make this move they adopt an extraordinarily abstract and ahistorical vision of the contemporary world. The density and intransigence of historical determinations are eclipsed, and there appears instead a concept of the social domain as "discourse," constantly available for deletion and recombination in ever-alterable "articulations," as fluid and malleable as words on a page. It is this ahistorical abstraction that makes it so easy for Laclau and Mouffe to sidestep the Marxist insistence on the dominative centrality of capital. Once one returns from the abstraction of discourse in general to the concrete specificities of the late twentieth century, the degree to which the logic of capital is busily "suturing" society—sewing up the planet in the net of the world market—becomes much more striking. To a greater extent than ever before, control over planetary resources, including the vital communicational and informational resource of "discourse" itself, is concentrated in the hands of a corporate order that now possesses truly global capacities of command and coordination, and whose organization increasingly subsumes and mediates other social hierarchies formed on the basis of gender and ethnicity. To skip over this point is to return—under the guise of postmodern sophistication—to a liberal, pluralistic view of an open society based on a multitude of freely competing interest groups. It is to evade, rather than surpass, the crucial point of Marx's analysis of "real subsumption"—the tendency of capital to impose its logic not just over the workplace, but over all areas of life.

This is the line of analysis that Negri and Guattari develop. In their analysis, capitalist totalization is a force that invades, permeates, and refracts every domain of social activity, and every other social antagonism. The market asserts its priorities over issues of gender equity or ecological preservation to a degree that it becomes impossible for feminist or green movements to succeed without coming into conflict with it. And it is the necessity of this challenge that provides the potential connecting point between the varied movements seeking to pursue other societal logics. From this point of view, there is no evading the issue of control over production—defined in its broadest social aspect: "Instead of new political alliances, we could say just as well: new productive co-operation. One always returns to the same point, that of production—production of useful goods, production of communication and of social solidarity, production of aesthetic universes, production of freedom."[126] Although Laclau and Mouffe's ideas have commanded an enormous academic interest, post-Marxism seems, a decade after its first enunciation, strangely dated. This is surely because analysis that has almost nothing to say about the international division of labor, new technologies of communication and exploitation, and changing conditions of labor misses some of the most dynamic aspects of contemporary social transformation.[127] In a massive failure of theoretical nerve, post-Marxism has shut its eyes to the approaching "big story" of the early twenty-first century—the consolidation of the world market. Moreover, in practice, "radical democratic" politics has proven peculiarly lackluster. It has been associated with a rejection of some of the most important actually occurring forms of militant struggle (such as the British miners' strikes and anti-poll-tax riots); with a fixation with electoral politics and reformist constitutional schemes; and with a recycling of that most exhausted shibboleth of social democracy—the mixed economy—at the very time when international capital has decisively signaled its lack of interest in such a settlement.[128]

Negri and Guattari's collaborative work lacks the enormous theoretical sophistication with which Laclau and Mouffe invest their proposals, and its sense of urgency sometimes translates into a purple, overblown rhetoric, and slapdash assembly. But in the years since they wrote, their analysis of "integrated global capitalism" grows in pertinence. Their discussion of new, technologically facilitated "machines of struggle" resonates with the actual paths being taken by a variety of coalitions and networks worldwide. And while their sketch of a revitalized communism is only rudimentary, it does begin to raise the pressing questions about the reorganization of work, income, and the allocation of social time that the general collapse of both state socialism and social democratic compromises necessitates. For these reasons, their postmodern/Marxism seems today a far more germane project than the eminently fashionable "post-Marxism."

Detotalizing Totalizations

As Harry Cleaver has observed, autonomist Marxism has "evolved in such a way as to answer the postmodern demand for the recognition of difference and the Marxist insistence on the totalizing character of capital."[129] Its project can be defined as a paradoxical "detotalizing totalization" that seeks to analyze the overarching social command of capital the better to dissolve it into a more multiplicitous and varied order. As Cleaver says, in spite of justifiable post-modern objections to master narratives, simple self-defense requires that for any social theory to be useful in the struggle for liberation, it must recognize and comprehend not only different forms of domination but the world-wide and totalizing character of the capitalist form. . . . what is required is an ability to grasp simultaneously: the nature of the totality/globality that capital has sought to impose, the diversity of self-activity which has resisted that totality and the evolution of each in terms of the other.[130]

Capital, in order to maintain its totalizing system, strives to prevent its variegated opponents from combining forces: dividing, splitting, and fracturing in order to maintain the systemic integrity of its world system. For the diverse anticapitalist movements, the problem is that in order to break out of capital's totalization they have to link their diversity, to ally across difference to circulate struggles.

I have suggested how, within this framework, we can recontextualize some of the important postmodern insights into contemporary conditions of communication. In introducing high technologies, a central aim of capital has been to reinforce its own circuits while paralyzing those of opposition movements through an increasingly intense regime of informational control. This decompositional, disintegrative, immobilizing tendency is recognized in the Baudrillardian school of postmodernism—which, however, completely fails to recognize the countervailing tendencies of oppositional groups. These groups have to some extent been able to reappropriate these same technologies capital has deployed and make them channels for new solidarities and alliances. This is the tendency partially recognized by Haraway and other "optimistic" postmodernists. In the work of Negri, Guattari, and Deleuze these two tendencies appear pitted against each other, as the collision of different "machines of struggle"—a conflict that might be characterized as "cyborgs versus the simulacrum."

However, while Negri, Guattari, and Deleuze envisage these struggles moving toward the constitution of a noncapitalist society, they offer only limited hints as to what this alternative might be. They clearly see it not as a state-socialist imposition of centralized uniformity but as an explosion of difference—a dissolution of the global command of profit that opens the way to alternatives that, like a volcanic magma, spread out in a "network of streams of enjoyments, of propositions, of inventions."[131] However it has to be said that these theorists have very little that is concrete to say about how such a self-organized society might operate—how the buses would arrive on time, the bread be on the shelf, or the AIDS vaccine be researched.

There are some good reasons for this reticence. Blueprints for a postrevolutionary society have too often had authoritarian implications. The stipulation of a preconceived set of ideal relations has resulted in "transitional programs" that repress anything deviating from their model. Postmodern/ Marxists emphasize that any project truly based on a belief in the self-determining capacities of people should avoid theoretical foreclosure of the paths this energy might take. Furthermore, if the aim of revolutionary activity is to break the "totalizing" logic of capital and shatter its homogenizing and systematizing tendencies, as Negri and Guattari suggest, any stipulation of a singular form of postrevolutionary society can be seen as self-contradictory; rather, the aim should be to create a space where a diversity of social, cultural, and economic ways of being can coexist.[132]

These are important points that nevertheless leave difficult problems unresolved. While a postcapitalist society definitely should encourage diversity of social organization and be open to evolving and unforeseen directions, this does not eliminate the need to think carefully about what arrangements, on a planet effectively unified by trade, transport, and communication, might enable such a coexistence, or of considering which within a plethora of possibly emergent noncapitalist ways of life are desirable and worth fighting for. So it is to these points that I turn in the next chapter.

ALTERNATIVES

Describing alternatives to capitalism has always troubled Marxists. Marx's early writings contain lyrical evocations of postcapitalist possibilities. But Marx and Engels were highly critical of "utopian socialisms"—many of them technocratic ancestors of today's information society theory—that drew-up elaborate pictures of ideal societies without recognizing the need for struggle and conflict to attain them.[1] Rejecting these "Comtist cookbooks about the future," they held that communism is "not a state of affairs which is to be established, an ideal to which reality [will] have to adjust itself" but rather "the real movement which abolishes the present state of things."[2]

Today, however, mere invocation of the "real movement" is not immediately encouraging. A vast block of despair and cynicism consolidates the dominance of the world market. The catastrophe of state socialism has left millions convinced that, however appalling the trajectory of capitalism may be, there is simply no alternative to it.

This resignation is reinforced by information capital's managers—those whom Pierre Bourdieu has called the "kings of technocracy"—in whose discourse any attempt to think beyond the "realities" of global competition and automating technology is instantly dismissed as tantamount to delirium.[3] As Massimo De Angelis observes, such "technicism" serves as the "ultimate legitimization" for capitalism, making its economic order into "a great Leviathan, the unchangeable and unquestionable constraint facing all political and cultural subjectivity, a constraint that subsumes everything."[4] De Angelis argues that in the face of this conceptual closure there is an urgent need to "recover a utopian discourse, in thought as well as in antagonistic and constitutive practice."[5] He observes that, "through an interesting play on words, the word utopia is defined in English as *nowhere*—no place. But this could also be read as *now* here—here and now."[6] De Angelis goes on to distinguish between "realizable" futures, that "presuppose a pre-conceived plan which must be realized (by subordinating to the plan all the people who don't like it)," and "actualizable" futures, where "whatever is actualizable is already existing in a virtual way, where virtuality is a dimension of reality."[7] He urges utopian invention, "not as *the* alternative model, not as a party program or a plan in search of subjects to subordinate" but as "an open and inclusive horizon of thought, antagonistic practice and communication" that can "show different possible horizons and contrast them to the poverty of the mainstream one."[8]

Against the Great Leviathan

It is in the spirit of De Angelis's proposal that I offer a sketch of an alternative future. I propose a series of measures—the institution of a guaranteed annual income, the creation of universal communications networks, the use of these networks in decentralized, participatory counterplanning, and the democratic control of decisions about technoscientific development. These elements would, in their full implementation and synergistic interaction, go a long way toward constituting a viable alternative to capitalism. Moreover, each of the separate elements proposed here, and each of the various gradients and steps in their realization, can be seen as delineating fronts of struggle. They are conceived of as invading beachheads that can be established on the shoreline of capital and advanced, up to the point where their combined effect overwhelms the logic of the entire system. The final section of this chapter briefly reflects on some conditions under which this might occur.

The ideas proposed here have not fallen from the sky. They extrapolate not only from a variety of theoretical sources but also from what is really being done, now, in what autonomists would term the "self-valorizing" practices of a multitude of activists.[9] The interweaving of elements and possibilities that are now commanding wide attention constitutes what might be considered a utopian future.[10]

This thought-experiment does, however, have some important limitations. It focuses only on those issues that relate to this book's major theme—the social uses of the new information technologies. Its basic orientation comes from Marx's observation in *Grundrisse*, that while machinery may be the "most appropriate form" of capital, capital is not necessarily the most appropriate social form for machines.[11] To illustrate this point, I assume a society in which high technologies are fairly readily available. Since currently these conditions obtain most strongly in a handful of advanced capitalist economies, the sketch is Eurocentric. There is a missing dimension, whose importance I acknowledge but do not address, one that involves issues such as the release of the South from an exterminatory debt burden, the reversal of the flows of value from South to North, the payment by the North for the preservation of the ecological resource vital to planetary survival, and the support of spaces for what is sometimes termed "autonomous development" freed from the economic and cultural constraints of neocolonialism.[12]

I think of this sketch as a proposal for "communism"—a continuation of the red thread that Marx and so many others have spun across centuries. But I also know that this name, "communism," has become so heavy, so sodden with blood and weighted with nightmarish history, and carries with it such a burden of explanation, repudiation and qualification, that many regard it as unspeakable, at least for this generation and probably several more.[13] What word might be used instead? I do not want to talk of "socialism," a concept profoundly tainted—in its authoritarian forms, by terror; and in its social democratic variants, by failed compromise.[14] I might follow the lead of Cornelius Castoriadis, who now speaks of an "autonomous society"—but this phrase also is freighted with its author's changing allegiances, and too rhetorically ponderous to be attractive.[15]

Therefore, I sometimes use another term: *commonwealth*. Some of the connotations of this word, too, are unappealing. But others are very appropriate. It designates exactly what I have in mind—a common-wealth of shared resources. It derives from a root around which cluster other concepts important to this study—like communism, communication, and commons. Commonwealth also recalls the energy of seventeenth-century revolutionary republicanism: if this proposal seems like a twenty-first-century version of the visions of Diggers and Ranters seeking a "world turned upside down," so be it.[16]

Zerowork: Guaranteed Income

Marx wrote that "the realm of freedom actually begins only where labour which is determined by necessity and mundane considerations ceases: thus in the very nature of things it lies beyond the sphere of actual material production. . . . Beyond it begins that development of human energy which is an end in itself, the true realm of freedom, which, however, can blossom forth only with this realm of necessity as its basis. The shortening of the working day is its basic prerequisite."[17] This is the prospect that the "information revolution" seems to bring in sight. Since the dawn of such computerized automation, people have been concerned about the consequences for employment. As early as 1949, Norbert Weiner, the father of cybernetics, raised the specter of a crisis of work resulting from robotization.[18] The classic reply to this anxiety was that labor displaced from the manufacturing sector would be reabsorbed in the service or information sector. For several decades, this optimistic prediction seemed to be borne out by the course of events. As I suggested in chapter 5, the diminishment of "direct labor" in production has been complemented by an expansion in "indirect" labor—both in the field of technoscientific work and in the myriad tasks of marketing, transportation, public service, cleaning, and caretaking that constitute the social matrix of a highly automated economy.

Today, however, there are signs that this logic may be exhausting itself. For the same types of technological systems that decimated manufacturing jobs are now being applied in the tertiary sectors meant to soak up the surplus labor displaced from industrial production. In the banking, insurance, wholesale, and retail industries, companies are using seamless, end-to-end information processing systems to eliminate whole layers of employees. Moreover, the acceleration of this process is an unacknowledged aspect of the "information highway." Teleshopping, video on demand, and virtual services mean the mass liquidation of clerks, salespeople, and other supernumeraries.[19] As the spate of layoffs in telecommunications demonstrates, those who are building the highway are the first to go. Capital is automating not just the factory but the entire social factory.

In many advanced capitalist economies—including those of Canada and much of the European Economic Union—unemployment rates are now at levels unthinkable fifty years ago. In the United States, visible joblessness is much lower. However, the relatively low U.S. official unemployment figures possibly disguise the scope of the job crisis behind a huge expansion of part-time and temporary work—the so-called "McJobs," which in effect institutionalize chronic underemployment.[20] This situation certainly can't all be laid at the door of automation. The global relocation of labor (capital's other major weapon against workers, itself made possible by technological advances in transportation and communication) is a factor. There are also further cyclical, organizational, and demographic elements in play. Nonetheless, attempts to deny the contribution of technological redundancy, along with all the negative multiplier effects of decreased consumer demand, seem increasingly obtuse. So serious is the consequent crisis of social disintegration that even some mainstream economists now concede that a serious problem exists.[21] And within the last few years several social theorists from a wide variety of perspectives—Stanley Aronowitz and William DiFazio in *The Jobless Future*, Jeremy Rifkin in *The End of Work*, Barrie Sherman and Phil Judkins in *Licensed to Work*, and various authors in the collection *Post-Work*, edited by Stanley Aronowitz and Jonathan Cutler—have acknowledged that we may be in view of the point foreseen by Marx, where the replacement of living labor by machines fatally undermines the wage relation.[22] Potentially, the extraordinary productivity increases created by high levels of automation could be realized in terms of general increases in income and/or supported leisure time. There emerges the potential for what Paolo Virno terms "the reduction of labor time to a virtually negligible part of life," making it possible to "conceive of wage labor as only one of the moments of existence, rather than as *hard* labor or as the source of a lasting identity."[23] However, because capital continues to impose the linkage of income to work (for all except the owners of the means of production) a diametrically opposite situation is produced: an intensified availability for work, enforced by the immiseration of unemployment. Thus, "the time of non-work, which is a potential richness, presents itself within the established system as a lack, as poverty."[24] Alongside practices of global relocation, computerization has in many sectors of the economy—and probably across the board—decreased the demand for socially necessary labor within the zones of advanced capitalism, thereby restoring what Marx identified as the central weapon of capitalist command over the working class—the maintenance of a permanent "reserve army" of the unemployed.[25]

The fear of joblessness promoted by accelerating high-technology automation is a sword held at the throat of labor. It undermines trade union strike power and allows management to coerce employee "cooperation," recruit desperate strike breakers, and drive down wages and working conditions. As workers compete among themselves for employment, capital sifts them into different strata—the declining core of permanent employees needed to run the new production systems, the periphery of temporary and part-time workers called up according to the fluctuations of the economy or the production cycle, the absolute rejects destined for the welfare lines or starvation. Labor is segmented into an increasingly vicious hierarchy whose rungs tend to correspond to and reinforce discriminations of gender, race, and age. Those at the top must work ever harder, faster, and more flexibly to save themselves from the immiseration below. Those at the bottom buy survival only at the price of superexploitation, pricing themselves into a job so cheaply it is not worth replacing them with machines.

Faced with this convulsion, the usual response of the socialist Left has been to call for the creation of "more jobs," engineered by a renewal of Keynesianism or an adjustment of interest rates. Not only does this response fly in the face of the actual capacities of technological innovation, but it forgets that, in origin, socialism was not a project for the extension of wage labor, but for the ending of what was understood as an exploitative and dominative institution: "wage slavery." The reduction of this aspiration to a call for full employment—a call, moreover, made more implausible by every advance in computer science—dramatically reveals the attachment of social democratic and trade union leaders to the basic structures of capitalist society, at the very moment when these walls are being breached. Putting the wage-form on an elaborate life-support system is a strategy of "making some people toil unnecessarily so that they can be paid without others complaining that they are hanging around with nothing to do."[26]

One sign of more creative thinking is the reemergence of an issue Marx saw as vital to the emancipation of labor, but which has since the end of World War II been largely abandoned by trade unions—the shortening of the working day.[27] Demands for the reduction of hours without loss of wage are now on the agenda of the most innovative sectors of labor revolt in North America, as in Europe, and even entertained by social democratic thinkers.[28] This strategy builds solidarity between the employed and the unemployed. Rather than dividing those impoverished by too little work and those exhausted by too much, it aims for a situation where "everyone works, but only a little."[29]

However, the real significance of such demands is that they point toward an even more radical possibility, namely, dissolving the link between work and income by the institution of guaranteed annual income. The case for this step is simple: capitalism has created a productive capacity so great that there is no necessity for anyone to suffer want because they cannot sell their labor time. Moreover, this productive capacity arises from an economic system so socialized—so much the product of a "combined effort" occurring not just in workplaces but in households, schools, and general social intercourse— that the allocation of income only to those who exert themselves at the immediate point of production is neither just nor even efficient. The social risks of people freeloading on a system of generalized income are now infinitely less than the problems created by consigning increasing masses to an income-less, because work-less, future.[30]

As Steve Wright notes, the institution of a universal guaranteed income "has long held an honored place within . . . autonomist discourse."[31] In the 1960s and 70s, theorists such as Negri were already suggesting that the automation and socialization of production had rendered labor theories of value anachronistic. They saw this as marking a crisis, not for Marxism, which has always seen wage labor as a historically transitory form of social organization, but for capital, which depends on upholding the necessity and rationality of the wage relation. Groups in the midst of militant shop-floor struggle argued that both rising technological productivity and the increasingly evident social nature of production should be recognized by the creation of a social wage, equal for all, tied to needs rather than performance, and available to those outside the traditional realms of paid work, such as houseworkers and students. This is sometimes known as the "zerowork" position.[32]

Such ideas were subsequently elaborated, popularized, and watered down by Andre Gorz, whose provocative writings are informed by a considerable familiarity with autonomist thought.[33] One of the few left optimists about computerization, Gorz in the mid-1980s suggested that the reductions in

labor-time made possible by microelectronics were opening "paths to paradise."[34] The realization of these prospects was, however, impeded by a "living dead" or "impossible" capitalism that preserved the wage and the market beyond the moment of their historical validity, retaining them merely as
techniques of domination.[35] Gorz rejected the traditional left focus on dignity in work, which he believed that rationalized and deskilled technological production made unattainable. Instead, he argued that the cutting edge
of social activism lay in the demand for freedom from work.

To this end, he proposed a program for a social income, distributed through
life, based on the requirement to perform a (low) minimum amount of socially
necessary labor; twenty thousand hours in a lifetime, or about ten years full-
time, twenty years part-time, or forty years of intermittent work.[36] If this was
implemented, Gorz suggested, work would no longer be a full-time occupation or the center of social existence. A wide variety of rhythms and styles
of activity would coexist, creating rich opportunities for citizens to exercise
their creative powers "autonomously," freed from the "heteronomous" constraints of work. "Let us work less," Gorz wrote, "so that we all may work
and do more things by ourselves in our free time."[37]

Gorz's work has had an ambiguous legacy. By developing the autonomists'
rather sketchy hints about a universal income, he pushed the frontiers of left
imagination beyond the boundaries of "a fair day's wage for a fair day's work."
But he also partially discredited the idea of liberation of work by associating
it with a sort of apolitical voluntarism. Whereas autonomists had always
emphasized that freedom from work was something that had to be fought for
against capital's tendency to reimpose the commodification of human activity, Gorz often seems to suggest that a general reduction of labor time could
be realized simply by dropping out from the wage economy. In his most notorious statement he suggested that we must say "farewell to the proletariat,"
as postindustrial socialism is quietly invented in do-it-yourself, backyard
experiments of the new "non-class of non-workers."[38] Because of this his work
has been widely criticized from the left as simply a recipe for what Wright
calls "self-managed poverty."[39]

An insistence on the *contested* nature of the guaranteed income project is
critical because versions of the idea have also been proposed from the right.
Indeed, its advocates include such free-market champions as Milton Friedman.[40] During the Nixon administration, a legislative proposal in the U.S.
Senate for a form of Guaranteed Annual Income (GAI) was only narrowly
defeated; in Canada in the 1980s a version of the idea was proposed by the
Liberal MacDonald Commission.[41] As De Angelis points out, these plans "to
separate access to income from the labour market" are in fact designed only
"to make the latter function effectively."[42] In such proposals, GAI is set low
(well below the poverty line) and delivered in terms of negative income tax;
the minimum wage is also low; and other social wage programs (unemploy-

ment insurance, welfare, family allowance) are abolished. The aim is to use the GAI to rationalize state expenses, to eliminate their universality, and to allow capital to pay inadequate wages, with the effect "not of eradicating poverty and unemployment, but of making them socially acceptable."[43] In the light of this "big business version" of a guaranteed annual income, some antipoverty activists are now intensely skeptical of the entire concept, believing it has been fatally coopted.[44]

However, at the same time, the intensifying crisis of unemployment and social disintegration precipitated by computerization and globalization has made others on the left increasingly interested in the concept. A new generation of autonomists has taken up the task of going, as Wright puts it, "beyond Gorz," developing schemes for a guaranteed income that "do not just coexist with capital, but can be used as a means to challenge it."[45] Their line of thought intersects with work on the same topic from a very wide variety of left and liberal orientations. Examples include the sustained theoretical arguments for a universal income offered by Philippe Van Parijs in the Netherlands; the campaigns waged by the Basic Income group in the United Kingdom; and proposals from political economists such as Diane Elson in England, Adam Przeworski in France, and Eric Shragge and Sally Lerner in Canada.[46]

Drawing on these sources, one can suggest some of the features of a guaranteed income scheme as it might figure in our commonwealth. Its level should be set high—well above the official poverty line. To the degree that such an income coexists with wage labor, as it might in the early stages of its introduction, it should be adequate to free people from the necessity of selling their labor power, even if the possibility of supplementation by this means continues. Its level should expand as and if the productivity of society grows, and accompany a generalized and egalitarian reduction in waged work time, to a point where guaranteed income eventually supersedes the wage as the main source of livelihood. Although receipt of such an income might initially be tied to some obligation to perform socially useful labor, this would not be construed in terms of participation in traditional paid productive employment (making it a "workfare") but of fulfilling responsibilities such as care for children, the sick, and the elderly. And it should be seen as an integral part of an expanding package of freely distributed services and use values, from housing and schooling to health, associated with the development of cooperative and collective forms of administration discussed later in this chapter, that would encourage forms of social solidarity going beyond the cash nexus.

Such an innovation would have multiple ramifications; I will comment on only three. First, the guaranteed income concept, while partly flowing from the technological crisis of paid jobs, also converges with feminist demands for the economic recognition of domestic labor. In the 1970s, Mariarosa Dalla Costa and Selma James integrated Marx's observations on the socialization of labor with the direct experience of millions of women, and pointed out the vast amount of monetarily unacknowledged, invisible but economically essential household labor done for free. Their proposal—immensely controversial within the women's movement—was "wages for housework."[47] Although this has been criticized as an attempt to commodify domestic work, it is clear that Dalla Costa and James intended "wages for housework" as a strategy to explode the wage form completely, undermining the attachment of income to a (male) job. Today, the drive to compensate domestic work is attracting widespread attention through the work of feminist economists such as Marilyn Waring.[48] A guaranteed annual income of the sort described here—perhaps tied to a requirement for men and women alike to participate in activities such as raising children, caring for the sick and elderly—would effectively annihilate the hierarchical division of waged and nonwaged labor that has so closely entwined capitalism and patriarchy.

Second, although the "zerowork" perspective focuses on reducing the overall amount of socially necessary labor, it should not be understood as precluding efforts to make what remains more enjoyable. Even in a society with a high level of technoscientific development, there will be tasks that, because of their inherent complexity, or their intrinsically satisfying nature, cannot or should not be automated. Although the commonwealth will abolish "work" as we know it—"work" as synonymous with "job," "boss," and "wage"—there will still be labor to be performed. Contrary to Gorz's gloomier statements, I do not believe that even highly technological tasks have to be alienating. There is now a vast literature on the enrichment and the qualitative improvement of such labor. Mike Cooley, from a trade unionist perspective, has written on ways in which computer systems can be designed to re-skill, rather than de-skill workers.[49] More recently, Shoshana Zuboff, from an enlightened managerial position, has discussed the ways in which high technologies can be used to "informate" workplaces rather than "automate" them, expanding workers' knowledge and control over operations rather than reducing and eliminating it.[50] The only (albeit very serious) problem with such analysis is that it usually represses the degree to which such humanizing innovations contradict capitalist imperatives of labor control and cost-reduction. Outside of this context, trade-offs between productivity and gratification could become a matter of social choice rather than profit-driven imperative. When a guaranteed income frees people from the necessity of enduring degrading, monotonous jobs there is every prospect for a creative remaking of labor. Thus, as Van Parijs suggests, the abolition of work should be seen as unfolding "along two converging routes: by giving work an ever-smaller place in life and by making it less and less like work."[51]

Third, freeing people from the compulsion to perform wage labor creates opportunities for more profound and creative involvement in other aspects of social life. One common and important objection to schemes for postcapitalist, self-organized societies is that they assume onerously high levels of political participation: Oscar Wilde's quip that "socialism is a good idea, but it requires too many evenings" springs to mind.[52] If one assumes a world like the present one, where most people are exhausted after eight-, ten-, or twelve-hour days of waged labor—plus the longer hours of unwaged domestic duties and "double shifts" that are the indispensable accompaniments of the current job system—this is a telling point. However, not the least important aspect of a guaranteed annual income and a drastically shortened and flexiblized work schedule is that they leave people with time and energy, some (though by no means all) of which can be devoted to collective discussion and decisions—and in ways that might even be rewarding and enjoyable. In other words, zerowork creates the communicative preconditions for other aspects of commonwealth. This potentiality can be enhanced by ensuring accessibility to the extraordinary communication systems that are, along with automation, the other major technological creation of information revolution. This is the prospect taken up in the next section.

Zero Commodity: Communication Commons

Under capital's direction, successive waves of electronic communication technologies—the radio, television, telecommunications, and computer networks whose networks now girdle the planet—have served mainly as the basis of vast, vertically and horizontally integrated commercial media empires. The consequences barely need rehearsal: an envelopment of society in corporate speech; market censorship of news and artistic expression; increasing privatization and stratification of access to information; and a relentless interpellation of audiences in the name, not of citizenship, but of consumerism.

The erosion of publicly owned media exacerbates these tendencies. Insofar as such institutions exist within advanced capital economies—and here conditions vary from the rudimentary services in the United States to the more developed institutions of Canada and Europe—the public ownership of media has largely centered around state-financed public broadcasting organizations. Always existing, like the other institutions of the welfare state, in an uneasy relationship to the market society that surrounds them, these organizations are now subjected to intensifying corporate encroachment. This proceeds under the watchwords of deregulation, the reduction of governmental limits on free enterprise activity, and privatization, the conversion of state institutions into corporate property. It is associated both with the use of new technologies to outflank and fragment the audiences of public broadcast systems and, even more important, with the ideological claim that the potential of new communicative technologies can only be realized by market forces. The net result is to deepen the communicative subsumption of society by capital.[53]

And yet, at the same time, electronic media display contrary tendencies that radically subvert the logic of the market. Because advanced communications networks can circulate information goods very fast and very widely, goods that are by their very nature dependent on extensive availability of appropriate machines, skills, and knowledge, imposing commodity exchange in this area has proven extraordinarily difficult. A wave of everyday media "piracy," including photocopying, home taping, bootlegged videos, unpaid reception of satellite signals, copying of computer software, and hacking is informally decommodifying information flows. These practices constitute a clandestine shadow-world that obstinately follows the attempt to enclose information in commodity form.[54] To give only one example, in the United States, where theft of satellite television signals was to be prevented by scrambling, it is estimated that half the descramblers are now used illegally.[55] Of course, much of this illicit activity is folded back into commodity form through black-market industries. However, what is remarkable is that so much corporate effort—both in terms of technological design and legal activity—is today being exercised to restrict what the media corporations ostensibly promote, that is, literally, the "free" flow of information.[56]

To understand this dynamic, we can elaborate on a hint of Marx's. He argued that a crucial motive behind the capitalist development of communications was its drive to shorten the circulation time of commodities—to speed the passage from commodity-form to money-form and back again. But Marx also observed that there was a limit to this acceleration. If a product passes instantly, without barrier or impediment, from producer to consumer, it destroys the moment of exchange. A commodity must remain in the owner's hands long enough to be sold. Capital might wish to maintain the continuity of circulation by passing through its different phases "as it does in the mind, where one concept turns into the next at the speed of thought."[57] But this dream cannot be realized. For the commodity to retain its essential attribute— that of being bought and sold—its passage must be interrupted: "it must spend some time as a cocoon before it can take off as a butterfly."[58] Today, electronic technologies are making a whole range of commodities central to the information economy—computer software, films, video, television programs, electronic music and games, and a proliferation of digital goods—into instant butterflies. Disseminated at virtually "the speed of thought" through electronic and digital channels, they take on aerial and evanescent forms difficult to contain within the commodity-form.

Nowhere is this more apparent than on the computer networks that capital hopes to make the central technology in its new wave of accumulation. The famous hacker slogan "information wants to be free" displays a naive technological determinism, but its mystification contains a kernel of truth— namely, that many people want information to be free, and they are finding in cyberspace the means to make it so. The Internet makes available a voluminous amount of information in uncommodified form. Vast databanks are available for free. Creators who prefer to see their work used rather than sold have dropped software into the Net gratis. Digital products have been electronically "liberated" from commercial owners and given instantaneous worldwide distribution.[59] Information society theorists have long pointed out that "ethereal goods" have qualities anomalous in a market economy: they can be used simultaneously by many people, be duplicated and transmitted cheaply and instantaneously, are not "consumed" or exhausted by use, and may grow in use-value the more widely they are shared. These features have become increasingly problematic to those concerned with policing digital commodity transactions. For what has emerged in cyberspace are collectivities of users who, rather than being subordinated to the laws of commodification, are rather characterized by a persistent, often gleefully overt, transgression of these rules. The massive confusion that now reigns over copyright and patent law in the electronic domain suggests that the enforcement of property rights in this arena will be extraordinarily vexed.[60]

Just as capital's introduction of new technologies, by potentially freeing huge surpluses of time, have unintentionally opened up prospects of liberation from work, so its expansion of new communication technologies inadvertently opens up a world of counterusage. As computerized automation, by reducing socially necessary labor time, makes possible either intensified exploitation or subversion of the wage form, so electronic communication, by reducing the necessary circulation time for information goods, opens onto two diametrically opposed options. It makes possible either a radical intensification of commodification—through pay-per services and consumer surveillance—or a fundamental attenuation of the commodity form, through the generalized transgression of electronic property rights.

Our commonwealth would build on and amplify this latter decommodifying tendency. Dorothy Kidd and others have referred to this process as the creation of a "communications commons"—a counterproject against capital's attempts to "enclose" the immaterial territories of airwaves, bandwidths, and cyberspaces in the same way it once enclosed the collective lands of the rural commons.[61] However, this project would advance along lines different from the state-operated public broadcasting systems favored by a previous generation of left media activists. While certain aspects of the public service, state-financed model remain valuable, these need to be revitalized and transformed by combining them with the more decentralized and diffuse practices of alternative media, from microwatt radio to community cable to the Internet.

Advocates of state-financed public media often find it difficult to marshal support against privatization, in part because of the frequent elitism, remoteness, overprofessionalization, and underaccountability of the institutions they defend. On the other hand, while the networks of autonomous media—the alternative press, community radio, public-access TV, microwatt broadcasting, and grassroots computer networking—have been the site of fertile experimentation in popular participation and public access, they have been stunted by the lack of resources that accompanies social and economic marginalization. Recently, however, analysts from a variety of backgrounds have begun to rethink the democratization of communication in terms that blend elements of the public service and alternative mode. They propose the public financing of a multiplicity of decentralized but collectively or cooperatively operated media outlets, licensed on the basis of commitment to encouraging participatory involvement in all levels of their activity.

Thus, for example, John Keane, writing from a liberal position, has argued that the undermining of "both arcane state power and market power" "requires the development of a dense network or 'heterarchy' of communications media that are controlled neither by the state nor by commercial markets."[62] Noting that "the new technologies strengthen the tendency whereby the element of rights to dispose of property privately becomes obsolete in the communications field," Keane argues for policies that would encourage the tendency for communication "to be seen as flows among publics rather than as an exchange among discrete commodities which can be owned and controlled privately as things."[63] This would involve a democratization of public broadcasting institutions, aimed at introducing greater accountability to and greater involvement of their various constituencies; creation of networks of leased-back broadcasting facilities made available for use by a wide variety of groups and collectivities; the support of cooperatively run publishers, community radio, and public access television; publicly funded faxes, videotext systems, and electronic mail facilities; and networks of media training and research institutions.

Somewhat similar suggestions have been made by Douglas Kellner. Drawing on his experience working on alternative television projects, Kellner suggests that the technological capacity for the multiplication of satellite and cable channels, often seen as a threat to public broadcasting, should be embraced as offering the potential for a diversified and decentralized version of such a service. He has urged the creation of a publicly funded satellite system, which, along with appropriate training and production facilities, would permit communities and movements from a wide variety of political and cultural orientations to broadcast their own programs.[64]

Popular support for decentralized and distributed public communications systems has been particularly strong in the field of computer networking. The development of the Internet arose, as we have seen, from a certain bizarre conjunction between publicly funded institutions—the original military-research ARPA Net—and the autonomous activity of a host of hackers, techno-hobbyists, and computer dissidents. In North America, the attempt to defend this unique experiment from commercial recolonization by the "information highway" has evoked a wealth of proposals for more fully releasing the democratic and participatory potential of digital technologies. Many of these come not from the usual centers of the Left but rather from technoscientific workers most familiar with the radical potentialities of the new technologies. Couched in idioms that combine liberalism, libertarianism, and undeniably communist impulses in an uncategorizable amalgam, the challenge of such initiatives to the prerogatives of corporate media empires is nevertheless unmistakable.

Thus, for example, a critique of the "information highway" put forward by Computer Professionals for Social Responsibility (CPSR) is predicated on "freedom to communicate," which it defines as having two essential features: first, freedom from censorship, and, second, "the opportunity to be heard in the first place."[65] This later is explicitly defined in terms of overcoming the condition so pithily defined by A. J. Liebling's aphorism that "the freedom of the press belongs to those who own one." CPSR suggests that the availability of increasingly cheap computer technology presents the possibility of breaking the corporate monopolies of communication established in print and broadcasting.[66] Recognizing the importance of the Internet in establishing a model of open, participatory computer communication, CPSR also notes its disadvantages—difficulties of navigation, technological complexities, and limitations of access.

It then makes the following proposals for a public network. There should be universality of access, defined not only in terms of availability of connections (with full service to homes, workplaces, and community centers), but also of low pricing, and the provision of subsidized hardware, software, and training. A basic feature of the network should be to enable all users to act as both producers and consumers: "every user . . . must have the option to generate new information as well as publish that information through the network." While CPSR concedes commercial interests a major role in the construction of the networks, it insists on preservation of "diversity of content."[67] Common carrier status—preventing the control of content by the owners of the channels—is crucial. A central aspect of any information infrastructure must, CPSR says, be the development of a "vital civic sector," constituting "public spaces" for discussion, governmental interaction, distribution of free software, and "the spontaneous development of communities of all kinds" among "groups . . . of people who want to discuss issues concerning their neighbourhood, worksite, nation or planet."[68]

Other local branches of CPSR have gone further. The Berkeley chapter calls for a national computer network infrastructure to be publicly built and maintained; for the creation of a "public information treasury" specifically aimed to "ensure that the widest possible kinds of social information are collected; and for the abolition of intellectual property laws."[69] On this last point, it notes that the ostensible and traditional rationale for such property rights is to promote progress and creativity. However, current patent and copyright systems do not perform this function but rather lead to secrecy, duplication, and litigation. As the CPSR activists observe, other models exist for organizing and rewarding intellectual work in ways that do not require proprietary title to the results—such as grants and peer or public recognition. They therefore call for a moratorium on computer software patents, accompanied by social funding of research and development, and the implementation of new systems, such as public competitions, to spur development of "socially needed technology."[70]

Even partial implementation of these ideas would represent a significant collective inroad on the capitalist information economy. But the significance of such a socialization of media goes well beyond the immediate reappropriation of resources from corporate conglomerates. Every communicational node and link established outside the control of capital diminishes its ability to naturalize commodification, to impose its "class-ifying" grids of surveillance, to suppress news of struggles, to censor, mystify, and deceive. Conversely, each instance of such countercommunication increases the possibility to explore variegated images of decommodified human identity, circulate struggles, and discuss the reorganization of society outside the parameters of the market. Because today's cultural industries take as their productive material forces basic to the constitution of individual and collective subjectivity, their liberation from capitalist control in turn enhances every other escape attempt.

Establishing a "communication commons" would both reinforce, and be reinforced by, the abolition of work proposed in the previous section. Diminishing the role of wage labor in society involves not just economic but cultural metamorphosis. This transformation would include lifting the cultural opprobrium attached to the sheer enjoyment of free time; validating the skill, difficulty, and worth of undervalued or nonmarket activities, such as collective decision-making or domestic labor; and constructing forms of subjectivity other than those revolving around the image of the "consumer." A diverse communication commons provides the matrix for such cultural experimentation, while the free time made available by the reduction of work creates the condition for the widespread involvement in cultural production necessary to give the new networks vivacity. Moreover, the establishment of such a commons creates unprecedented opportunities for cooperative organization—not least in the sphere of social governance, to which I now turn.

Zero State: Computerized Counterplanning

To pose an alternative to advanced capital is, necessarily and centrally, to raise the issue of planning. In the socialist tradition, centralized state planning has been the alternative to the market. The years of the high-technology revolution have also, and not coincidentally, been a period during which both the necessity and viability of the nation-state as a central unit of social organization has been seriously challenged. This challenge has, however, appeared simultaneously in two different and antagonistic forms: privatization and socialization.

Marx glimpsed both these tendencies a century ago. Writing of the roads, railways, and canals of his age, he described "the production of the means of communication, of the physical conditions of circulation," as part of "*communal, general conditions of social production* as distinct from the conditions of *particular capital* and *particular production* process."[71] As capital expands in scope and scale, such systems become increasingly necessary for individuals to reproduce themselves as members of a social collectivity "and hence to reproduce the community, which is itself a general condition of productive activity."[72] Marx noted that the enormous cost of investment in such infrastructures usually resulted in capital's leaving their initial development to the state: only subsequently does business reclaim them from the realm of "public works" as sources of private profit—precisely what we know as "privatization."[73] This takeover of the means of communication and other public infrastructures represents "the highest development of capital" and "indicates the degree to which the real community has constituted itself in the form of capital."[74]

But Marx also saw a contrary tendency. For example, in *Grundrisse* he describes how institutions of information, such as the mails and telegraph, are established by capital in an attempt to overcome the "crises, etc." that arise from the contradiction between increasing global "interdependence" and the "indifference" of privatized production.[75] The new means of communication are instruments in the "autonomization of the world market"—the alienation of human powers to a vast transpersonal apparatus of monetary exchange.[76] Yet at the same time, they open "relations and connections" with the potential to overcome this alienation. They introduce the possibility of "suspending the old standpoint" and replacing it with a "real communality and generality" that affirms the "general bond" of planetary humanity.[77]

Today, of course, the privatizing tendency is actualized in the neoliberal program of marketization and deregulation. Its essence is the reversion of the apparatus of government, which the era of the welfare state had (as a result of pressures from labor and other social movements) attained a certain "relative autonomy" from the immediate imperatives of business, back into direct instruments of capital accumulation. In some respects this involves a diminution in state functions: the erosion of welfare expenditures, reduction in social services, and sale of public industries. In others, it expands these functions—most notably in the intensification of the state's security, surveillance, and coercive role. Privatization abolishes the state only insofar as it presses the interdependence of capital and state to the point of identity, making the latter, in effect, the direct administrative and coercive arm of the former. As Gilles Deleuze and Felix Guattari put it, "Never before has a State lost so much of its power in order to enter with so much force into the service of the signs of economic power."[78]

This fusion of capital and state relates to the issue of information technology in several ways. It is increasingly through the state, by means of government-industry consortia, university-business partnerships, training and education schemes, military contracts, and business subsidies that capital mobilizes the range of cooperative social activities necessary to generate the technological innovations on which it depends. Moreover, much of the drive to privatization is aimed at expropriating technoscientific systems first developed as public utilities and now sufficiently advanced to become profitable for private operation; hence the selling off of telephone systems, research institutes, library resources, and so on. At the same time, high technologies allow corporate power to exercise both the carrot and the stick in compelling privatization and deregulation. The stick is the threat of capital flight into the global webs of investment and speculation. The carrot is the promise (to compliant regimes) of instrumentation for reducing costs—automating public service jobs, intensifying surveillance of welfare "cheats," deploying Robocop-like security forces to mop up social disintegration, and so on. And this technologically aided reduction in social expenditures is itself one of several avenues to reduce the so-called tax burden on corporations, thus freeing funds for the gigantic investments required by new high-technology systems. The emergent conditions of technoscientific production are thus profoundly connected—both as end and means—to the dynamic of privatization.

Confronted by this onslaught, the usual response of social democratic parties and trade unions has been a defensive cry for the maintenance of the welfare state. But calls for a return to the era of Keynesian "big government" are as inadequate as the demand that unemployment be solved by "more jobs." They forget the important critique of the welfare state mounted by workers, feminists, and antipoverty movements during the 1960s and 1970s, which addressed not only the quantitative limits of social expenditures and programs but also the qualitative problems arising from their frequently demeaning and invasive administration.[79] It is important to recognize that neoliberal success in deregulation and privatization rests in part on mobilizing these real popular resentments against remote, bureaucratic and hierarchical forms of state power. Moreover, a purely defensive response to privatization neglects the real possibilities for more responsive and participatory "governmentality" than that of the old Planner State.[80]

The "withering away of the state" was once viewed on the left as an occasion for jubilation rather than dismay. This perspective can be maintained without lapsing in to any sort of anarchist romanticism.[81] The response to neoliberal privatization should not simply be a plea for return to the welfare state, but rather a project for destatification of a different kind—one that restores and increases social expenditures but devolves administrative power toward a multiplicity of collective, democratic projects and agencies.[82] This project of "destatification downward" or "socializing without statifying," a longstanding element in the autonomist tradition, has recently been voiced from many other sections of the European Left.[83] Broadly speaking, such proposals aim to relay financial and administrative control over publicly funded governmental services away from the state apparatus toward a variety of other social loci—housing and medical cooperatives, social and cultural movements, research and innovation centers. The role of government is redefined as supporting collective initiatives rather than substituting for them, diffusing rather than concentrating control, nurturing social transformation from the bottom up rather than engineering it from the top down.

The potentiality for this diffusion arises from the proliferation of ecological, feminist, labor, educational, housing, and public transport activism that has been such a marked feature of capitalist societies over the last twenty-five years. Such activism constitutes an already-existing tissue of agencies and organizations, many operating at sophisticated levels of administrative, technological, and communicative practice. This can be seen as an arena of "counterplanning"—a term autonomists have used to designate the ability of socialized labor to run things according to priorities different from those of capital, either on shop floor or in the social factory as a whole.[84] Destatification downward rests on reinforcing and amplifying this nascent network of counterplanning agencies and institutions, so that they play an increasing role in the conception and administration of governmental regulation and spending in the workplace, welfare, education, health, and environment. Where privatization dissolves the state into capital with the aim of better subordinating society to corporate will, "socializing without statifying" reabsorbs the functions of the state within myriad noncommercial collectivities with the aim of surrounding and encroaching on capital from a variety of directions.

The products of the information revolution can be put to serve this alternative at least as effectively as they are now being marshalled in the service of privatization. Within the context of "communication commons" of the sort outlined in the previous section, computerization and telecommunication could provide the channels for access to data and analysis, cooperative assistance, and easy-to-use accounting and administrative systems necessary for complex and decentralized systems of social self-organization. In fact we are now witnessing, in embryonic form, the emergence of such capacities.

For example, in the United States, agitation by green groups has resulted in the establishment of the Right-to-Know Computer Network (RTK Net). This offers free, on-line access to the U.S. government's Toxics Release Inventory (TRI), with information on industrial releases of toxic chemicals from some twenty-four thousand U.S. industrial facilities. Grassroots groups around the country have used TRI information to produce dozens of reports on pollution, garnering public attention and compelling industry cleanup efforts in a number of states.[85] In Canada, the Ottawa-based Rural Advancement Foundation International, which serves as a clearing house and information source for movements of indigenous peoples and First and Third World farmers fighting biotechnological enclosures, uses electronic data-base searches to identify pending corporate patent claims. It disseminates its analysis via the World Wide Web.[86] To these examples can be added others—feminists coordinating proposals for international conferences by email; unions establishing in-house electronic data-bases on health and safety practice; community networkers making available public information on health or recreational activities on free-nets. All these experiments are in various ways using the networks to accumulate and distribute knowledge and coordinate activities on a scope and scale that was previously the prerogative of state and business organizations.[87] Limited as these instances are, one can extrapolate from them to envisage the potential role of computers in providing the fibers for destatification from below.

Here it is possible that information technologies may help resolve a major dilemma of the Left—that of large-scale economic coordination. It is widely held today that on this issue there exist only two options—the Free Market, or the Command State—and that the latter of these has been decisively discredited.[88] Neither, in my view, offers a desirable prospect, the former because it drives inexorably toward the commodification of human lifetime, the latter because of its tendencies—tragically demonstrated in previously existing socialism—to official despotism. Reformist combinations of state and market in a mixed economy have revealed their extreme instability.[89] In this situation, attempts to envisage an emancipatory social order seem stymied between two unacceptable choices—command by money or bureaucracy: *non tertium datur.*

There *is,* however, a third way, periodically proposed by the antiauthoritarian Left: decentralized democratic planning, sometimes known as participatory economics. The classic riposte to this suggestion is that the volume and complexity of information required to coordinate a modern economy could never be processed in time to allow any exercise of democracy or participation. However, the emergence of highly distributed, very fast information systems throws this rebuttal into question. Some radical economists are now asking whether the extreme sophistication of contemporary communications technologies does not make feasible highly decentralized forms of planning previously considered unwieldy, eliminating the need to chose between the "single brain" of the centralized state or the blind exchanges of the market.[90]

Proposals along these lines encompass varied, perhaps contradictory, possibilities. For example, the socialist-feminist Diane Elson forecasts a crucial role for communication systems in her vision of a "socialized market."[91]

Elson's economy assumes a guaranteed income—along the lines discussed earlier—and a situation where production is predominantly in the hands not of corporations, but cooperatives, the self-employed, or publicly owned but worker-managed companies. Centralized economic planning would be limited to the setting of a guiding strategy by means of fiscal and monetary policy, with the daily coordination of supply and demand left to the market. However, the market would be "socialized" by rendering it *transparent*. Enterprises would be obliged to divulge information about the design, production processes, price formation, wage conditions, and environmental consequences of the goods that they make. Publicly supported collectives—"consumers unions"—analyze this data and propose norms to govern various aspects of these practices. Information about actual production processes and proposed norms would then be disseminated via universal communication networks—something like the Internet or the information highway—publicly supported so that every individual, or at least every household, had easy access to telephones, photocopiers, fax machines, computers, and modems.

In this way, Elson says, people could know what enterprises offered, not merely in terms of price but of social and environmental costs of what was consumed. In a situation where it would be immediately apparent what goods had been produced in low-wage or environmentally dubious conditions, shopping would, she suggests, become a series of decisions about the collective, as well as individual, costs and benefits of goods selected. Collective control over information is thus interpreted in terms of democratization rather than centralization.[92] Arguing that "open access to information is the key to conscious control of the economy," Elson concludes by arguing for a strategy that aims to "attack capital's prerogatives over information, and to begin to develop networks which prefigure those a socialist economy would need."[93] Issues ranging through environmental and consumer protection, industrial democracy, and open government should be woven into a coherent campaign around access to information "appealing to a wide range of non-socialists as well as to socialists, while going to the heart of capital's ability to exploit labour."[94]

Michael Albert and Robin Hahnel propose an even more comprehensive model of decentralized planning.[95] They conceive a society in which production and consumption are entirely organized by decisions of workers' and consumers' cooperatives. Initial statements of needs, in the case of consumer councils, and capacities, in the case of workers councils, are matched and then adjusted one to another according to what emerges about the overall situation. This process proceeds by several rounds of discussion or "iterations," ascending and descending through various levels of neighborhood, regional, national, and international organization. Now, this is of course the sort of scheme that might be suspected of taking so long nothing would ever get produced or consumed. However, Albert and Hahnel argue strongly that the rapidity of information processing, the speed and scope of networked communication, and the relative ease-of-use of contemporary computer technology would make involvement in the process no more complex or time consuming than the daily processes we take for granted in a market economy

These models are, as their authors admit, necessarily abstract and schematic. But the possibilities they raise of linking high-technology communications to nonstatist planning models are important. If we consider the incredible sophistication of the electronic networks now used by global stock exchanges, or corporate just-in-time production, or military Star Wars systems, the prospect that these might be used to facilitate highly decentralized forms of collective negotiation, decision making, and resource management does not seem farfetched. By facilitating economic coordination without commodity exchange *or* dependence on centralized state bureaucracies, the information technologies capital has created dissolve a major barrier to actualizing a noncapitalist alternative.

Zero Technology? The Reconstitution of Machines

Writing of technology, in a statement to which I will return in chapter 9, Marx observed: "Nature builds no machines, no locomotives, railways, electric telegraphs, self-acting mules etc. These are products of human industry: natural material transformed into organs of the human will over nature, or of human participation in nature."[96] How far the author recognized the significance of this apparently casual distinction must remain unsure.[97] What is certain is that today the issue of whether technology is conceived as an organ of "will over" or "participation in" nature marks a momentous line of struggle.

For capitalism, the use of machines as organs of "will over nature" is an imperative. The great insight of the Frankfurt School—an insight subsequently improved and amplified by feminists and ecologists—was that capital's dual project of dominating both humanity and nature was intimately tied to the cultivation of "instrumental reason" that systematically objectifies, reduces, quantifies, and fragments the world for the purposes of technological control.[98] Business's systemic need to cheapen labor, cut the costs of raw materials, and expand consumer markets gives it an inherent bias toward the piling-up of technological power. This priority—enshrined in phrases such as "progress," "efficiency," "productivity," "modernization," and "growth"—assumes an automatism that is used to override any objection or alternative, regardless of the environmental and social consequences. Today, we witness global vistas of toxification, deforestation, desertification, dying oceans, disappearing ozone layers, and disintegrating immune systems, all interacting in ways that perhaps threaten the very existence of humanity and are undeniably inflicting social collapse, disease, and immiseration across the planet. The degree to which this project of mastery has backfired is all too obvious.

Confronting this catastrophic scene, one understandable response is an outright refusal of technoscience. This, for example, is the position of the eco-feminist Maria Mies. Writing primarily in the context of a discussion of bio-technologies, but referring also to computerization, Mies argues that high technology is so implacably stamped with a capitalist/patriarchal logic of domination that it can only be met by an act of absolute refusal. Marxism, because of its attachment to technological development, is rejected. Any left-ist who uses a computer is "schizophrenic."[99] The project of oppositional politics is defined as the construction of a society based on "subsistence pro-duction" that largely repudiates machine production and happily accepts voluntary frugality.[100] This type of perspective is now widespread in ecologi-cal, feminist, and anarchist movements.

Contrary to the celebrants of preindustrial conditions, I would argue that a return to such relative impoverishment sets the likely conditions for the re-imposition of all the most unpleasant forms of parochial and patriarchal tyr-anny. Notwithstanding the enormous problems of environmental degradation that have accompanied their development, machines are a prerequisite for cre-ating the surpluses that support human freedom. Moreover, the technologi-cal changes that have already been wrought on the natural and social habitat are often irreversible. Short of accepting the need for mass extinction of sur-plus peoples (as some misanthropic sections of the ecology movement do), the sustainability of human society can no longer be predicated on reversion to a supposedly natural, preindustrial condition. Rather, it will require continu-ous levels of intervention and management even in order to contain or undo the dangers already set in motion by damage to the planetary ecology.[101]

This interpenetrating of "first" and "second" natures is not *necessarily* terrible. As capital has been compelled by labor struggles to develop technolo-gies that could *potentially* end the need for wage work, so it has been spurred by green activism to create machines that *potentially* diminish the depletion of the natural world. Computer and communications networks could (if used in conjunction with electricity sources other than catastrophic megaprojects) be elements in a benign and careful planetary metabolism that, rather than pillaging and defiling ecological systems, repaired and protected them. The experiments of many ecological movements—for example, in the satellite mapping of endangered resources—demonstrate this capacity. However, just as capital makes of automation a means to increase people's availability for work, so it deforms resource-saving technologies into means to extend and intensify the reduction of nature to raw materials. Undoing this paradox re-quires a governance of technology free from capital's compulsion to convert the world into commodity-form.

Thus, rather than rejecting technological development *tout court* it seems more useful to reconsider whether there is some possibility of breaking with the capitalist project of technology as "will over nature" and of developing Marx's hint that machines might instead be developed as organs of "participation in nature." This of course was the issue raised by Herbert Marcuse nearly fifty years ago when he called for the possibility of an alternative technology based on active partnership with nature rather than Promethean conquest.[102] His suggestion was stingingly attacked by Jürgen Habermas, who, in a highly influential article, accused Marcuse of a romanticism that confused the proper domains of "communicative" and "instrumental" reason.[103] The natural world was mute and never could become a coparticipant and interlocutor in the development of technology, but must always remain an object of human control.

In my view, however, Habermas's refutation is not definitive. Marcuse does not have to be understood as proposing a conversation with dolphins, owls, and rain forests, but a dialogue among humans who perceive a more reflexive and participatory relationship with such creatures than instrumental rationality acknowledges. The development of machines as "organs of participation in nature" means recognizing that the human wielders of technology are embedded in and dependent on the world they transform, and intervening with an awareness of the limits and uncertainties that flow from this recursive situation.[104]

Moreover, as Andrew Feenberg has argued, since the time Marcuse issued his call, the project of developing a new science and technology has taken concrete social form. Social movements in conflict with the technoscientific agenda of capital—feminists, ecologists, community health, and worker movements—have, at both theoretical and practical levels, challenged the characteristic methods, preoccupations, and institutional structures of corporate technoscience.[105] Such movements have attempted to develop modes of investigation and experimentation that do not align themselves with the assumptions of capitalist progress. In a field of workplace, medical, and environmental settings they have challenged the rigid instrumental division between subjects and objects of knowledge, and investigated research practices emphasizing holism, interaction, complexity, and self-reflexivity. They have questioned the privileging of certain forms of theoretical inquiry over others—for example, the adoption of physics rather than biology as a model of scientific inquiry—and disputed the automatic dismissal of alternative knowledge-systems, such of those of indigenous peoples. They have experimented with using the machines designed by capital in ways different from what was intended, and with intentionally designing machines in ways different from those employed by capitalism.[106]

Technoscientific innovation is a collective, social process. It is not so much something that capital creates as appropriates—an activity it must forcibly shape and twist to its purposes, by acts of exclusion, repression, and marginalization. Moreover, scientific practices are manifold rather than monolithic. Thus, although reductionism, fragmentation, and "will over nature" are elements in technoscientific endeavor to which the path of capitalist development has given precedence and emphasis, they are not the whole story. As Evelyn Fox Keller has argued from a feminist perspective, they are only part of a more complex and variegated bundle of impulses and approaches associated with scientific activity, which also includes very different tendencies toward holistic perspectives, reverence, and curiosity.[107] If these aspects have been devalued in capital's expropriation of social knowledge, they have never been completely extinguished, and they can be revived.

The commonwealth would create space for these emergent forms of counterknowledge and alternative ways of doing. It would not reject technological development, but broaden its scope, opening and creating institutions to allow the emergence of experiments, innovations, and logics other than those that have hitherto been admitted, and assessing them not according to the needs and priorities of capital, but by far more widely determined criteria. Adherents of many movements are now arguing for a "democratization of technology" . . . "a democracy deep enough to function even at the level at which the machines are shaped—from the uses to which those machines are applied to their design and construction and use."[108] Theorists such as Andrew Feenberg, Richard Sclove, Michael Goldhaber, and Hilary Wainwright have done valuable work in suggesting noncapitalist criteria that might be applied in evaluating technologies for collective adoption—for example, the degree to which they support ecological sustainability, local economic self-reliance, satisfying work experiences, flexible life scheduling, and equalitarian and diverse social relations.[109]

They have also suggested the array of new institutions necessary to make application of these criteria feasible. These include the creation of extensive opportunities for citizen involvement in technological research, development, and design; publicly funded organizations to assist communities to research and develop technologies shaped to their needs; public programs to overcome traditional patterns of marginalization and exclusion in the institutions of science and technology; and a wide array of collective bodies to monitor, test, evaluate, and debate the consequences of specific lines of research and determine the level of funding for their development, possible redirections, or termination. As Douglas Schuler observes, while these approaches could not and should not *control* technoscientific innovation, which depends on the surprising and unpredictable, it could *shape* its trajectory—just as capitalist control today channels it, but in very different directions.[110]

The only shortfall of this approach is the apparent reluctance of many of its advocates to recognize that the adoption of such arrangements, on any large scale, is incompatible with capitalism. For the liberal-sounding slogan "democratization of technology" is, if taken seriously, tantamount to a call for the reappropriation of the means of production, and will be resisted by established power accordingly. Such a "democratization" is, however, consistent with a noncapitalist commonwealth characterized by decentralized, networked collective planning and an abundance of free time. Moreover, the advance of such initiatives for the collective control of machine development is itself a way of struggling for the institution of such a commonwealth.

The commonwealth outlined here is clearly not a primitivist one based on the abolition of machines. But it does imply a very different relation between machines and people from that which exists under capital—to a degree that perhaps subverts commonly accepted notions of "technology." Historically, machines have incarnated expropriation from the means of production. In their fixity of design, industrial technologies embodied—or metalized—the alien will of their owner, so much so that sayings like a "cog in the machine" summon up a world of dispossession and powerlessness. Indeed, as Marx often pointed out, in a certain sense this association with dominative power became definitive of what a machine *is*.

There are in play today, in social struggles and in everyday experimentations like hacking, tendencies to erode this situation. The commonwealth envisaged here would accelerate this dissolution. In particular, it would undo the capitalist line of machine development (dynamic in some respects, sluggish and constrained in others) that in the name of progress and efficiency assumes the status of a natural law, repressing question or deviation and canceling the autonomy of the humans it ostensibly serves. Instead, the selection or refusal of particular paths of innovation would be the outcome of collective reflection and discussion.

This collective decision-making might well lead to the phasing out of certain machines that the capitalist structuring of everyday life has made indispensable (such as the private automobile) or the rapid development of others (such as the universal provision of adequate cooking and clean drinking-water facilities on a global scale) to which it has paid little or no attention. In the absence of capital's compulsion to accumulate, any number of more, less, or differently technologized futures, currently ruled out of play as inefficient or noneconomic, become available. This would not be because of any magical translation to some realm of infinite abundance, but because a self-organized society is empowered to make the difficult decisions as to how to allocate its resources.

If the commonwealth itself has any technological imperative, it is a paradoxical and self-reflexive one, namely, that there will be enough machines to permit choice about whether to develop more machines. Sufficient automation to free ample time from work, a communications infrastructure capable of acting as an organ of democratic debate and planning, enables collective decision and reflection. The aim is to subordinate the imperative aspects of technology to the collective, communicative determination of social directions.

The Future in the Present

I have pointed to various ingredients for the creation of a social order different from capital. The elements for this alternative are at hand, but not combined. They exist, here and now, but only here and there, just as at a certain point in the prehistory of capital its various ingredients—wage labor, market exchange, new machinery—all existed in scattered form but had not cohered, or been violently welded, into a new order.

Under what conditions, and through what pressures, the new ensemble might come into being is uncertain. I do not believe its emergence is inevitable. It is, however, obvious that capitalism is experiencing serious difficulties in managing the world-transforming technologies it has itself brought into being. The problems of sustaining employment in the face of blisteringly fast automation; the consequent contrast between restricted consumption power and endlessly expanding production; the tendencies of social spending cuts to erode the very public infrastructures on which technological development depends; the repeated failures to restrain the depredation of the planet's ecology; and the manifest instabilities introduced by the lightning-fast transactions of global financial markets (recently dramatically revealed by the meltdown of the South East Asian economies) all suggest that maintenance of the existing order may be a project no less utopian (in the negative sense of inviting incredulity) than the creation of an alternative.

What is offered here is not so much a blueprint as a battlefield map. It does not identify an agenda to be implemented "after the revolution," but a series of initiatives whose advancement would contaminate and overload the circuitry of capital with demands and requirements contradictory to the imperatives of profit. Pursuit of these interrelated measures would cumulatively undermine the logic that binds society around market exchange and increasingly require the reassembly of everyday activities into a new configuration.

The actualization of such an alternative will, however, be contested. While the recent disintegration of Soviet state-socialism presents the historically unusual case of a system so demoralized and undermined that it collapsed without major exercise of force, a repetition of this pattern should not be assumed; as a wall poster on the streets of Vancouver recently warned, "present policies are not accidental: capital will put up a fight." Insurrectionary concepts of revolution—the storming of the Winter Palace—are today a dead letter. But capital's capacity to unleash violence against any serious challenge is undiminished. To agitate for social change while ignoring this would be to act in bad faith. I can imagine a commonwealth born in extreme tumult. It could come out of mounting civil disorder arising from intensifying unemployment and social disintegration, accompanied by increased activity of protofascist militias and extreme-right parties and resistance against them. A social democratic government elected to implement part of the commonwealth program—say, a guaranteed annual income—might face a reactionary coup, whose defeat in turn propels deeper social transformation. A region or nation attempting to secede from the world market by debt-repudiation might actualize some parts of the program, at the risk of invasion or intervention. At worst, the alternative may emerge in the wake of ecological catastrophe or the devastation of intercapitalist war.

Whatever path their actualization might take, the measures suggested here, combined in some concerted societywide ensemble, would make up a world very different from that which we today accept as normal. It would be a world where wage-work would have a steadily decreasing importance or vanish entirely; where, although there would be work to be done, livelihood would not be dependent on a job; where, consequently, people would have more time to think about and participate in decisions about organizing life in association with others; where they would have access to a very wide variety of communication channels, with a very wide diversity of representations and images about different possibilities of being; where these channels served also as routes for a flow of participatory decision making about the production and distribution of goods—and also about the directions taken and not taken in technological development. Distant as these prospects may seem, they are potentialities germinating in the soil of our everyday lives, today.

INTELLECTS

At the beginning of this work I described class conflicts within high-technology capitalism as a "contest for general intellect." This final chapter returns to that phrase. After describing Marx's original use of the term "general intellect," I examine the recent reworking of his concept by a group of theorists clustered around the French journal *Futur Antérieur*, and I suggest how their perspective helps frame some of the issues discussed in the preceding pages. I then conclude with some reflections on the implications of this analysis of "general intellect" for those who teach and study in universities.

General Intellect

Marx introduces the concept of "general intellect" in a passage of the *Grundrisse* known as the "Fragment on Machines."[1] In these pages he departs from his customary emphasis on the role of work in creating the surpluses needed for social progress. Rather, he suggests that at a certain point in the development of capital the creation of real wealth will come to depend not on the direct expenditure of labor time in production but on two interrelated factors: technological expertise, that is, "scientific labor," and organization, or "social combination."[2] The crucial factor in production will become the "de-

velopment of the general powers of the human head"; "general social knowledge"; "social intellect"; or, in a striking metaphor, "the general productive
forces of the social brain."[3]

The main expression of the power of "general intellect" is the increasing
importance of machinery—"fixed capital"—in social organization:

> Nature builds no machines, no locomotives, railways, electric telegraphs,
> self-acting mules etc. These are products of human industry: natural
> material transformed into organs of the human will over nature, or of hu-
> man participation in nature. They are organs of the human brain, created
> by the human hand: the power of knowledge, objectified. The develop-
> ment of fixed capital indicates to what degree general social knowledge
> has become a direct force of production, and to what degree, hence, the
> conditions of the process of social life itself have come under the con-
> trol of the general intellect and been transformed in accordance with it.[4]

There are two forms of technology Marx particularly notes as signaling
capitalism's mobilization of "general intellect." One is the development of
production systems based on "an automatic system of machinery . . . consisting of numerous mechanical and intellectual organs, so that the workers
themselves are cast merely as its conscious linkages."[5] The other, to which
his allusions are more scattered but equally persistent, is the network of transport and communication integrating "the world market." The development
of human-eliminating, globe-spanning machines indicates the degree to which
"general intellect" has been successfully mobilized and mastered by business,
and "the accumulation of knowledge and skill, of the general productive
forces of the social brain . . . absorbed into capital."[6]

However—and this is the whole point of Marx's analysis—such a level of
technological advance, which seems at first a capitalist utopia, contains
within itself the seeds of a capitalist nightmare. By setting in motion the
powers of scientific knowledge and social cooperation, capital ultimately
undermines itself. This occurs for two reasons. First, as advances in machinery and organization reduce the requirement for direct labor in production,
the need for people to sell their labor power—the very basis of capitalism's
social order—is systematically eroded. There arises a monstrous dispropor-
tion between individual labor time and the forces set in motion by organized
science.

This is reinforced by a second tendency, the increasingly social nature of
activity required for technoscientific development, which unfolds not on the
basis of individual effort but as a vast cooperative endeavor. As this becomes
more and more apparent, highlighted by the diffusion and integration of communication and transport networks, both private ownership and payment for
isolated quanta of work-time appear increasingly as irrelevant impediments
to the full use of social resources. Automation and socialization together

create the possibility of—and necessity for—dispensing with wage labor and private ownership. In the era of general intellect "capital thus works towards its own dissolution as the form dominating production."[7]

Today, "The Fragment on Machines" seems simultaneously astoundingly prescient and sadly anachronistic. In its extrapolation of capital's techno-scientific trajectory it is surely prophetic. What Marx describes is eminently recognizable as a portrait of what is now commonly termed an "information society" or "knowledge economy," in which the entire intellectual resources of society, from shop-floor production teams, to university-industry partnerships, to the regional "innovation milieux" of microelectronic and biotechnology companies, is mobilized to produce the technological wonders of robotic factories, gene splicing, and global computer networks. Yet any suggestion that this development of the productive forces leads automatically to the advent of socialism appears definitively refuted. Instead, we seem to be witnessing a triumphant reorganization of capitalism that is deploying the new technological innovations to solidify an unprecedented level of global domination. What—if anything—can now be made of the revolutionary optimism of Marx's account of "general intellect"?

Futur Antérieur

It is this question that is addressed by the work of a group of theorists associated with the French journal *Futur Antérieur*. This group includes veterans of the Italian *autonomia* movement whose earlier course was charted in chapter 4, such as Antonio Negri and Paolo Virno; younger scholars making new departures within this tradition, such as Michael Hardt and Maurizio Lazzarato; and others with roots in different lines of Marxism, such as Jean-Marie Vincent.[8] The central points of their analysis can be summarized as follows.

The "mass worker" struggles of the 1960s and 1970s and the consequent crisis of Fordism compelled capital toward extraordinary levels of high-technology automation and global mobility. These post-Fordist experiments have now brought capital to a point corresponding to Marx's account of "general intellect." However, rather than generating the ordained demise of capitalism, these developments are resulting in something much more ambiguous. Paradoxically, the revolutionary tendencies Marx identified—the erosion of wage labor, the increasingly "social" nature of production—are occurring, but in forms prescribed by an order that continues to organize itself on the basis of the wage and private ownership. As Virno remarks, these processes remind one of what Marx wrote about joint-stock companies; that in such institutions "one witnesses the disappearance of private property on the very ground of private property."[9] Today post-Fordist capital displays a similar transformation of communist potentialities into capitalist actualities. As Virno puts it, "the displacement is real, but the ground on which it is accomplished is no less real. To think these two aspects jointly, without reducing the first to a mere virtuality and the second to an external 'rind': such is the difficulty that cannot be avoided."[10] In this situation it is not enough to focus, as Marx did, on the objectification of social knowledge in new technologies. Rather, the critical issue is that of the nature of the human activity required to create, support, and enable this technoscientific apparatus. Here, *Futur Antérieur* suggests, we encounter another paradox. While capital has developed machines to subordinate and reduce labor at the point of production, this development itself demands the emergence a new range of social competencies and cooperations—the cultivation of "general social knowledge." This subjective component of general intellect the *Futur Antérieur* group explores under the label of "mass intellectuality" (*intellectualité de masse*).

"Mass intellectuality" is the ensemble of "know-hows" that supports the operation of the high-tech economy. It is "the social body" as a "repository of knowledges indivisible from living subjects and from their linguistic co-operation."[11] It comprises a "whole gamut of qualifications, modes of communication, local knowledges, informal 'language games' and even certain ethical preoccupations."[12] Negri says that "mass intellectuality" is the activity of a "post-Fordist proletariat . . . increasingly directly involved in computer-related, communicative and formative work . . . shot through and constituted by the continuous interweaving of technoscientific activity and the hard work of production of commodities, by the territoriality of the networks within which this interweaving is distributed, by the increasingly intimate combination of the recomposition of times of labour and of forms of life."[13] Mass intellect appears not just in production but throughout a whole network of educational and cultural relations. It is present in industrial and service workers, laboring at the interface with digital technologies, in students keeping pace with technological innovation through "lifelong learning," and in the various technocultural literacies on which new markets for electronic and entertainment goods depend. Mass intellectuality is intimately bound up with the new prominence of what Negri and Lazzarato term "immaterial labour"—the "distinctive quality and mark" of work in "the epoch in which information and communication play an essential role in each stage of the process of production."[14] Overflowing and surpassing previous Marxist distinctions between base and superstructure, economics and culture, mass intellectuality is "difficult to describe in economic terms" but is "for that very reason (and not despite it) the fundamental ingredient of today's capitalist production."[15]

The crucial question thus becomes how far capital can contain what Vincent terms "this plural, multiform, constantly mutating intelligence" of mass intellect within its structures.[16] As he observes, it "appears to domesticate general intellect without too much difficulty."[17] But this absorption demands an extraordinary exercise of "supervision and surveillance," involving "complex procedures of attributing rights to know and/or rights of access to knowledge which are at the same time procedures of exclusion":

> Good "management" of the processes of knowledge consists of polariz-ing them, of producing success and failure, of integrating legitimating knowledges and disqualifying illegitimate knowledges, that is, ones con-trary to the reproduction of capital. It needs individuals who know what they are doing, but only up to a certain point. Capitalist "management" and a whole series of institutions (particularly of education) are trying to limit the usage of knowledges produced and transmitted. In the name of profitability and immediate results, they are prohibiting connections and relationships that could profoundly modify the structure of the field of knowledge.[18]

The *Futur Antérieur* group suggests that these structures of exclusion and limitation can become the occasion for new forms of social conflict.

Team Concept

Perhaps the most detailed analysis of the new antagonisms that *Futur Antérieur* sees as characterizing the era of "general intellect" is Negri and Lazzarato's discussion of "participative management."[19] As they point out, in many post-Fordist industries the quantitative elimination of labor by com-puterized automation has paradoxically been accompanied with increasing managerial concern about the quality of the remaining workers. To prevent or fix the many breakdowns of new production systems, to run them at peak capacity, requires operators who are creative, cooperative, and alert. This requirement has resulted in innumerable post-Taylorist experiments in work organization—"quality circles," "team concept," "Japanese management techniques," "Total Quality Management"—in which the intellectual and intersubjective aspects of labor previously suppressed by Taylorism are mo-bilized for problem solving and participation. Such systems demand that workers "become 'active subjects' in the co-ordination of the different func-tions of production, instead of being subjected to it as simple command. As the new management prescribes, today it is 'the soul of the worker' which must come down into the factory."[20] If new production systems are the ob-jective side of capitalized "general intellect," the work team represents its subjective side, in cellular form.

Such participative management schemes are, Lazzarato and Negri say, "techniques of power."[21] In an apparent contrast to Taylorism, capital grants its labor power a certain fusion of conception and execution. Despotic man-agement seems to retreat from the shop floor. Capital continues, however, to dominate the overall process from the heights of the enterprise, retaining control of finance, investment, marketing, and, of course, profit. Problem solving is predicated on accepting these predetermined parameters.[22] Although management exhorts dialogue and interaction, communication is actually reduced to "a simple relay of codification and decodification, within the context . . . that has been completely normalized by the firm."[23] In this context, the exhortation to participate is, as Lazzarato observes, authoritarian: "one must express oneself, one must speak, one must communicate, one must co-operate."[24] Indeed, the new team organization is even more totalitarian than the old assembly line, precisely because it seeks to involve the very subjectivity and will of workers, making them "control" themselves so that command "arises from the subject itself, and from the communicative process."[25]

However, Negri and Lazzarato suggest there is another side to this process. In delegating—even nominally—certain managerial responsibilities to workers, capital is partially relinquishing its claim to act as the mediator and coordinator of production. There is a potential tension between capital's control of enterprises and the increasingly self-directed nature of work. Drawing on Negri and Lazzarato's work in the context of the South African auto industry, Franco Barchiesi observes that

> a massive contradiction arises for capital: it has to stimulate and harness subjectivity by encouraging increasing worker responsibilization, even creativity, in order to grasp a social and communicational surplus value in the workplace. This . . . comes to constitute a competitive edge in the global fight for shrinking and specialized markets. But in doing so, capital has to be careful in depriving worker subjectivity of any implication in terms of power and control. . . . In this way, capital silences subjectivity just at the same time it calls it into life. Capital has not found, yet, the ways to deal with this contradiction.[26]

Such tension becomes increasingly pronounced as business uses the knowledge squeezed from team production to intensify automation, speed up work, and increase layoffs. In this sense, Negri and Lazzarato suggest, post-Fordist production methods, although devised as a means of circumventing and coopting workers organizations, contain the seeds of an aggravated conflict.

Lazzarato has examined some of these dynamics in strikes at Peugeot car factories in France in 1989.[27] These strikes were significant because they broke a relatively long period of industrial peace in the French automobile industry. They involved a new generation of employees, supposedly distanced from the militancy of the older "mass" assembly-line workers, including many immigrants, and trained for work in a highly automated environment. Lazzarato argues that the company's rhetoric about "involvement," "participation," and "dignity," although at first quite attractive to workers, gradually became mired in contradiction. There emerged an increasing discrepancy between the company's supposed willingness to entertain all and any "suggestions" and its evident determination to implement only those that enhanced productivity. The alleged ethic of cooperation was riddled with actual grievances about pay and pace of work. This led to mounting tension on the shop floor, which eventually exploded. In the strikes, one of the workers' demands was for the company to live up to its own rhetoric about respect and cooperation. Moreover, in this strike, Lazzarato argues, new forms of shop-floor and community organization could be seen emerging, in some ways supplanting the more conventional and rigid forms of trade union hierarchy. This, he suggests, shows that the "cooperative" aspects of the new work organization were being mobilized, but in the form of counterpower. Moreover, these isolated strikes can now been seen as anticipations of the societywide explosions of the 1995/1996 general strike in France, with its remarkable spontaneous organization by myriad popular assemblies.

Although North America has not seen concerted unrest on the scale of the French strikes, *Futur Antérieur*'s claim that forms of "participative management" are generating new flash-points for industrial conflict receives some confirmation if we examine workplace tendencies in the United States and Canada. A survey of fifteen hundred workers and managers on the topic of team organization, conducted by the U.S. consultant firm Kepner-Tregoe, produced findings so shocking that the researchers had them checked by another company. The verified results clearly showed that every aspect of participative management elicited cynicism among workers. In the words of Kepner-Tregoe's president: "The vitriolic response was amazing. . . . Workers don't like their companies, and there is a fundamental social change going on in this country regarding workplace relations. The workers hear the verbiage about how 'our people are the most important asset we have' and they want to throw up."[28]

In at least one sector where capital's drive for teamwork and other new management techniques has been very intense, the automobile industry, the mid-1990s saw a series of significant strikes. Tactically, car workers discovered the susceptibility of highly integrated, technologically sophisticated "just-in-time" production to work stoppages. Strategically, they made demands that responded aggressively to new technological conditions by challenging traditional managerial prerogatives. In Flint, Michigan, car workers struck to compel hiring new workers rather than increasing overtime, linked these demands to the need to reduce unemployment, and won partial successes.[29] In 1996 Canadian autoworkers responded to contracting-out by General Motors by striking in support of "job ownership"—shorter work time, restrictions on outsourcing, and guaranteed job levels for the communities in which plants were located.[30] The strike, which lasted twenty-one days and included the workers' occupation of a plant from which GM was attempting to remove dies to start production elsewhere, won considerable public support and was victorious. In both these cases, workers fought for a real, not token, voice in production decisions and linked workplace demands to a social agenda aimed at counteracting the destructive consequences of capitalism's post-Fordist restructuring.

As we saw in chapter 5, some North American factory workers, faced with drastic downsizing, have gone so far as to introduce alternative production proposals. Auto workers in Ontario and Los Angeles have entered into alliances with environmental and community groups to introduce "green work" plans, and they have even begun to connect these projects internationally. In other sectors, especially in the manufacturing and defense industries, there have been instances where workers faced with automation and relocation have challenged capital's right to shut down.[31] Plant closures have been met with plant occupations and picket lines aimed not only at stopping strikebreakers' getting in but at preventing machines' being taken out. Facing the withdrawal of waged work, labor has deployed its invention power not so much to stop production (as in classic strike strategies), but to keep it going— and, sometimes, to transform it, converting military or ecologically damaging industries toward "socially useful production."[32] Sometimes symbolic, sometimes sustained, such actions have occasionally either forced capital to continue operations contrary to its intentions or transferred management entirely into the hands of the workers. Repeatedly they have involved the creation of alliances with wider community groups negatively affected by capital flight. Initially defensive and local, usually limited in their aims, most of such efforts are painlessly reabsorbed within the overall logic of market relations. Nevertheless, in their proliferation, they constitute a multiplicity of subversive question marks about the priorities of capitalist production.

Alongside these projects for what be might called "autonomous production" is a whole series of struggles over the allocation of time.[33] The objectives workers seek are diverse: resistance to layoffs, rollbacks, speedups, and contracting out; demands for redundancy compensation; support for retraining; better pay and conditions for contingent workers; protection of health and other benefits. These goals are not, per se, new. But they are set in a new context—that of the vast potential surpluses of labor time produced by automation. Underlying the new wave of struggles is a rejection of capital's prerogative to plan and manage these surpluses to its own advantage. Most important, there are, as discussed in chapter 8, a number of initiatives to address the crisis of employment by shortening the working day and by introducing new, general forms of income distribution separate from the wage. These movements often emerge at the intersections of labor, feminist, green, and poor people's movements. Currently tentative and in a state of flux, they have wide potential in terms of weakening the wage relation, reorganizing household labor, and stopping the environmental destruction resulting from capital's drive for "production for productions sake." In such movements social labor has mobilized the same intellectual and cooperative capacities that capital tries to harness through teamwork, but in different directions, and with a vastly expanded horizon of collective responsibility. These movements establish networks of counterresearch and pools of shared experience, new connections and alliances; they build a capacity for counterplanning from below. In short, they represent a nascent alternative to capitalism's organization of "general intellect."

Interactive Networks

The other field where *Futur Antérieur* has investigated the contradictions of "general intellect" is that of media and communication. As Vincent puts it, "general intellect" is in fact "a labour of networks and communicative discourse": "In effect, it is not possible to have a 'general intellect' without a great variety of polymorphous communications, sequences of communication in the teams and collectivities work, communications to use in a creative fashion the knowledges already accumulated, communications to elaborate and record new knowledges."[34] Capital has developed technologies of information—mass media, telecommunications, and computer networks—to consolidate markets and ideological control. But here too it has been unable to develop the objective, fixed, machine side of "general intellect" without also involving the subjective, variable, human aspect. Negri specifically rejects media critiques framed only in terms of "manipulation."[35] Although we now inhabit a world where corporate media seem to constitute a vast "machine" that dominates society, there are, he says, spaces on the "inside" of this machine within which new individual and collective subjectivities can emerge.[36]

The *Futur Antérieur* authors have studied a number of movements in France where groups opposing neoliberal policies have shown great dexterity in using media and information technologies to publicize their causes. These include strikes by cultural workers—film crews and audio-visual technicians—fighting for improvements in the conditions of contingent work; the movements of nurses opposing cutbacks and privatization of health care; and the student revolts of 1986, which I discussed in chapter 4.[37] In Italy, Lazzarato has analyzed the media practices of the "Panther" student movement, which in the late 1990s closed some one hundred fifty Italian colleges and universities in protests against privatization.[38] These movements, Lazzarato says, were characterized by their extreme sophistication in countermanagement of the media. The students exercised careful control of how, and under what conditions, journalists covered their actions. They refused to subscribe to conventions damaging to the political integrity of the movement (e.g., focus on leaders). And they made constant use of information technologies—particularly fax—to generate their own coverage and bulletins. Lazzarato argues that the Panthers' careful orchestration of refusals and reappropriations displays the generational characteristics of subjects who, having come of age in a media environment, are capable of shaping this terrain for their own political purposes, rather than merely being passively exploited as objects of spectacular display.

These tendencies are, again, manifest in a North America context, where alternative and subversive media channels—political film and video networks, community TV, microwatt and community radio—have spread like fireweed. Even as corporate media consolidate more massive, vertically integrated empires, "mass intellect" seems engaged in a proliferating counterusage of information technology, springing hundreds of leaks and counterflows within capital's communication apparatus. Such grassroots media experiments are not only plaguing capital with an epidemic of transgressions against "intellectual property" but are playing a crucial role in circulating news and analysis of struggle across sectorially diverse and geographically distant movements.

Nowhere has this been more apparent than in the field of computer-mediated communications. As we have seen, in the development of this extraordinarily powerful technology capital has depended on a mass of informal, innovative, intellectual activity—"hacking"—on whose creativity commerce constantly draws even as it criminalizes it. It was out of capital's inability to contain such activity that there emerged the astounding growth of the Internet. This is surely the quintessential institution of "general intellect." For, despite all the admitted banalities and exclusivities of Internet practice, one at moments glimpses in its global exchanges what seems like the formation of a polycentric, communicatively connected, collective intelligence.[39]

Today, of course, capital is trying to recuperate this collective intelligence by channeling it along the information highway, forcing its traffic into the commodified pathways of video-on-demand, teleshopping, telegambling, and personalized advertising. It is funneling network interactions into a commercial "interactivity." As Chris Carlsonn observes, there is an interesting parallel between such media interactivity and participative management techniques.[40] The control that corporate interactivity offers media audiences "mimics the false control offered by workers participation schemes, wherein workers decide how to accomplish the businesses mission, but, crucially, not what the mission is."[41] If the work team is the microcosmic, cellular, shopfloor form of capitalism's "general intellect," the media interactivity of the corporate information highway is its macrocosmic mode, expanding through the entire social metabolism in an attempt to integrate subjects into a seamless circuit of labor and consumption.

On the Internet, however, mass intellect has spectacularly refused to be corralled. In earlier chapters, I have already given several instances of the dissident cyber-communication that has become a significant part of oppositional politics in the 1990s. To these examples I can do no better than add one more example, originally reported by the conservative Canadian newspaper the *Globe and Mail* but, appropriately enough, emailed to me by a telecommunications worker who tirelessly relays anticapitalist news to an array of electronic contacts. The report, dated May 1998, concerns the stalling of the Multilateral Agreement on Investment (MAI).

The MAI was intended as the latest in the round of international agreements assuring the untrammeled activity of the world market, in this case by effectively removing governments' ability to regulate corporate direct investment. Officials of the OECD pursued preliminary discussions in high secrecy for eighteen months, apparently pursuing a stealth strategy designed to circumvent opposition. In 1997, however, activists from the Malaysian-based Third World Network alerted international social movements to these proceedings. The Council for Canadians, a movement fighting the destruction of the Canadian welfare state, obtained a draft of the treaty and, as the *Globe and Mail* puts it, "immediately posted it on Web site and made sure allies around the world knew it was there through e-mail."[42] As news of the planned agreement leaked, it was met by a wave of protest:

> High-powered politicians had reams of statistics and analysis on why a set of international investing rules would make the world a better place. They were no match, however, for a global band of grassroots organisers, which, with little more than computers and access to the Internet, helped derail a deal. . . . Using the Internet's capability to broadcast information instantly worldwide [these] groups . . . have been able to keep each other informed of the latest developments and supply information gleaned in one country that may prove embarrassing to a government in another. By pooling their information they have broken through the wall of secrecy that traditionally surrounds international negotiations, forcing governments to deal with their complaints. "We are in constant contact with our allies in other countries," said Maude Barlow, the Council of Canadians chairwoman. "If a negotiator says something to someone over a glass of wine, we'll have it on the Internet within an hour, all over the world."[43]

The *Globe and Mail* goes on to remark that "the OECD's efforts to harness the Internet have not caught up in colour, content and consumer friendliness to those of the advocacy groups" and reports an official's rueful comment that it had failed on a "strategy on information, communication and explication." The extent of the failure was clear when disagreements among the negotiators, disagreements at least partly reflecting the pressures brought to bear on their domestic governments by popular movements, resulted in their missing their initial deadline for negotiating the MAI. "This is the first successful Internet campaign by non-governmental organisations," said one diplomat. "It's been very effective."[44]

It would be foolish to exaggerate the significance of what may prove only a temporary and tactical victory. Nor should the *Globe and Mail*'s account of "How the Net Killed the MAI" be uncritically accepted, for it overlooks the amount of very traditional, on-the-ground, in-person meetings, marches, demonstrations, and pickets involved in the anti-MAI mobilization. However, if its analysis is even partially true—and my own experiences of the campaign suggest that it is—then this is a striking vindication of Negri's thesis about the capacities of mass intellect to reclaim advanced capital's means of communication. Moreover, this example suggests that social movements are beginning to consider using this capacity in more than a reactive way. Spokespeople for the Council of Canadians are cited as suggesting that the next stage of the anti-MAI campaign will activate the global communications network to circulate plans for alternatives to capitalism's globalizing project. They stress that while anti-MAI groups are "against this model of economic globalisation" their use of the Internet shows their own commitment to the "idea of coming together and working together" across international boundaries.[45] Such global, electronically facilitated counterplanning is precisely what might be expected of a movement of "mass intellect."

How General Is "General Intellect"?

Looking at the contradictions appearing both in the workplace and in the larger societal networks of the post-Fordist economy, the *Futur Antérieur* group argues that mass intellect is potentially explosive for capital. This volatility arises not only from a dynamic of immiseration—with more and more people being expelled from production by automation—but also from a reappropriative process in which "mass intellect" begins to fold back into itself the organizational and technological knowledge necessary for the running of society. Negri now calls this capacity "constituent power" and describes the task of radical politics as the creation of a "republic" that dissolves both capitalist command and state authority.[46] Virno speaks of an "exodus" from the "society of work" made possible by a radical redisposal of the surplus time arising from automation.[47] It is these potentialities of "mass intellect" that *Futur Antérieur* now sees pulsing through a wave of social protest in advanced capitalist societies of the 1990s—in France, but also in the large-scale strikes and protests in Italy, Germany, and Belgium, and to a lesser degree in North America.

The argument that subversive potentialities exist at the very heart of the technological armature that seems to make contemporary capital so impregnable is an attractive one. Although only starting to be translated and discussed in North America, *Futur Antérieur*'s revival and reworking of the category of "general intellect" has already sparked some debate in Europe. This has, however, included some substantial criticisms.[48]

Perhaps the most serious of these objections is that, in its capitalist form, "general intellect" is not "general" at all, but rather structured by an intensely hierarchical division of labor. This restricts crucial types of knowledge to a narrow stratum of privileged, and hence loyal, employees, leaving the rest to suffer the effects of technological deskilling. The edifice of scientific-technological power depends not just on scientists, engineers, programmers, and various "symbolic analysts" but on a mass of janitors, homeworkers, fast-food cooks, and other service workers. But the crucial point, the critics say, is that these latter are excluded from the intellectual functions of the capitalist economy. The whole capitalist organization of work is, in fact, predicated on dividing the "head" of the collective worker from the "arms," "feet," "digestive," "excretory," and "reproductive" organs. Given this, the capacities *Futur Antérieur* focuses on would seem to be very unevenly distributed.

Associated with this criticism is a suspicion about some of *Futur Antérieur*'s terminology—particularly its references to "immaterial labor." For this can easily be read as obscuring the continued importance of a vast mass of all-too-physical and material work in the post-Fordist economy—domestically, in the service sector, and internationally, in everything from labor on coffee plantations to the trade in body organs. To speak of "immaterial labor" can also easily occlude some very corporeal components of high-tech work, such as the epidemic of repetitive-strain injuries associated with computer use.

These problems are clearly related to the relatively cursory analysis of the gendered or international dimensions of "general intellect" offered by *Futur Antérieur*—omissions that might be pointedly related to the fact that most of its authors are men, located in Europe or North America. The new circuits of capital, it could be argued, look a lot less "immaterial" and "intellectual" to the female and Southern workers who do so much of the grueling physical toil demanded by a capitalist "general intellect" whose headquarters remain preponderantly male and Northern. Indeed, the *Futur Antérieur* analysis has been accused of an all-too-familiar sort of Marxist vanguardism, whose protagonist is now not the "industrial" but the "intellectual" proletariat—a vanguardism that is, however, made peculiarly implausible by the relatively privileged conditions that its chosen protagonist enjoys.

Futur Antérieur authors have replied to these objections. Hardt and Negri go to some pains to designate "mass intellectuality" as a general propensity of the post-Fordist proletariat, and not as some "recomposed vanguard or leading sector."[49] Technoscientific labor, they say, is a "massified quality of laboring intelligentsia, of cyborgs and hackers."[50] It is "a quality of subjectivity that extends through the various sectors of production."[51] Discussing the category of "immaterial" work, Hardt and Negri underline that "however immaterial this labor might be it still involves both brains and bodies."[52] "Mass intellectuality" has to be understood as including the affective, emotional work performed inside and outside the home by women—for example, the labor of nurses, which is both "both highly technical and affective."[53] Moreover, the *Futur Antérieur* focus on advanced, post-Fordist production methods is not, Hardt and Negri say, meant to deny the existence of other Fordist or even more archaic techniques, particularly in the South. Rather, it only suggests that high-technology practices furnish the command, control, and communication capacity through which the whole system operates, and that these practices, in both obvious and subtle ways, bathe the whole arena of struggle in their influence.

As Ed Emery has suggested, resolution of the debate about "general intellect" really calls for a project along the lines of what Marx called "a workers' inquiry," involving a network of researchers engaged in participatory study of emergent forms of struggle.[54] In the meantime, my own view is that while the *Futur Antérieur* analysis has to be very seriously qualified, it is also "onto" something important. Although the initial propositions of Negri, Lazzarato, and Hardt need extensive revision to take fuller account of capital's tendency to polarize the allocation of skills and competencies along lines of gender and race, such a reworking need not invalidate the concept of "mass intellect."

As we saw in chapter 5, while capital has found in computers and other forms of informatics the weapons to assault the old factory-centered fortifications of working class, this attack is rebounding in unpredicted ways. The creation and operation of such technologies depends on widespread scientific and organizational competencies. It presumes the very development of so-called human capital that neoliberalism is now eroding through its curtailment of the educational, medical, and communicational infrastructures of the welfare state. The paradoxical result is that a technologically armed corporate order finds itself confronting the trade unionist with an autonomous production plan and a computer network of international contacts; the anti-poverty activist with a microwatt transmitter and an alternative budget; the reproductive rights worker trained in medical science, scanning the data banks for genetic patents; the anti-AIDS organizer with camcorder and pharmaceutical expertise; the rioter connected to the Internet—in short, by a force of social labor that will not acquiesce to technological revolution commanded from above but rather demands the right to direct this epochal transformation from below.

Moreover, although capital clearly attempts to limit and divide access to the social knowledge vital to technoscientific power, it should not be assumed that this division and fragmentation always succeeds. The globalization process described in chapter 6 is in large part unfolding as a story of capital's *failure* to maintain such segregation. In an era when Silicon Valley janitors can access email to embarrass the computer companies they are striking against, and the World Wide Web carries the messages of Zapatistas and East Timorese resistance fighters, it is clear that the wretched of the earth are neither entirely outside the mechanisms of high-technology production or (more important) completely powerless to appropriate them.[55] The question of whether capital will successfully segment post-Fordist labor power, or if, on the contrary, rebellious subjects will break down these barriers to establish new alliances, lies at the core of what I call "the contest for general intellect." In this contest the contemporary proletariat fights to actualize "general intellect," not according to the privatizing, appropriative logic of capital, but in ways that are deeply democratic and collective, and hence truly "general."

Virtual Universities

This account of the cycles and circuits of struggle in high-technology capitalism was written in an academic context. It is therefore only appropriate to end by considering what the analysis of "general intellect" might mean for those specifically and particularly intellectual laborers who teach and study at universities. For, as Negri and Lazzarato observe, no site could be more vital to capital's harnessing of collective intelligence than academia.[56] Over the last twenty-five years, it has been reshaped by an inexorable dialectic. Capitalist industry, mutating into its informational phase, has become more intellectual; Microsoft calls its central production facilities a "campus." Simultaneously, universities have become more industrial, acting as ancillary research and training facilities for capital's overall project of high-technology development: Academia, Inc.

This advancement of "corporate-university partnership" has as its aim what David Noble, North America's most trenchant critic of this union, terms "the systematic conversion of intellectual activity into intellectual capital, and, hence, intellectual property."[57] As Noble points out, this process has passed through a series of phases. In the first stage, unfolding through the 1970s and 1980s, the research activities of the university were effectively commercialized. This was accomplished partly through the fostering of industry-sponsored or -targeted programs at the expense of basic research, partly through the installation of research parks and other entrepreneurial experiments on campus sites, and partly by legal changes that give post-secondary institutions an interest in merchandising patents resulting from faculty research. The second stage in the university-corporate merger, however, has appeared only during the 1990s, with the drive toward the "virtual university," based on large-scale, computer-assisted, telelearning—a development that, Noble says, has as its aim nothing less than the commodification of the university's teaching function.

Virtual university experiments, now widespread in both the United States and Canada, are promoted under the banner of accessibility, innovation, and inevitable technological progress. But, Noble argues, they are really concerned with "transforming courses into courseware, [and] the activity of instruction itself into commercially viable proprietary products that can be owned and bought and sold in the market."[58] At the core of this process is a classic industrial strategy of deskilling and automation, downloading instructors' courses into reusable software packages over whose use they surrender all pedagogical control. Experiments in this direction typically involve the universities in complex partnerships with computer corporations, carrier companies, and "edutainment" providers. These commercial interests look to the virtual university as a market for hardware and software products, and at educational software as a saleable on-line commodity. For university administrators, virtual universities offer a dramatic way of cutting labor costs and centralizing managerial control. This is accomplished either by the simple elimination of the faculty whose knowledge has been extracted in digital form, or, in the case of the remaining live instructors, through an envelopment in multiple on-line teaching requirements, complete with endless email solicitations, Web-site preparations, and monitored electronic activities. Although there is an almost complete lack of substantial evidence as to the pedagogical benefits of computerized education, this does not deter the rush to convert universities into what Noble scathingly terms "digital diploma mills."[59]

Those who have followed the cycle of capitalist restructuring and class recomposition outlined in this book will not be surprised to hear that the "virtualizing" of universities has already provoked resistance. At both the University of California Los Angeles and the University of British Columbia, students have opposed the additional fees that universities impose to implement such high-tech schemes. And at Canadian universities, such as York and Acadia, faculty have struck to maintain control over teaching methods in the face of mounting administrative attempts to technologically control their work.[60]

These resistances should be supported and extended. But it is also important to ask whether there are any aspects of the "virtual university" agenda, and the larger process of academic-corporate fusion of which it is part, that offer not just threats, but opportunities for counterattack. Writing in a European context, Negri and Lazzarato suggest that this might be the case. In the era of the "ivory tower," they say, when universities were only partially integrated into capitalism, or marginal to its central functions, academics appeared (however much this actually mystified real interconnections) to be removed from industrial activity and its attendant class conflicts. It was from this position of apparent exteriority that intellectuals could commit to or engage themselves with political movements. From the end of the Second World War, however, this distance began to diminish rapidly. Today, when the distance separating the university from business has shrunk to virtually nothing, university teachers find themselves unequivocally involved in capital's appropriation of "general intellect."

These changed conditions, Negri and Lazzarato suggest, create the grounds for a new relation between dissenting academics and oppositional social movements. Rather than descending from the heights of the university to involve themselves in causes largely external to their daily experience, possibilities emerge for academics to make more "transverse" connections.[61] Academics perhaps lose some pretensions as the bearer of great truths and grand analysis, but they become the carriers of particular skills, knowledge, and accesses useful to movements in which they participate on the basis of increasing commonalties with other members of post-Fordist "mass intellect."[62]

I would add that the matrix for these connections is formed by the new movements of social unrest. Participation in these movements pulls academics and students into contact with other public service workers protesting cutbacks, wider labor and trade unionist organizations, and the many diverse constituencies surging against capital's agenda of high-technology austerity. Out of such contacts comes a corporate-university interaction very different from that which capital intends—one that disseminates opposition to corporate rule from the streets back onto the campuses, and again from the campuses out onto the streets.

The possibility of such a counterflow exists because, to effectively harness mass intellect to accumulation, capital must maintain a certain degree of openness within the universities. Part of what business seeks in its invasion of academia is the creativity and experimentation of social labor-power, qualities vital to a high-technology economy based on perpetual innovation. But if industry is to benefit from such invention-power, it cannot entirely regiment the institutions of education. However carefully it circumscribes the budgets and mission statements of academia, capital's incessant search for competitive advantage requires chances for unforeseen synthesis, opportunities for the unpredicted but really profitable idea or invention to emerge. And this unavoidable condition of an economic order based on general intellect gives a limited but real porosity to universities. This porosity can be exploited by dissident academics—to research and teach on topics of value to social movements in opposition to capital; to invite activists and analysts from these movements onto campuses and into lectures and seminars; and to use the university's resources, including its easy access to the great communication networks of our age, to circulate news and analysis that are otherwise marginalized.

The administrative imposition of computers and other high-technology on the classroom should sometimes simply be opposed as pedagogically destructive. But in other cases—and sometimes simultaneously—it is possible to recapture the virtual apparatus for alternative purposes. It should be remembered that students and academics played a major part in the unauthorized creation of the Internet that took the nascent technology of computer-mediated communication beyond the unilateral control of the military-industrial complex and opened it to a popular use that blindsided corporate planning. Subsequently, students have continued to use the networks for subversive purposes.

To give just one example, in spring 1994, Latino and Chicano students at the universities of Michigan, Colorado, Nebraska, and numerous sites in California staged hunger strikes and occupations. They demanded new programs, antiracist initiatives, grape boycotts in support of farmworkers, and the naming of buildings in memory of Cesar Chavez. Their protests were extensively connected and coordinated by computer-communications facilitated by sympathetic librarians, faculty, and union organizers.[63] Similarly, 1995 and 1996 saw the email coordination of multicampus protests against reductions in student aid and rising tuition fees in Canada and the United States.[64]

More broadly, it is clear that much of the circulation of oppositional content on the Internet today is conducted from academic centers by students and teachers, acting as relays to wider constituencies. Noble would be the first to appreciate the irony that his denunciations of "digital diploma mills" have been primarily distributed by way of email and electronic publications. In the era of mass intellect, a purely Luddite stance is not enough. To grasp the tactical and strategic chances presented by capital's failure to control the technological dynamics it has set in motion, activists must be, as Dorothy Kidd and I once put it, "Luddites on Monday and Friday, cyberpunks the rest of the week."[65]

Conclusion

In academia, as elsewhere, labor power is never completely controllable. To the degree that capital uses the university to harness general intellect, insisting its work force engage in lifelong learning as the price of employability, it runs the risk that people will teach and learn something other than what it intends. In my own practice, a crucial aspect of teaching that "something other" is to address critically the utopian promises of information revolution examined at the beginning of this book.

It is both very important and relatively easy to demonstrate how hollow these promises have proven over the last three decades: how they have brought the majority of people in Europe and North America not new, technologically generated wealth, but declining or stagnant real wages; how the mirage of increased, enriched leisure has evaporated into rates of unemployment and poverty unimaginable twenty years ago; how the "knowledge class" that was to humanize capital has found itself pink-slipped by its corporate masters, sharing the welfare line with millions of others; how the high-skill, high-tech service jobs are fractional compared to the burgeoning mass of poorly paid and precarious "McJobs"; how the "cooperative" workplace is terrorized by downsizing, closures, and concessionary rollbacks; how the heralded multiplication of media channels masks an intensifying concentration of ownership; how promises of "all information everywhere" translate into a vast extension of property rights and corporate power. From this point of view, the utopian announced by information revolutionaries is mere fraud.

However, to teach this, unalloyed, can simply reinforce despair and cynicism. Demystification, practiced alone, leads to a dead end—to the assertion of monolithic and unbreakable capitalist power that characterizes so much of what passes for Marxism today. The more difficult task is to identify the possibilities of things being other than they are. As Raymond Williams wrote, the crucial challenge is "making hope practical, rather than despair convincing."[66]

For this purpose, I have found the analysis offered by autonomist Marxism, with its emphasis on the constantly changing and renewed cycle of struggles between capital and labor, particularly valuable. This perspective shows how the information revolution came into being as a result of a social contest, as part of a vast restructuring by capital intended to evade and suppress working-class opposition. More important, it suggests that this informational restructuring has *failed*. Rather than pacifying class conflict, digitalization and genetic engineering only displace capital's constant internal war—so that the lines of contestation now run along the inside of the very technological systems deployed to overcome them.

To contain crisis, capital has been compelled to set in play agents and subjects whose capacities outrun its control. Now, more than ever before, it has "conjured up such gigantic means of production and of exchange" that it becomes like "a sorcerer, who is no longer able to control the powers of the nether world which he has called up by his spells."[67] If workers' refusal of work has resulted in extraordinary levels of automation, the new machine-systems now threaten the viability of the wage economy itself. If local militancies have provoked capital to seek global mobility, the very communication and transportation networks down which it flees provide the threads of new, transnational solidarities. If people's desires for education and self-development have been made the stuff of a knowledge-for-profit market, collective intelligence turns to criticize the human and environmental costs of this trajectory—and to devise alternatives.

At its present very high level of technoscientific development, corporate power finds itself dependent on levels of cooperative activity, unimpeded communication, and free circulation of knowledge that, far from being easily integrated into its hierarchies, exist in persistent tension with its command. Thus the possibilities that information revolutionaries speak of cannot *just* be written off as false promises. Rather, they are a refracted and distorted version of real potentialities for a new social order, liberated from the despotic constraints of constant work, denied wealth and destructive accumulation.

However, the actualization of these hopes demands breaking through the limits that capital currently imposes on human development. I have argued that there are now visible signs of an emergent collectivity refusing the logic of commodification, uprising at the very moment that the world market seems to have swallowed the entire planet. Deepening and expanding this process of recomposition depends on interconnection between many and disparate movements at different points along capitalism's circuits. Ironically, the conversations necessary for creation of the new combination are now being conducted across the world-spanning communication networks that information-age capital has itself created. It is as a contribution to such a circulation of struggles that this book is offered.

NOTES

I DIFFERENCES

1. See William Gibson and Bruce Sterling, *The Difference Engine* (New York: Bantam, 1991).

2. Ibid., 293.

3. Simon Schaffer, "Babbage's Intelligence: Calculating Engines and the Factory System," *Critical Inquiry* 21 (1994): 203–27.

4. "The Babbage principle" asserted that the benefit of the division of labor is not solely increased productivity, but the cheapening of labor that arises from the establishment of a hierarchy of wages. On this topic, see Harry Braverman, *Labor and Monopoly Capital: The Degradation of Work in the Twentieth Century* (New York: Monthly Review Press, 1974), 81, 85–88.

5. Schaffer, "Babbage's Intelligence," 210.

6. Ibid., 225.

7. Karl Marx, *Capital: A Critique of Political Economy*, vol. 1 (New York: Vintage Books, 1977), 470, 553.

8. Andrew Ure, *The Philosophy of Manufacture* (London, 1835), 367–70; quoted by Marx, *Capital*, vol. 1, 564.

9. Marx, *Capital*, vol. 1, 563.

10. Karl Marx, *Capital: A Critique of Political Economy*, vol. 3 (New York: Vintage Books, 1981), 198–99.

11. Ibid., 199.

12. Ibid.

13. Marx, *Grundrisse: Foundations of a Critique of Political Economy* (Harmondsworth, Eng.: Penguin, 1973), 690.

14. Ibid., 705.

15. Ibid., 706.

16. Ibid., 700.

17. See Francis Fukuyama, *The End of History and the Last Man* (New York: Macmillan, 1992).

18. See Jacques Derrida, *Specters of Marx: The State of the Debt, the Work of Mourning, and the New International* (London: Routledge, 1994).

19. See Karl Heinz Roth, *L'Autre mouvement ouvrier en Allemagne, 1945–1978* (Paris: Christian Bourgeois, 1979).

20. For a general critique of the Bolshevik record see Maurice Brinton, *The Bolsheviks and Workers' Control, 1917 to 1921: The State and Counter Revolution* (Montreal: Black Rose, 1975); for a specific discussion of Lenin's attitude to technology see Langdon Winner, *Autonomous Technology: Technics-Out-of-Control as a Theme in Political Thought* (Cambridge, Mass.: MIT Press, 1977), 268–76; and for broader views of the Soviet state's attitudes to technological development, see Steve Smith, "Taylorism Rules OK? Bolshevism, Taylorism and the Technical Intelligentsia in the Soviet Union, 1917–41," *Radical Science Journal* 13 (1983): 3–27, and the essays in *Technology and Communist Culture: The Socio-Cultural Impact of Technology under Socialism*, ed. Frederic J. Fleron Jr. (New York: Praeger, 1977). On the USSR and Fordism, see Robin Murray, "Fordism and Post-Fordism," in *New Times: The Changing Face of Politics in the 1990s*, ed. Stuart Hall and Martin Jacques (London: Lawrence and Wishart, 1989), 38–54.

21. Marx, *Capital*, vol. 1, 872–76.

22. This point is made by Fredric Jameson, "Actually Existing Marxism," in *Marxism beyond Marxism*, ed. Saree Makdisi, Cesare Casarino, and Rebecca E. Karl (London: Routledge, 1996), 14–54, and by Michael Hardt and Antonio Negri, *The Labor of Dionysus: A Critique of the State-Form* (Minneapolis: University of Minnesota Press, 1994).

23. Jameson, "Actually Existing Marxism," 54.

24. See Alain Touraine, *The Post-Industrial Society: Tomorrow's Social History: Classes, Conflicts and Culture in the Programmed Society* (New York: Random House, 1971); Alberto Melucci, *Nomads of the Present: Social Movements and Individual Needs in Contemporary Society* (Philadelphia: Temple University Press, 1989); Timothy W. Luke, *Screens of Power: Ideology, Domination, and Resistance in an Informational Society* (Urbana: University of Illinois Press, 1989).

25. See Ernesto Laclau and Chantal Mouffe, *Hegemony and Socialist Strategy: Towards a Radical Democratic Politics* (London: Verso, 1985).

26. Ibid., 176.

27. John McMurtry, "The Cancer Stage of Capitalism," *Social Justice*, on-line, Internet, PEN-L@anthrax.ecst.csuchico.edu, 24 July 1996 [accessed 24 July 1996]. The remaining quotations from McMurtry in this chapter are also from this source. An extended version of McMurtry's argument appears in his book *Unequal Freedoms: The Global Market as an Ethical System* (Toronto: Garamond, 1998).

2   REVOLUTIONS

1. See, for example, Frank Webster, *Theories of the Information Society* (London: Routledge, 1995). Also useful, though now dated, is David Lyon, *The Information Society: Issues and Illusions* (Cambridge: Polity, 1988).

2. See chap. 1, and also the discussion of the Saint Simonian tradition in Krishan Kumar, *Prophecy and Progress* (Harmondsworth, Eng.: Penguin, 1978).

3. For discussion of this concept, see Daniel Bell, *The End of Ideology* (New York: Free Press, 1961).

4. On Bell's early interest in Marxism see his account in "First Love and Early Sorrow," *Times Higher Education Supplement*, 16 Jan. 1981, 9–11.

5. See G. Ross, "The Second Coming of Daniel Bell," in *Socialist Register*, ed. Ralph Miliband and John Saville (London: Merlin, 1974), 56–84.

6. See E. G. Mesthene, ed., *Technological Change* (Cambridge, Mass.: Harvard University Press, 1970); Peter Drucker, *The Age of Discontinuity* (New York: Harper and Row, 1968); Zbignew Brzezinski, *Between Two Ages: America's Role in the Technotronic Era* (New York: Viking, 1970); Herman Kahn and Anthony J. Weiner, *The Year 2000: A Framework for Speculation on the Next Thirty-three Years* (New York: Macmillan, 1967); Daniel Bell, *The Coming of Post-Industrial Society* (New York: Basic, 1973). Although *The Coming of Post-Industrial Society* appeared in 1973, Bell published "Notes on the Post-Industrial Society" in *The Public Interest* 6/7 (1967). He competes with Alain Touraine as the first to introduce the term "postindustrial" in its contemporary sense.

7. Bell, *The Coming of Post-Industrial Society*, x.

8. Ibid., 27–30.

9. Ibid., 20.

10. Ibid., 374.

11. Ibid., 49.

12. Ibid., x.

13. On this point see Kevin Robins and Frank Webster, "Information as Capital: A Critique of Daniel Bell," in *The Ideology of the Information Age*, ed. Jennifer D. Slack and Fred Fejes (Norwood, N.J.: Ablex, 1987), 95–117.

14. Bell, *The Coming of Post-Industrial Society*, 374.

15. Bell's technocratic vision retains a connection to his early brushes with Trotskyism. His idea of the "new class" is adapted from the analysis of bureaucratized socialism offered by a maverick Trotskyite, James Burnham, in *The Managerial Revolution* (1941; reprint, Bloomington: Indiana University Press, 1966). Burnham's thesis of a "managerial revolution" establishing bureaucrats and administrators as the dominant class in capitalist and socialist countries alike was in turn inspired by Trotsky's critique of the "bureaucratic degeneration" of the USSR.

16. Bell, *The Coming of Post-Industrial Society*, 288, 291.

17. Ibid., 40.

18. At one point Bell speaks of corporate power's being "subordinated" to the new class (*The Coming of Post-Industrial Society*, 269); elsewhere he says it remains "dominant" (372). If at times Bell suggests that the role of organized knowledge supersedes the logic of the market—allowing economic planning—at others he sees such knowl-

edge as itself subject to commodification, defined and prioritized by its marketability.

19. Kumar, *Prophecy and Progress*, 190–92.

20. Bell, *The Coming of Post-Industrial Society*, 480. For his later elaboration on the topic of the "adversary culture" see his *The Cultural Contradictions of Capitalism* (New York: Basic, 1976).

21. Only Masuda's work is generally available in English. On the other authors, see Tessa Morris-Suzuki, *Beyond Computopia: Information, Automation and Democracy in Japan* (London: Kegan Paul International, 1988).

22. Ibid., 8–13.

23. See Yoneji Masuda, *The Information Society as Post-Industrial Society* (Washington, D.C.: World Future Society, 1981).

24. See Alex S. Edelstein, John E. Bass, Sheldon M. Hasel, eds., *Information Societies: Comparing the Japanese and American Experiences* (Seattle: University of Washington Press, 1978).

25. See Marc Uri Porat, *The Information Economy: Definition and Measurement*, vol. 1 (Washington D.C.: U.S. Department of Commerce, 1977).

26. Published in English, with an introduction by Daniel Bell, as Simon Nora and Alain Minc, *The Computerization of Society* (Cambridge, Mass.: MIT Press, 1981).

27. Ibid., 3.

28. Daniel Bell, "The Social Framework of the Information Society," in *The Computer Age: A Twenty Year View*, ed. Michael L. Dertouzous and Joel Moses (Cambridge, Mass.: MIT Press, 1979); Bell's essay is reprinted in *The Complete Guide to the New Technology and Its Impact on Society*, ed. Tom Forester (Cambridge, Mass.: MIT Press, 1980), 500–549; the quotation is from p. 526.

29. Representative titles include James Martin, *The Wired Society* (Englewood Cliffs, N.J.: Prentice Hall, 1978); Kimon Valaskakis, *The Information Society* (Montreal: Gamma, 1979); Anthony Oettinger, "Information Resources: Knowledge and Power in the Twenty-first Century," *Science* 209 (1980): 191–98; Barry Jones, *Sleepers Wake!* (Melbourne: Oxford University Press, 1982); W. P. Dizard, *The Coming Information Age* (New York: Longman, 1982); Tom Stonier, *The Wealth of Information* (London: Methuen, 1983); James Beniger, *The Control Revolution* (Cambridge, Mass.: Harvard University Press, 1986); John Naisbitt, *Megatrends* (New York: Warner, 1982); Alvin Toffler, *The Third Wave* (New York: Morrow, 1980).

30. It is symptomatic of this new tone that when Bell recast his earlier arguments in terms of the information society theory, he retained three of his original five descriptors of postindustrialism but dropped two—those dealing with the professional and technical class and the future orientation that he had previously related to the enlarged scope of governmental planning and public policy. He does not explain this change. It can be speculated that it reflects not only the heightened prominence automated information systems had gained since the early 1970s, but also a rightward political shift, creating a "free-enterprise" climate hostile to planning and bureaucracy in which computerization was frequently hailed as a means of eliminating professional and technical jobs—technology turned against technocracy.

31. For a useful account, see Kees Brants, "The Social Construction of the Information Revolution," *European Journal of Communication* 4 (1989): 79–87.

32. See William Leiss, "The Myth of the Information Society," in *Cultural Politics in Contemporary America*, ed. Ian Angus and Sut Jhally (New York: Routledge, 1989), 282–99.

33. See, for example, Kevin Robins and Frank Webster's distinction between "administrative" and "apocalyptic" versions of informational futurism in their "Athens without Slaves . . . or Slaves without Athens? The Neurosis of Technology," *Science as Culture* 3 (1988): 19.

34. In making this summary I have found very useful the similar synopses by Manuel Castells, *The Informational City: Information Technology, Economic Restructuring and the Urban-Regional Process* (Oxford: Blackwell, 1989), and Jorge Reina Schement and Leah A. Lievrouw, "Introduction: The Fundamental Assumptions of Information Society Research," in *Competing Visions, Complex Realities: Social Aspects of the Information Society*, ed. Schement and Lievrouw (Norwood, N.J.: Ablex, 1987), 1–10.

35. Masuda, *Information Society*, 46.

36. Bell, *The Coming of Post-Industrial Society*, 25–26.

37. Drucker, *Age of Discontinuity*, 271; Alvin Toffler, *Powershift* (New York: Bantam, 1990), 17.

38. Examples of this fusion include the use of computers in genetic analysis, the search for the "biochip," the modeling of electronic automata on cellular phenomena, and research into molecular robotics or "nanotechnologies."

39. See Castells, *The Informational City*, on this point.

40. In his postindustrial thesis Bell drew on the economic model of the "march through the sectors" developed by Colin Clark in *The Conditions of Economic Progress* (London: Macmillan, 1940). Dividing the economy into three sectors—primary (agricultural), secondary (manufacturing), and tertiary (services)—Clark had posited a historical process in which, as productivity rose successively in each sector, the bulk of the labor force migrated to the next. Postindustrial society, Bell claimed, marked the point where the majority of the labor force moved into the service sector—a proposition whose definitional clarity and historical accuracy has since been the subject of hot debate. Subsequently, some information society theorists have built on this contested foundation to posit a distinct, delimited "quaternary" information sector constituted by high-technology industry and succeeding agriculture, manufacture, and services at the leading edge of economic growth. However, such models have been strongly criticized for obscuring the actual interconnection and overlap of allegedly discrete sectors. What studies such as Porat's show is the ubiquity of symbol-manipulating tasks across a wide range of occupations. Many theorists have therefore abandoned the notion of a distinct information "sector" and now favor an analysis that stresses the increased importance of informational activity as a component within *all* aspects of the economy. Information processing is seen not so much as a successor to manufacturing, services, or agriculture, but rather as a "superordinate" function whose productivity-improving powers span each of these areas.

41. Oettinger, "Information Resources," 191.

42. Jorge Reina Schement, "The Origins of the Information Society in the United States: Competing Visions," in *The Information Society: Economic, Social and Structural Issues*, ed. Jerry Salvaggio (Hillsdale, N.J.: Lawrence Erlbaum, 1989), 21.

43. Toffler, *Powershift*, 232–34.

44. Bell, "The Social Framework of the Information Society," 545.

45. Dizard, *The Coming Information Age*, 11, 23.

46. See Drucker, *Age of Discontinuity*, and also Marc Porat, "Global Implications of the Information Society," *Journal of Communication* 28.1 (1978): 70–80.

47. Marshall McLuhan, *Understanding Media: The Extensions of Man* (New York: McGraw-Hill, 1964), 87.

48. Toffler, *The Third Wave*, 328.

49. Beniger, *The Control Revolution*, 104.

50. Hans Moravec, *Mind Children: The Future of Robot and Human Intelligence* (Cambridge, Mass.: Harvard University Press, 1988), 5.

51. Ibid., 5–6.

52. Karl Marx, *The Poverty of Philosophy* (New York: International Publishers, 1971), 109.

53. Alvin Toffler, *Previews and Premises* (London: Pan, 1984), 195.

54. Toffler, *Future Shock*, 361.

55. Toffler, *The Third Wave*, n.p.

56. Hendrick Hertzberg, "Marxism: The Sequel," *New Yorker*, Feb. 1995, 7.

57. Quoted by Hertzberg, "Marxism," 7.

58. Toffler, *The Third Wave*, 440.

59. Toffler, *Powershift*, 421.

60. Ibid., 421–22.

61. Ibid., 421.

62. See Robert Blauner, *Alienation and Freedom: The Factory Worker and His World* (Chicago: University of Chicago Press, 1964).

63. Toffler, *Powershift*, 240.

64. Ibid.

65. Toffler, *Previews and Premises*, 103–5.

66. For a sophisticated version of this position, one can do no better than the account of the "control revolution" offered by James Beniger, a former official with the U.S. Food and Drug Administration (Beniger, *The Control Revolution*). Beniger argues that information technologies represent the consummation of a century-long quest by capitalism to develop instruments of technobureaucratic control adequate to overcome the repeated crises in production, consumption, and distribution of goods encountered during industrialization. As the culmination of this "control revolution," microelectronic methods of automation, coordination, inventory, and advertisement are revolutionary in a double sense. On the one hand, they mark a dramatic advance in human progress, offering means to overcome the protracted tumults and disorder of industrialism. But they are also revolutionary in a "cyclical," "astronomical," sense, implying the "restoration of levels of economic and political control" exercised in the preindustrial era (433). The information revolution is thus both a progressive and a conservative revolution. Like so many information revolutionaries, Beniger tips his cap to Marx, citing the famous aphorism about the way "men make their own history, but not under circumstances of their own choosing," and observing that because these circumstances have shifted "from land and capital to information," social theory inherited from the nineteenth century is now challenged "much as the Industrial Revolution challenged Marx . . . to reconsider preindustrial theories" (32). But the end

to which Beniger's reconsideration of social theory looks is precisely antithetical to Marx's. For what it envisages is a market society in which the development of cybernetic programming and feedback techniques—including, Beniger believes, the eventual creation of synthetic life forms—enables the removal of all blockages and interruptions from the circuits of the commodity. The result is a seamlessly integrated totality, a perfect structural-functionalist harmony in which every moment of social activity is organically connected with and responsive to the other in fulfillment of the needs of the capitalist whole.

67. Yoneji Masuda, *Managing in the Information Society: Releasing Synergy Japanese Style* (Oxford: Blackwell, 1990), 130–31.

68. William E. Halal, *The New Capitalism: Democratic Free Enterprise in Post-Industrial Society* (New York: Wiley, 1986), 7.

69. See Ithiel de Sola Pool, *Technologies of Freedom* (Cambridge, Mass.: Harvard University Press, 1983).

70. F. A. Hayek, "The Use of Knowledge in Society," *American Economic Review* 35 (1945): 519–30.

71. See Zbignew Brzezinski, *The Grand Failure: The Birth and Death of Communism in the Twentieth Century* (New York: Macmillan, 1988).

72. Kenichi Ohmae, "Global Consumers Want Sony, Not Soil," *New Perspectives Quarterly* 8.4 (1991): 72.

73. Toffler, *Powershift*, 411.

74. Francis Fukuyama, *The End of History and the Last Man* (New York: Macmillan, 1992). Fukuyama first expressed his ideas in "The End of History," *The National Interest* 16 (1989): 3–18. He subsequently developed them in the book. All citations here are from the book.

75. Ibid., xii.

76. Ibid.

77. Ibid., xv.

78. Ibid.

79. Ibid., 93.

80. Ibid., 98.

81. ibid., xxiii.

82. See, for example, the excellent discussion of the highway metaphor by Robert Adrian X, "Infobahn Blues," in *Digital Delirium*, ed. Arthur and Marilouise Kroker (New York: St. Martin's, 1997), 84–88.

83. For a fuller discussion of Fordism, see chap. 3.

84. Albert Gore, "The National Information Infrastructure: Information Conduits, Providers, Appliances and Consumers," *Vital Speeches of the Day*, Feb. 1994, 229. See also his "Information Superhighways: The Next Information Revolution," *The Futurist* (Jan.–Feb. 1991): 21–23.

85. See Christopher Scheer, "The Pursuit of Techno-Happiness: Third Wavers and Tekkie Cults," *Nation*, 8 May 1995, 632–34; Thomas M. Disch, "Newt's Futurist Brain Trust" *Nation*, 27 Feb. 1985, 266–70; David Corn, "CyberNewt," *Nation*, 6 Feb. 1995, 154–55.

86. Progress and Freedom Foundation, "Cyberspace and the American Dream: A Magna Carta for the Knowledge Age," on-line, Internet, 1994. Available at http://

www.pff.org/position.html. Subsequent quotations from the document are from this source.

87. On the Telecommunications Bill, see Robert McChesney, "The Global Struggle for Democratic Communication," *Monthly Review* 48.3 (1996): 1–20.

88. Bill Gates, *The Road Ahead* (New York: Norton, 1995), 171, quoted in Michael Dawson and John Bellamy Foster, "Virtual Capitalism: The Political Economy of the Information Society," *Monthly Review* 48.3 (1995): 51.

89. See Herbert J. Schiller, "The Global Information Highway: Project for an Ungovernable World," in *Resisting the Virtual Life: The Culture and Politics of Information*, ed. James Brook and Iain A. Boal (San Francisco: City Lights, 1995), 17–33.

90. Michael Rothschild, *Bionomics: The Inevitability of Capitalism* (New York: Henry Holt, 1990), xiii, xv.

91. Hans Moravec, "Pigs in Cyberspace," in *Thinking Robots, An Aware Internet, and Cyberpunk Librarians*, ed. R. Bruce Miller and Milton T. Wolf (Chicago: Library and Information Technology Association, 1992), 15–21.

92. Ibid., 19.

93. Ibid.

94. Ibid.

95. Karl Marx, "Economic and Philosophic Manuscripts," in *Early Works* (Harmondsworth, Eng.: Penguin, 1973), 366, quoted in McKenzie Wark, *Virtual Geography: Living with Global Media Events* (Bloomington: Indiana University Press, 1994), 175.

3   MARXISMS

1. Karl Marx, *The Poverty of Philosophy* (New York: International Publishers, 1971), 109.

2. For a valuable discussion, see Donald McKenzie, "Marx and the Machine," *Technology and Culture* 25.3 (1984): 473–502.

3. Karl Marx, *Preface to the Contribution to a Critique of Political Economy* (London: International Publishers, 1971), 20–21.

4. Ibid., 21.

5. On this debate see Laurence Harris, "Forces and Relations of Production," in *A Dictionary of Marxist Thought*, ed. Tom Bottomore et al. (Cambridge, Mass.: Harvard University Press, 1983), 178–80.

6. Frederick Engels, *Socialism: Utopian and Scientific* (Peking: Foreign Languages Press, 1975), 88.

7. See Nikolai Bukharin, *Historical Materialism: A System of Sociology* (New York: International Publishers, 1925); J. D. Bernal, *The Social Function of Science* (Cambridge, Mass.: MIT Press, 1939); Gerald Cohen, *Karl Marx's Theory of History: A Defence* (Oxford: Clarendon, 1978).

8. Karl Marx, *Capital: A Critique of Political Economy*, vol. 1 (New York: Vintage Books, 1977), 1026–40.

9. Ibid., 1037.

10. Georg Lukács, "Technology and Social Relations," *New Left Review* 39 (1966), 27–34; Harry Braverman, *Labor and Monopoly Capital: The Degradation of Work in*

the *Twentieth Century* (New York: Monthly Review Press, 1974); David Noble, *Forces of Production* (New York: Knopf, 1984).

11. Marx, *Capital*, vol. 1, 590–614.

12. Marx, "Speech at the Anniversary of the *People's Paper*" (14 Apr. 1856), in *Karl Marx: Selected Works*, vol. 1, ed. V. Adoratsky and C. P. Dutt (London, 1942), 428, quoted in Simon Schaffer, "Babbage's Intelligence: Calculating Engines and the Factory System," *Critical Inquiry* 21 (1994): 206.

13. Marx, *Capital*, vol. 1, 618.

14. Ibid.

15. Ibid.

16. Ibid.

17. Ibid.

18. Ibid.

19. For such an account see Paul S. Adler, "Marx, Machines, and Skill," *Technology and Culture* 3.4 (1990): 780–812.

20. An excellent collection of Marx's writings in this sphere is Yves de la Haye, *Marx and Engels on the Means of Communication: The Movement of Commodities, People, Information and Capital* (New York: International General, 1979). See also Armand Mattelart and Seth Siegelaub, eds., *Communication and Class Struggle*, vol. 1, *Capitalism and Imperialism* (New York: International General, 1979).

21. Karl Marx and Friedrich Engels, *The Communist Manifesto* (New York: Washington Square, 1969), 63.

22. Karl Marx, *Wage Labor and Capital: Value, Price and Profit* (New York: International Publishers, 1976), quoted in de la Haye, *Marx and Engels on the Means of Communication*, 52.

23. Marx, *Grundrisse*, quoted in de la Haye, *Marx and Engels on the Means of Communication*, 102–3.

24. Marx and Engels, *The Communist Manifesto*, 73.

25. Ibid., 68.

26. James Billington, *Fire in the Minds of Men: Origins of the Revolutionary Faith* (New York: Basic, 1980), 309.

27. Peter Waterman, *International Labour Communication by Computer: The Fifth International?* Working Paper Series 129 (The Hague: Institute of Social Studies, 1992), 9.

28. On these definitions see James O'Connor, *The Meaning of Crisis* (Oxford: Blackwell, 1987), 49–108.

29. See Ernest Mandel, *Late Capitalism* (London: New Left Books, 1975).

30. Ibid., 195–97.

31. Ibid., 191.

32. Ibid., 501.

33. The following summary draws on Mandel's own account in his *An Introduction to Marxist Economic Theory* (New York: Pathfinder, 1969).

34. Mandel, *Late Capitalism*, 111.

35. Karl Marx, *Capital: A Critique of Political Economy*, vol. 3 (New York: Vintage Books, 1981), 339–49. For the immensely complex debate on this topic see Michael A. Lebowitz, "Marx's Falling Rate of Profit: A Dialectical View," *Canadian Journal*

of Economics 9.2 (1976): 233–54; Harry Cleaver, "Karl Marx: Economist or Revolutionary?" in *Marx, Schumpter and Keynes: A Centenary Celebration of Dissent*, ed.
Suzanne W. Helburn and David F. Bramhall (New York: M. E. Sharpe, 1986), 121–48,
and O'Connor, *The Meaning of Crisis*.

36. O'Connor, *The Meaning of Crisis*, 76.

37. Mandel, *Introduction*, 50.

38. Mandel, *Late Capitalism*, 207.

39. Ibid., 110.

40. Ibid., 405–7.

41. Ibid., 208.

42. The phrase is from Russell Jacoby, "Towards a Critique of Automatic Marxism:
The Politics of Philosophy from Lukács to the Frankfurt School," *Telos* 10 (1971): 119–
46, and "The Politics of Crisis Theory: Toward a Critique of Automatic Marxism II,"
Telos 23 (1975): 3–52.

43. This resemblance is tellingly revealed by the facility with which Mandel's theories, stripped of political implication, have been reabsorbed by postindustrial long-wave
theorists, with their visions of ineffable economic cycles pulsing tidally across the
centuries. "Long wave theorists" such as Christopher Freeman (*Technology, Policy and
Economic Performance* [London: Frances Pinter, 1987]) and Peter Hall and Paschal
Preston (*The Carrier Wave: New Information Technology and the Geography of Innovation, 1846–2003* [London: Unwin Hyman, 1988]) base their work on Kondratieff's
theory of inexorable—albeit mysterious—economic cycles linked to clusters of technological innovation and situate the emergence of computers and communications
as part of the new "techno-economic" paradigm marking the onset of the "fifth
Kondratieff" or "carrier wave" about to unfold at the turn of the century. While this
construction moderates claims of the scope of epochal change made by postindustrialists—situating the microchip and fiber-optic cable within a sequence of technological revolutions whose predecessors include steam, steel processing, electricity—
it retains the idea of momentous technological change overtaking the contemporary
world with inhuman irresistibility.

44. See note 35, above.

45. Ernest Mandel and George Novak, *The Revolutionary Potential of the Working
Class* (New York: Pathfinder, 1974), 6.

46. For discussion of the "STR" see Frederic J. Fleron Jr., ed., *Technology and Communist Culture: The Socio-Cultural Impact of Technology under Socialism* (New York:
Praeger, 1977); Krishan Kumar, "Futurology: The View from Eastern Europe," *Futures*
4.1 (1972): 90–95; and Dallas Smythe, *Dependency Road: Communications, Capitalism, Consciousness and Canada* (Norwood, N.J.: Ablex, 1981).

47. See, for example, Bell's comments on Trotsky's concepts of economic planning
in "The Social Framework of the Information Society," in *The Computer Age: A
Twenty Year View*, ed. Michael L. Dertouzos and Joel Moses (Cambridge, Mass.: MIT
Press, 1979); Bell's essay is reprinted in *The Complete Guide to the New Technology
and Its Impact on Society*, ed. Tom Forester (Cambridge, Mass.: MIT Press, 1980), 500–
549. Bell derives his model of "technocracy" in part from Soviet practice, and scientific
socialists model the "scientific technological revolution" from American futurists.

48. See Theodor Adorno and Max Horkheimer, *Dialectic of Enlightenment* (1947; New York: Herder and Herder, 1972), 120–68.

49. See Herbert Marcuse, *One-Dimensional Man: Studies in the Ideology of Advanced Industrial Society* (Boston: Beacon, 1964).

50. See Braverman, *Labor and Monopoly Capital.*

51. Ibid., 114.

52. Ibid., 193.

53. See the essays in Andrew Zimbalist, ed., *Case Studies in the Labor Process* (New York: Monthly Review Press, 1979).

54. Noble, *Forces of Production.*

55. See David Noble, *Progress without People: New Technology, Unemployment, and the Message of Resistance* (Toronto: Between the Lines, 1995).

56. See Adorno and Horkheimer, *Dialectic of the Enlightenment,* 120–68.

57. Representative works include Vincent Mosco, *Pushbutton Fantasies: Critical Perspectives on Videotex and Information Technology* (Norwood, N.J.: Ablex, 1982) and *The Pay-Per Society: Computers and Communication in the Information Age: Essays in Critical Theory and Public Policy* (Toronto: Garamond, 1989); Nicholas Garnham, *Capitalism and Communication: Global Culture and the Economics of Information* (London: Sage, 1990); Smythe, *Dependency Road.* For works by Schiller see note 59.

58. See in particular Mosco's *The Political Economy of Communication* (London: Sage, 1996).

59. Schiller's work includes *The Mind Managers* (Boston: Beacon, 1973); *Communication and Cultural Domination* (New York: International Arts and Sciences Press, 1976); *Who Knows: Information in the Age of the Fortune 500* (Norwood, N.J.: Ablex, 1981); *Information in the Crisis Economy* (Norwood, N.J.: Ablex, 1984); *Culture, Inc.: The Corporate Takeover of Public Expression* (Oxford: Oxford University Press, 1989).

60. Schiller, *Who Knows,* xi–xiii.

61. Schiller, *Culture, Inc.,* 165.

62. Ibid., 45.

63. Schiller, *Information in the Crisis Economy,* 22.

64. See Schiller, *The Mind Managers.*

65. See Kevin Robins and Frank Webster, "Cybernetic Capitalism: Information Technology, Everyday Life," in *The Political Economy of Information,* ed. Vincent Mosco and Janet Wasko (Madison: University of Wisconsin Press, 1988), 44–75.

66. Frank Webster and Kevin Robins, *Information Technology: A Luddite Analysis* (Norwood, N.J.: Ablex, 1986), 311.

67. Ibid., 329.

68. Ibid., 347.

69. David Noble, "Social Choice in Machine Design: The Case of Automatically Controlled Machine Tools," in Zimbalist, *Case Studies in the Labor Process,* 19.

70. For statements on the pathological nature of capitalist technology, see Schiller, *Information in the Crisis Economy,* 25; Kevin Robins and Frank Webster, "Athens without Slaves . . . or Slaves without Athens? The Neurosis of Technology," *Science as Culture* 3 (1988): 7–53; and Noble, *Forces of Production.*

71. For classic statements of this position see Kevin Robins and Frank Webster, "Luddism: New Technology and the Critique of Political Economy," in *Science, Technology and the Labor Process*, vol. 2, ed. Les Levidow and B. Young (Atlantic Highlands, N.J.: Humanities Press, 1983), 9–48, and the series of articles by David Noble, "Present Tense Technology," parts 1–3, *Democracy* (Spring, Summer, Fall 1983): 8–24, 70–82, 71–93. Noble's articles are reprinted in his *Progress without People*. Although Schiller does not adopt the Luddite label, in *Who Knows* he concurs with Noble's call to halt the rapid diffusion of IT, urging "a maximum effort directed at slowing down, and postponing wherever possible, the rush to computerisation" (149).

72. Noble, "Present Tense Technology: Part 1," 8.

73. Noble, "Present Tense Technology: Part 3," 87.

74. Stuart Ewen, review of *The Mind Managers*, by Herbert Schiller, *Telos* 17 (1973): 186.

75. See the articles mentioned in note 71.

76. Marx, *Capital*, vol. 1, 554–55.

77. See David Harvey, *The Condition of Postmodernity: An Enquiry into the Origins of Cultural Change* (Oxford: Blackwell, 1989). For an interesting anthology containing several of the more "radical" strains of post-Fordist analysis, see Ash Amin, ed., *Post-Fordism: A Reader* (Oxford: Blackwell, 1994).

78. Two foundational works of Regulation School analysis are Michel Aglietta, *A Theory of Capitalist Regulation: The US Experience* (London: New Left Books, 1979), and Alain Lipietz, *Mirages and Miracles: The Crisis of Global Fordism* (London: Verso, 1987).

79. Lipietz, *Mirages and Miracles*, 32–33.

80. D. Leborgne and Alain Lipietz, "New Technologies, New Modes of Regulation: Some Spatial Implications," *Environment and Planning D: Society and Space* 6 (1988): 264.

81. Antonio Gramsci, *Selections from the Prison Notebooks*, ed. Q. Hoare and G. Nowell-Smith (New York: International Publishers, 1971), 217–316.

82. Aglietta, *Theory of Capitalist Regulation*, 123–24.

83. Ibid., 124, 385.

84. See Michael J. Piore and Charles Sabel, *The Second Industrial Divide: Possibilities for Prosperity* (New York: Basic, 1984).

85. They discuss workplaces based on "flexible specialization" as places where "Proudhon might have taken Marx to show him where cooperation and competition meet." Ibid., 287.

86. Ibid., 258–80.

87. Ibid., 305.

88. Representative works include Martin Kenney and Richard Florida, "Beyond Mass Production: Production and the Labor Process in Japan," *Politics and Society* 16.1 (1988): 121–58; John Mathews, *Age of Democracy: The Politics of Post-Fordism* (Melbourne: Oxford University Press, 1989), and *Tools of Change: New Technology and the Democratisation of Work* (Sydney: Pluto, 1989); Robin Murray, "Fordism and Post-Fordism," in *New Times: The Changing Face of Politics in the 1990s*, ed. Stuart Hall and Martin Jacques (London: Lawrence and Wishart, 1989), 38–54. Although it

does not use the "post-Fordist" terminology, the work of Larry Hirschhorn is very similar in themes and in its trajectory away from Marxism. See Fred Block and Larry Hirschhorn, "New Productive Forces and the Contradictions of Contemporary Capitalism: A Post-Industrial Perspective," *Theory and Society* 7.3 (1979): 363–95; Hirschhorn, "The Post-Industrial Labor Process," *New Political Science* 2.3 (1981): 11–32; idem, *Beyond Mechanization: Work and Technology in a Post-Industrial Age* (Cambridge, Mass.: MIT Press, 1984).

89. On this point see Chris Smith, "From the 1960s Automation to Flexible Specialization: A Déjà Vu of Technological Panaceas," in *Farewell to Flexibility*, ed. A. Pollert (Oxford: Blackwell, 1991), 138–57, and Stephen Wood, "The Transformation of Work?" in *The Transformation of Work? Skill, Flexibility and the Labour Process*, ed. Stephen Wood (London: Unwin Hyman, 1989), 1–43.

90. See Eloina Pelaez and John Holloway, "Learning to Bow: Post-Fordism and Technological Determinism," *Science as Culture* 8 (1990): 15–27.

91. Leborgne and Lipietz, "New Technologies, New Modes," 263, 272.

92. Les Levidow, "Foreclosing the Future," *Science and Society* 8 (1990): 59–79.

93. Mike Parker and Jane Slaughter, "Management by Stress," *Science and Society* 8 (1990): 27–58.

94. Many of these articles are collected in Hall and Jacques, *New Times.*

95. See Dick Hebdige, "After the Masses," in Hall and Jacques, *New Times*, 76–93.

96. Stuart Hall, "The Meaning of New Times," in Hall and Jacques, *New Times*, 121, 129.

97. A. Sivanandan, "'All that Melts into Air Is Solid': The Hokum of New Times," *Race and Class* 31.3 (1989), 1–23.

98. See Michael Rustin, "The Trouble with 'New Times,'" in Hall and Jacques, *New Times*, 303–20, and John Clarke, *New Times and Old Enemies: Essays on Cultural Studies and America* (London: Harper Collins, 1991).

99. See Nick Witheford and Richard Gruneau, "Between the Politics of Production and the Politics of the Sign: Post-Marxism, Postmodernism, and New Times," *Current Perspectives in Social Theory* 13 (1993): 69–92.

100. Julie Graham, "Fordism/Post-Fordism, Marxism/Post-Marxism," *Rethinking Marxism* 4.1 (1991): 39–58.

101. Ibid., 48.

102. Ibid.

103. Ibid., 49.

104. See Levidow, "Foreclosing the Future."

105. This tendency appears even where supporters of the Regulation School believe they are arguing most strongly for the "radicalism" of their outlook. Thus Mark Ellam ("Puzzling Out the Post-Fordist Debate: Technology, Markets and Institutions," in Amin, *Post-Fordism*, 66) contrasts the Regulation School with "neo-Schumpeterean" and "neo-Smithian" approaches and approves it because it "sees the new rulebook of capitalist life as only partially written with room for many more coauthors." In support of this, he asserts the insight that "in fact, *strategic resistance against new technology and the market may well be essential if a new period of stable capitalist growth is to be secured.*" This strikes me as the apogee of functionalist recuperation: "strategic resistance" is defined by its role in consolidating accumulation. .

4   CYCLES

1. On "class-struggle" Marxism see James O'Connor, *The Meaning of Crisis: A Theoretical Introduction* (Oxford: Blackwell, 1987), 52–54; on "subjective" Marxism, Russell Jacoby, *Dialectic of Defeat: Contours of Western Marxism* (Cambridge: Cambridge University Press, 1981); on "open Marxism," the essays in Werner Bonefeld, Richard Gunn, and Kosmas Psychopedis, eds., *Open Marxism*, vol. 1, *Dialectics and History* (London: Pluto, 1992), especially John Holloway, "The Relevance of Marxism Today."

2. O'Connor, *The Meaning of Crisis*, 52.

3. Ibid., 53.

4. See Michael Lebowitz, *Beyond Capital: Marx's Political Economy of the Working Class* (New York: St. Martin's, 1992). Lebowitz argues that this focus on the activity of capital, rather than workers, began with Marx himself, who completed *Capital* but never his projected book on wage labor. The result is a perspective in which the worker appears primarily as a passive object ground under the wheels of capital's exploitative machine. This machine is, to be sure, a self-destructive one—driven toward disaster by inexorable internal laws. But it runs toward breakdown on its own—until eventually, in a moment of massive reversal, the immiserated proletariat revolts. While the consequences of such a view have varied, they have been almost uniformly catastrophic, and indeed largely justify the many criticisms of Marx made by new social movement theorists. On the one hand, that view has generated a teleological—and fatally misplaced—confidence in the inevitability of revolution. On the other, when it is suspected that the "laws" of economic collapse are not manifesting on schedule, it fosters the vision of capital as an invincible juggernaut capable of assimilating every opposition within its one-dimensional order. Further, insofar as such an account can see workers only, as it were, through the eyes of capital—as so much labor power— it does, as so many anti-Marxists have claimed, tend toward a reductionism in which people are regarded only as the bearers of economic categories, emptied of sexuality, culture, pleasure—everything, indeed, that makes life worth living. And finally, since the only historical project visible in such an account is that of capital, it makes it difficult to imagine revolution as anything but the extension or completion of capitalist development, albeit in either faster or fairer terms, that is, "socialism." For the argument that at a certain point in his project of immanent critique Marx "fell into a trap baited by political economy," see E. P. Thompson, *The Poverty of Theory* (New York: Monthly Review Press, 1978), 62.

5. On this period see Peter Rachleff, *Marxism and Council Communism* (Brooklyn: Revisionist Press, 1976); Russell Jacoby, "Towards a Critique of Automatic Marxism: The Politics of Philosophy from Lukács to the Frankfurt School," *Telos* 10 (1971): 119–46; idem, "The Politics of Crisis Theory: Toward a Critique of Automatic Marxism II," *Telos* 23 (1975): 3–52; and also Dick Howard, *The Unknown Dimension: European Marxism since Lenin* (New York: Basic, 1972).

6. For discussion of the Johnson-Forest tendency, see William Santiago-Valles, "Memories of the Future: Maroon Intellectuals from the Caribbean and the Sources of Their Communication Strategies, 1925–1940" (Ph.D. diss., Simon Fraser University, Vancouver, Canada, 1997), and Harry Cleaver, *Reading Capital Politically*

(Brighton, Eng.: Harvester, 1979), 45–49. There are now a large number of anthologies of C. L. R. James writings, including *The Future in the Present* (London: Alison and Busby, 1977); *Spheres of Existence* (London: Alison and Busby, 1980); *At the Rendez-vous of Victory* (London: Alison and Busby, 1984); and *The C. L. R. James Reader*, ed. A. Grimshaw (Oxford: Blackwell, 1992).

7. For materials from Socialisme ou Barbarie, see Cornelius Castoriadis, *Political and Social Writings*, vols. 1–3, trans. David Ames Curtis (Minneapolis: University of Minnesota Press, 1988–1993); and for analysis of the group, Cleaver, *Reading Capital*, 49–51.

8. See E. P. Thompson, *The Making of the English Working Class* (London: Gollancz, 1963).

9. See Karl Heinz Roth, *L'Autre mouvement ouvrier en Allemagne, 1945–1978* (Paris: Christian Bourgeois, 1979).

10. Thompson, *The Making of the English Working Class*; C. L. R. James, *The Black Jacobins: Toussaint L'Ouveture and the San Domingo Revolution* (New York: Vintage Books, 1989).

11. The early work of Armand Mattelart, particularly his "Introduction," with Seth Siegelaub, to *Communication and Class Struggle*, vol. 2, *Liberation and Socialism* (New York: International General, 1983), which sets out valuable analytic schemas of popular resistances to capitalism's informatic control, is an important exception to this criticism. But even this analysis today seems stamped with categories and concepts overtaken by the massive global restructuring that has reconfigured both capital and its subversion over the last two decades.

12. In adopting this term I follow Cleaver, *Reading Capital*. This work constitutes the major English definition and mapping of the theoretical positions and historical unfolding of autonomist Marxism.

13. See Cleaver, *Reading Capital*, and also *The Texas Archives of Autonomist Marxism*, ed. Harry Cleaver and Conrad Herold, available from the Department of Economics, University of Texas, Austin.

14. On the Italian autonomists see Cleaver, *Reading Capital*; Yves Moulier, "Introduction," in *The Politics of Subversion: A Manifesto for the Twenty-first Century*, by Antonio Negri (Cambridge: Polity, 1989), 1–46; and Michael Ryan, *Politics and Culture: Working Hypotheses for a Post-Revolutionary Society* (London: Macmillan, 1989). The articles in Marie Blanche Tahon and Andre Corten, eds., *L'Italie: Le Philosophe et le gendarme*, Actes du Colloque de Montréal (Montreal: VLB Editeur, 1986), provide a valuable retrospective assessment. Important anthologies of autonomist Marxist writings are Red Notes Collective, ed., *Working Class Autonomy and the Crisis* (London: Red Notes, 1979), and Sylvere Lotringer and Christian Marazzi, eds., *Italy: Autonomia—Post-Political Politics* (New York: Semiotext(e), 1980). Robert Lumley, *States of Emergency: Cultures of Revolt in Italy from 1968 to 1978* (London: Verso, 1990), gives a rather hostile account of autonomia in its historical context, while Nanni Balestrini, *The Unseen* (London: Verso, 1989), gives a vivid novelistic description of its rise and fall. My synoptic account of Italian autonomist Marxism necessarily distorts its complex history: in particular it scants the relationship of the earlier Italian *operaismo*, or "workerism," focused around factory struggles of the industrial proletariat, to the later currents merged in the broad social movement of autonomia. Tronti

and Panzieri belong to the former, not the latter. Indeed, Tronti split politically with theorists of autonomia such as Negri who built substantially on his work. Nonetheless, I find sufficient continuity in their line of thought to classify all as "autonomist Marxists." A key English-language analysis of the Italian New Left is Steven Wright, "Forcing the Lock: The Problem of Class Composition in Italian Workerism" (Ph.D. diss., Monash University, Australia, 1988), which emphasizes the difference between *operaismo* and *autonomia* and gives a fascinating analysis of the debates and struggles within the movement.

15. Readers will look in vain for any mention of autonomist Marxism in Perry Anderson's periodic reports on the state of Western Marxism in his *Considerations on Western Marxism* (London: New Left Books, 1976) or his *In the Tracks of Historical Materialism* (London: Verso, 1983), or in the *Dictionary of Marxist Thought*, ed. Tom Bottomore et al. (Cambridge, Mass.: Harvard University Press, 1983).

16. The main vehicle for the current work of Negri and his colleagues is the French journal *Futur Antérieur*. Some indication of its direction can be found in Michael Hardt and Antonio Negri, *Labor of Dionysus: A Critique of the State-Form* (Minneapolis: University of Minnesota Press, 1994), and in chap. 9 of the present book. Other strands of autonomist Marxism include, in the United States, the work of Harry Cleaver, and of George Caffentzis and the Midnight Notes Collective—a selection of whose work can be found in Midnight Notes Collective, *Midnight Oil: Work, Energy, War, 1973–1992* (New York: Autonomedia, 1992); in Britain, by the publications of the Red Notes Collective; in Italy, by a new generation of autonomist activists, largely focused around the formation of "social centers"; and internationally by Selma James's organization of the "wages for housework" campaign. During the final stages of my research for this book, Negri returned to Italy from exile, and, as the book goes to press, is in prison, while an international movement seeks a revocation of the charges against him.

17. See Antonio Negri, *Marx beyond Marx: Lessons on the Grundrisse* (New York: Bergin and Garvey, 1984).

18. Moulier, "Introduction," *The Politics of Subversion*, 19; Karl Marx, *Grundrisse* (Harmondsworth, Eng.: Penguin, 1973), 361. This definition of "labor" is quoted in the preface to Hardt and Negri, *Labor of Dionysus*, 1.

19. Mario Tronti, "Lenin in England," in Red Notes Collective, *Working Class Autonomy*, 1. Tronti's major work, *Operai e Capitale* (Turin: Einaudi, 1966), from which most of the essays cited here are extracts, has never been completely translated into English. It is available in French as *Ouvriers et capital* (Paris: Christian Bourgeois, 1977).

20. See also Mario Tronti, "The Strategy of Refusal," in Red Notes Collective, *Working Class Autonomy*, 8–9: "The existence of a class of capitalists is based on the productive power of labour. . . . It is productive labour that produces capital. . . . The worker is the *provider of capital*. In reality, he is the possessor of that unique, particular commodity which is the condition of all the other conditions of production. Because, as we have seen, all these other conditions of production are, from the start, capital in themselves—a dead capital which, in order to come to life and into play in the social relations of production, needs to subsume under itself labour power, as the subject and activity of capital."

21. Karl Marx, *Capital: A Critique of Political Economy*, vol. 1 (New York: Vintage Books, 1977), 719. On this "theoretical juxtaposition of labor power to working class" see also Cleaver, *Reading Capital*, 53.

22. Zerowork Collective, "Introduction," *Zerowork: Political Materials* 1 (1975): 3. This piece from a now unfortunately defunct journal, though perhaps somewhat dated, remains the best single, short, English-language introduction to autonomist perspectives. It is reprinted in Midnight Notes Collective, *Midnight Oil*, 108-14. On defining the working class by struggle see also Karl Marx, *Selected Correspondence* (Moscow: Progress Publishers, 1965), 165: "The working class is revolutionary or it is nothing."

23. Harry Cleaver, "Secular Crisis in Capitalism: The Insurpassability of Class Antagonism," paper presented at the Rethinking Marxism Conference, University of Massachusetts, Amherst, 1992.

24. For the concept of class composition, see Cleaver, *Reading Capital*, 112, and "The Inversion of Class Perspective in Marxian Theory: From Valorisation to Self-Valorisation," in *Open Marxism*, vol. 2, *Theory and Practice*, ed. Werner Bonefeld, Richard Gunn, and Kosmas Psychopedis (London: Pluto, 1992), 106-44; Moulier, "L'Operaisme italien: Organisation/représentation/idéologie: Ou la composition de classe revisitée," and Bruno Ramirez, "Notes sur la recomposition des classes en Amerique du Nord," both in Tahon and Corten, *L'Italie*, and Zerowork Collective, "Introduction."

25. Cleaver, "The Inversion of Class Perspective," 113.

26. Marx, *Capital*, vol. 1, 461.

27. As Cleaver puts it, both the organic composition of capital and class composition refer to the organization of the production process, but where Marx's original concept focuses on the "aggregate domination of variable by constant capital," "class composition" involves a "disaggregated picture of the structure of class power existing within the division of labor associated with a particular organisation of constant and variable capital" ("The Inversion of Class Perspective," 113). Note that such a perspective does not dismiss the interminably debated falling-rate-of-profit tendency. But it sees this tendency as defining a line of combat. In this view, capital's drive to increase constant over variable capital derives not primarily from the exigencies of competition, but in the first instance from its need to control with automation the threat of class conflict. Equally, its ability to avoid being choked by the weight of technological innovation depends on how inventively and on what terms it can impose countertendencies to offset this self-asphyxiation—perhaps most importantly through manufacturing new work to replace that being eliminated by machines. In each case, what is critical to the organic composition of capital is the class composition of the labor it confronts—the degree of resistance or compliance that capital encounters in its attempts to exercise command.

28. Zerowork Collective, "Introduction," 3-4.

29. Antonio Negri, *Revolution Retrieved: Selected Writings on Marx, Keynes, Capitalist Crisis and New Social Subjects* (London: Red Notes, 1988), 209.

30. As Cleaver notes, in using the concept "cycle of struggles" it is important to avoid suggesting a foreordained, quasi-Spenglarian natural "law" of history which determines that workers' struggles spring up, die down, and are renewed ("The Inversion of Class

Perspective," 6). If these conflicts do indeed recur, even after the most crushing defeats, it is because (1) capital has not, to date, freed itself from the need for workers, and (2) its logic imposes on the lives of these workers constraints and limitations against which they rebel. This view does not preclude the possibility either of false starts—workers' initiatives that are suppressed within the existing structures of capital without compelling it to restructure—or, equally, worker's victories—in which capital fails to reattain control.

31. Negri, *Revolution Retrieved,* 209.

32. Tronti, "The Strategy of Refusal," 10.

33. "Subsumption" designates the degree to which labor is absorbed into capital's processes of value extraction. In the "Unpublished Sixth Chapter" of *Capital,* Marx described this process in terms of successive stages. In "formal subsumption"—roughly the early stages of the industrial revolution—capital simply imposes the form of wage labor on preexisting modes of artisanal production. But in the subsequent phase, "real subsumption," it undertakes a wholesale reorganization of work. Science is systematically applied to industry; technological innovation becomes perpetual; exploitation focuses on "relative" intensification of productivity rather than "absolute" extension of hours; economies of scale and cooperation are systematically sought out; and consumption is organized by the cultivation of new needs that beckon new industries in an orgy of "productions for production's sake." In this phase, Marx observes, the agent of production ceases to be the individual laborer and becomes the "collective worker," made up of labor power "socially combined" in all the manifold differentiated yet interdependent tasks "which together form the entire production machine." The "limbs" of this collective subject encompass both physical and intellectual activity: "Some work better with their hands, others with their heads, one as a manager, engineer, technologist, etc., the other as overseer, the third as manual labourer or even drudge.... And here it is quite immaterial whether the job of a particular worker, who is merely a limb of this aggregate worker, is at a greater or smaller distance from the actual manual labour." See *Capital,* vol. 1, 1026–40. Autonomist Marxism's concept of the "social factory" is the theory of this "collective worker" grown beyond a point even Marx could anticipate.

34. Quoted in translation in Mario Tronti, "Social Capital," *Telos* 17 (1973): 105.

35. See Mariarosa Dalla Costa and Selma James, *The Power of Women and the Subversion of the Community* (Bristol, Eng.: Falling Wall Press, 1972).

36. The concept of "unwaged" labor is particularly associated with Selma James and has been developed by her in a series of writings, from *Sex, Race and Class* (Bristol, Eng.: Falling Wall Press, 1975) to "Women's Unwaged Work—The Heart of the Informal Sector," *Women: A Cultural Review* 2.3 (1991): 267–71. Other works drawing on this analysis include Nicole Cox and Silvia Federici, *Counterplanning from the Kitchen—Wages for Housework: A Perspective on Capital and the Left* (Bristol, Eng.: Falling Wall Press, 1975), and, more recently, Leopoldina Fortunati, *The Arcane of Reproduction: Housework, Prostitution, Labor and Capital* (New York: Autonomedia, 1995).

37. Selma James, "Marx and Feminism," *Third World Book Review* 1.6 (1986): 2.

38. Ibid.

39. See Sergio Bologna, "The Tribe of Moles," in Lotringer and Marazzi, *Italy,* 36–61.

40. Tronti, *Ouvriers et capital*, 305.

41. On self-valorization see Cleaver, "The Inversion of Class Perspective."

42. Werner Bonefeld, "Human Practice and Perversion: Beyond Autonomy and Structure," *Common Sense* 15 (1994): 43-42.

43. Thus in a review of Midnight Notes Collective, *Midnight Oil*, a writer for the journal *Aufheben* criticizes the book for a tendency to see "capital as an undifferentiated unity imposing an agreed structure on the working class," which verges on conspiracy theory (review of *Midnight Oil*, by Midnight Notes Collective, *Aufheben* 3 [1994]; 35-41). For other interesting critical analyses of the autonomist tradition, see Aufheben Collective, "Decadence: The Theory of Decline or the Decline of Theory— Part II," *Aufheben* 3 (1994); 24-34; and Radical Chains, "'Autonomist' and 'Trotskyist' Views: Harry Cleaver Debates Hillel Ticktin," *Radical Chains* 4 (1994): 9-17.

44. See Lebowitz, *Beyond Capital*.

45. Raniero Panzieri, "Surplus Value and Planning: Notes on the Reading of Capital," in *The Labour Process and Class Strategies*, ed. Conference of Socialist Economists (London: Conference of Socialist Economists, 1976), 4-25, and "The Capitalist Use of Machinery: Marx versus the Objectivists," in *Outlines of a Critique of Technology*, ed. Phil Slater (Atlantic Highlands, N.J.: Humanities Press, 1980), 44-69.

46. Marx, *Capital*, vol. 1, 563.

47. Panzieri, "Surplus Value and Planning," 12.

48. Ibid.

49. Ibid. In "The Capitalist Use of Machinery," Panzieri observed that "the relationship of revolutionary action to technological 'rationality' is to 'comprehend' it, but not in order to acknowledge and exalt it, rather in order to subject it to a new use: to the socialist use of machines" (57).

50. Harry Cleaver, "Technology as Political Weaponry," in *Science, Politics and the Agricultural Revolution in Asia*, ed. Robert Anderson (Boulder, Colo.: Westview, 1981), 261-76.

51. My account here rewrites in terms familiar to English-speaking audiences the distinction Negri makes ("Domination and Sabotage," in Red Notes Collective, *Working Class Autonomy*, 93-138) between "sabotage" and "invention power."

52. Extracts are reprinted in Red Notes Collective, *Working Class Autonomy*.

53. By "sabotage" Negri was designating a very broad concept of refusing and undermining capitalist development. But there is no doubt that this included its most concrete and pointed applications: strikes, "direct action," and the destruction of machinery—tactics for which *autonomia* was to develop a formidable reputation.

54. See, for example, Frank Webster and Kevin Robins, *Information Technology: A Luddite Analysis* (Norwood, N.J.: Ablex, 1986).

55. Cleaver, "Technology as Political Weaponry," 264.

56. For an account of the most famous of these radio stations, which broadcast out of Bologna, see Collectif A/Traverso, *Radio Alice, Radio Libre* (Paris: J.-P. Delarge, 1977). Other autonomist-influenced radio stations included Radios Red Wave and City of the Future in Rome, and Radio City of the Future in Venice.

57. Franco Berardi, *Le Ciel est enfin tombé sur la terre* (Paris: Seuil, 1978), 27 (my trans.).

58. Ibid.

59. Berardi, quoted in Collectif A/Traverso, *Radio Alice, Radio Libre,* 107.

60. *Détournement* is a term deriving from the Situationists, with whom the Italian autonomia had a distinct affinity. It describes the reassemblage of elements torn out of their original context in order to make a subversive political statement; see Ken Knabb, ed., *Situationist International Anthology* (Berkeley, Calif.: Bureau of Public Secrets, 1981), 8–14, 55–56; Guy Debord, *Society of the Spectacle* (Detroit: Black and Red, 1977); and, for useful commentary, Cleaver, "The Inversion of Class Perspective."

61. Cleaver, "Technology as Political Weaponry," 264; Panzieri, ""Surplus Value and Planning," 12.

62. On the fluidity of technology, see Cleaver, "Technology as Political Weaponry," 268: "A given technology is never the same when it is implemented in different historical and political contexts. As an organization of social production, technology organizes the existing social relations, and those shift and change according to the changing composition of political power."

63. Moulier, "L'operaisme italien." He stresses that the periodization should not be taken to mean that the workers of one era vanish at the commencement of the next: professional workers persist into the era of the mass worker, and mass workers into the epoch of the socialized worker: the point is that the innovative and strategic center of struggle shifts. For a compact summary of the three cycles described here, see Antonio Negri, "Interpretation of the Class Situation Today: Methodological Aspects," in Bonefeld, Gunn, and Psychopedis, *Open Marxism,* 2:69–105.

64. See Sergio Bologna, "Class Composition and the Theory of the Party at the Origin of the Workers Councils Movement," in *The Labour Process and Class Strategies,* ed. Conference of Socialist Economists (London: Conference of Socialist Economists, 1976), 68–91.

65. On the social wage and the factory wage see Tronti, "Social Capital," and Negri, "The State and Public Spending," the latter in *Labor of Dionysus,* 179–216.

66. See Tronti, "Social Capital."

67. John Merrington, notes to Negri, *Revolution Retrieved,* 200.

68. Witness Henry Ford's obsessive concern with workers' morality and marital status. An excellent account of these developments is Stuart Ewen, *Captains of Consciousness* (New York: McGraw-Hill, 1976).

69. Tronti, "The Strategy of Refusal." For autonomist accounts of this crisis see Cleaver, *Reading Capital,* and many of the articles in Midnight Notes Collective, *Midnight Oil.* For a collection of nonautonomist articles on the same topic, see Colin Crouch and Alessandro Pizzorno, eds., *The Resurgence of Class Conflict in Western Europe since 1968,* vols. 1 and 2 (London: Macmillan, 1978).

70. Negri, "Interpretation of the Class Situation Today," dates it specifically to the 1971 Nixon/Kissinger departure from the gold standard.

71. Ibid., 87.

72. George Caffentzis, "The Work/Energy Crisis and the Apocalypse," in Midnight Notes Collective, *Midnight Oil,* 221–22 (orig. *Midnight Notes* 3 [1980]).

73. See Michael J. Crozier, Samuel P. Huntington, and Joji Watanuki, *The Crisis of Democracy* (New York: New York University Press, 1975), 115.

74. Collettivo Strategie, "The 'Technetronic Society' According to Brzezinski," in

Compulsive Technology, ed. Tony Solominides and Les Levidow (London: Free Association Books, 1985), 128.

75. Ibid.

76. Ibid.

77. Ibid.

78. Ibid., 129. It should be underlined that this analysis was no abstract exercise: Brzezinski, an originator of information revolution theory, was also, at the time of the Collettivo Strategie's writing, advising the Italian government on how to dispose of *autonomia* and other dissident groups, a process that was to end in arrest and imprisonment for hundreds of activists. See Zbigniew Brzezinski, *Power and Principle: Memoirs of the National Security Adviser* (New York: Farrar, Strauss, Giroux, 1983).

79. See Paul Virilio, *Popular Defense and Ecological Struggles* (New York: Semiotext(e), 1990). Although Virilio has now departed on a very different trajectory, he is a theorist at one time loosely connected to *autonomia*—see his "Popular Defense and Popular Assault," in Lotringer and Marazzi, *Italy,* 266–72.

80. See David Noble, *Forces of Production* (New York: Knopf, 1984); Les Levidow and Kevin Robins, "Towards a Military Information Society?" in *Cyborg Worlds: The Military Information Society,* ed. Levidow and Robins (London: Free Association Books, 1989), 159–77.

81. See Starr Roxanne Hiltz and Maurice Turoff, *The Network Nation: Human Communication via Computer* (Reading, Mass.: Addison-Wesley, 1978).

82. On this process, see, in addition to the sources cited above, Manuel Castells, *The Informational City: Information Technology, Economic Restructuring and the Urban-Regional Process* (Oxford: Blackwell, 1989); Jim Pomeroy, "Black Box S-Thetix: Labor, Research, and Survival in the (Art) of the Beast," in *Technoculture,* ed. Constance Penley and Andrew Ross (Minneapolis: University of Minnesota Press, 1991), 271–94; Manuel De Landa, *War in the Age of Intelligent Machines* (New York, 1991).

83. Negri, *Revolution Retrieved,* 181, borrows this phrase from Herbert Marcuse and applies it specifically to the transition from Planner State to Crisis State.

84. For ratification of this point from a nonautonomist source see Castells, *The Informational City,* 29: "while informationalism has by now been decisively shaped by the restructuring process, restructuring could never have been accomplished, even in a contradictory manner, without the unleashing of the technological and organisational potential of informationalism."

85. See Dave Feikert, "Britain's Miners and New Technology," *Issues in Radical Science,* Radical Science 17 (London: Free Association Books, 1985), 22–30; G. Santilli, "Peau de leopard: L'Automatisation comme forme de controle social," *Travail* 8 (1985): 20–28; Cynthia Cockburn, *Machinery of Dominance: Women, Men and Technical Know How* (London: Pluto, 1985).

86. Caffentzis, "The Work/Energy Crisis," 234.

87. Fergus Murray, "The Decentralisation of Production—the Decline of the Mass-Collective Worker?" *Capital and Class* 19 (1983): 74–99.

88. Ibid., 95.

89. Ibid.

90. Negri, *Revolution Retrieved,* 209.

91. Negri first uses the term in *La Classe ouvrière contre l'etat* (Paris: Edition Galilee, 1978) and *Del Obrero-Masa al Obrero Social* (Barcelona: Editorial Anagrama, 1980). His fullest English-language statements of the position are "Archaeology and Project: The Mass Worker and the Social Worker," in his *Revolution Retrieved*, 199–228; *The Politics of Subversion: A Manifesto for the Twenty-first Century* (Cambridge: Polity, 1989); and "Interpretation of the Class Situation Today." For the context of the theory, and the controversy surrounding it, see Wright, "Forcing the Lock," 287–339.

92. Negri, *Politics of Subversion*, 89.

93. Ibid.

94. Negri, *La Classe ouvrière*, 254.

95. For an autonomist analysis of this in the context of Fiat, see Mariella Bierra and Marco Revelli, "Absentéisme et conflictualité: L'Usine reniée. Crise de la centralité de l'usine et nouveaux comportements ouvriers," in *Usines et ouvriers: Figures du nouvel ordre productif*, ed. Jean Paul de Gaudemar (Paris: François Maspero, 1980), 105–36.

96. Negri, *Politics of Subversion*, 79.

97. Negri, "Interpretation of the Class Situation Today," 79.

98. Negri, *Revolution Retrieved*, 219.

99. Negri, *Politics of Subversion*, 204.

100. Ibid., 84.

101. Ibid., 94–95.

102. Negri, *Del Obrero-Masa al Obrero Social*, 36–37 (trans. Santiago Valles).

103. Negri, *Politics of Subversion*, 87.

104. Negri, "Gauche et coordinations ouvrières," *Lignes* 5 (1989): 94 (my trans.).

105. Ibid., 93.

106. Ibid., 94.

107. Ibid.

108. Felix Guattari and Toni Negri, *Communists Like Us: New Spaces of Liberty, New Lines of Alliance* (New York: Semiotext(e), 1990), 128.

109. Negri, "Luttes sociales et control systemique," *Futur Antérieur* 9 (1992): 15 (my trans.).

110. Ibid.

111. Ibid., 18.

112. Negri, *Revolution Retrieved*, 239.

113. Negri, *Politics of Subversion*, 93.

114. Ibid.

115. Ibid.

116. Ibid., 85–86. Referring specifically to computerization, Negri observes that the more "abstract" and "immaterial" the "instrumentation of production" becomes, the more it is itself "implicated in the struggle that traverses the social" and reveals "sectors which are vulnerable, and ever more vulnerable to the autonomy of social cooperation and the auto-valorisation of proletarian subjects." Negri, "Interpretation of the Class Situation Today," 89.

117. See Serge Mallet, *Essays on the New Working Class* (St. Louis: Telos, 1975).

118. Maurizio Lazzarato and Toni Negri, "Travail immaterial and subjectivité," *Futur Antérieur* 6 (1991): 87.

119. Negri, *Politics of Subversion*, 52.

120. Ibid., 116.

121. Ibid.

122. Ibid., 118.

123. Ibid., 119. He notes that the distinction is "imprecise"; in practice information and communication are not easily separable: computerization represents an attempt by capital to enhance its informational powers, but may in practice allow communicational opportunities for workers. For this reason, "One must . . . be very careful in the use we make of the distinction, occasionally using it, if one wishes, as an abstract, definitional distinction, but bearing in mind that it is quite inadequate for analysis of the concrete."

124. Ibid., 117-18.

125. Ibid., 118, 58.

126. For information on these events, see Marie Marchand, *The Minitel Saga* (Paris: Larousse, 1988).

127. "There was also detailed information on the demonstrations (starting points, routes, buttons to wear, slogans to shout) and statements rejecting any attempts by political parties and their allies to coopt the student movement. They were truly dead set on maintaining their independence and they said as much on display page after display page" (ibid., 153).

128. Negri, *Politics of Subversion*, 137.

129. Alain Lipietz, *The Enchanted World: Inflation, Credit and the World Crisis* (London: Verso, 1985), 141. The resemblances between the autonomists' theory of "cycles of struggle" and the Regulation School's concept of successive "regimes of accumulation"—with the era of the mass worker corresponding to Fordism, and the socialized worker to post-Fordism—will be apparent. In fact, both groups have influenced each other, while taking very different orientations, the Regulation School theorists preoccupying themselves with the requirements for successful capitalist accumulation, the autonomists searching for possibilities to explode that process. Perhaps predictably, they arrive at different conclusions, with autonomists, or at least Negri, perceiving the onset of a new era of struggle, and Regulation School theorists settling for accommodation. Negri, although sometimes using the Fordist/post-Fordist terminology, has criticized the Regulationists as an "academic school" who have abandoned the "critique of political economy" in favor of a "functionalist and programmatic schema" ("Interpretation of the Class Situation Today," 104-5).

130. For example, Sergio Bologna was intensely critical of Negri's attempt to contain the complexities arising from the restructuring of labor power within a single grand theoretical construct. For an exciting and informative summary of the criticism of Negri's "socialized worker" thesis by Bologna and other of his Italian comrades see Wright, "Forcing the Lock," 287-339. An important, scathing critique of Negri's work from the perspective of the German "autonomen" movements appears in George Katsiaficas, *The Subversion of Politics: European Autonomous Social Movements and the Decolonization of Everyday Life* (Atlantic Highlands, N.J.: Humanities Press, 1997). It should, however, be noted that Negri's account of the "socialized worker" has developed over the course of time, and its most recent versions are more substantial than its initial enunciation.

131. Leon Trotsky, *History of the Russian Revolution* (London: Pluto, 1977), 26–27.

132. Caffentzis, "The Work/Energy Crisis," 235.

133. Ibid.

134. For a selection from the large literature on this issue, see Cockburn, *Machinery of Dominance*; idem, *Brothers: Male Dominance and Technological Change* (London: Pluto, 1991); Margaret Lowe Benston, "For Women, the Chips Are Down," in *The Technological Woman: Interfacing With Tomorrow*, ed. Jan Zimmerman (New York: Praeger, 1983), 44–54; Heather Menzies, *Fast Forward and Out of Control: How Technology Is Changing Your Life* (Toronto: Macmillan, 1989); Sally Hacker, *"Doing It the Hard Way": Investigations of Gender and Technology* (Boston: Unwin Hyman, 1990).

135. See the discussion between Guido Baldi, "Negri beyond Marx," and Bartleby the Scrivener, "Marx beyond Midnight," both in *Midnight Notes* 8 (1985): 32–36.

136. Thus in his book *The Informational City*, which provides a striking contrast to Negri's account of the socialized worker, Manuel Castells argues that a high-technology economy—unlike smokestack industry with its massed blue-collar work force—tends to polarize employment. Computerization results in the elimination of jobs insufficiently skilled to escape automation but expensive enough to be worth replacement. Of the remainder a substantial number are "upgraded" to provide "intellective" tasks for a new echelon of technicians and programmers. A larger portion are downgraded, "recycled in low-skill, low-pay activities in the miscellaneous service sector, or integrated in the booming informal economy in both manufacturing and services" with lower wages and little or no social protection. This generates a dualized occupational pattern whose divisions follow predictable lines of gender and ethnicity, and are reinforced by self-perpetuating residential enclaves, educational chances, and differential exposure to media and information flows. What results is "a highly differentiated social structure, both polarised and fragmented" (205). Professional and managerial classes identify with capital. The remainder of the working population is divided into "socially discriminated communities that *cannot constitute a class*" (228, emphasis added).

137. Negri, *Politics of Subversion*, 145–46.

138. Ibid., 133.

139. Negri, "Interpretation of the Class Situation Today," 87.

5   CIRCUITS

1. This concept of the circuit of capital recurs throughout Marx's work, but perhaps finds its most systematic exposition in vol. 2 of *Capital* (London: Vintage Books, 1978) and, in a somewhat different form, in the Introduction to *Grundrisse* (Harmondsworth, Eng.: Penguin, 1973), 81–114.

2. The autonomist development of the concept can be found in Mario Tronti, "Social Capital," *Telos* 17 (1973): 105. Tronti elsewhere writes: "The more capitalist development advances, that is to say the more the production of relative surplus value penetrates everywhere, the more the circuit production-distribution-exchange-consumption inevitably develops; that is to say that the relationship between capitalist production and bourgeois society, between the factory and society, between society and the state, become more and more organic. At the highest level of capitalist development social relations become moments of the relations of production, and the whole

society becomes an articulation of production. In short, all of society lives as a function of the factory and the factory extends its exclusive domination over all of society." (Tronti, *Operai e Capitale*, quoted and trans. Harry Cleaver, "The Inversion of Class Perspective in Marxian Theory: From Valorisation to Self-Valorisation," in *Open Marxism*, vol. 2, *Theory and Practice*, ed. Werner Bonefeld, Richard Gunn, and Kosmas Psychopedis [London: Pluto, 1992], 137.) See also Raniero Panzieri: "The factory is becoming generalised. The factory is tending to pervade, to permeate the entire arena of civil society" ("Lotte Operaie nello Sviluppo Capitalistico," *Quaderni Piacentini* (1967), quoted by James O'Connor, *Accumulation Crisis* [Blackwell: Oxford, 1984], 151). For a very clear exposition of the concept of the social factory, see Harry Cleaver, "Malaria, the Politics of Public Health and the International Crisis," *Review of Radical Political Economics* 9.1 (1977): 81–103, and Peter F. Bell and Harry Cleaver, "Marx's Crisis Theory as a Theory of Class Struggle" *Research in Political Economy* 5 (1982): 189–261, to which this chapter owes a considerable debt.

3. See Mariarosa Dalla Costa and Selma James, *The Power of Women and the Subversion of the Community* (Bristol, Eng.: Falling Wall Press, 1972).

4. For important recent Marxist theoretical perspectives on ecological issues see the journal *Capitalism, Nature, Socialism*.

5. On new production systems see Stanley Aronowitz and William DiFazio, *The Jobless Future: Sci-Tech and the Dogma of Work* (Minneapolis: University of Minnesota Press, 1994); Benjamin Coriat, *L'Atelier et le robot* (Paris: Christian Bourgeois, 1990); Jean Paul de Gaudemar, ed., *Usines et ouvriers: Figures du nouvel ordre productif* (Paris: François Maspero, 1980); David Noble, *Forces of Production* (New York: Knopf, 1984); Stephen Wood, ed., *The Transformation of Work? Skill, Flexibility and the Labour Process* (London: Unwin Hyman, 1989).

6. See G. Santilli, "Peau de leopard: L'Automatisation comme forme de controle social," *Travail* 8 (1985): 20–28.

7. Paul Kennedy, *Preparing for the Twenty-first Century* (New York: Random House, 1993), 88; Jim Davis and Michael Stack, "Knowledge in Production." *Race and Class*, 34.3 (1992): 8.

8. Marx, *Grundrisse*, 692.

9. On Hormel, see Peter Rachleef, *Hard Pressed in the Heartland: The Hormel Strike and the Future of Labor* (Boston: South End, 1992).

10. An early, and brilliant, critique of Bell's concept of the service sector is to be found in Krishan Kumar, *Prophecy and Progress* (Harmondsworth, Eng.: Penguin, 1978).

11. Marx, *Grundrisse*, 705.

12. Ibid.

13. On this point, see Doug Henwood, "Info Fetishism," in *Resisting the Virtual Life: The Culture and Politics of Information*, ed. James Brook and Iain A. Boal (San Francisco: City Lights, 1995), 163–72.

14. See Kim Moody, *Workers in a Lean World: Unions in the International Economy* (London: Verso, 1997), 193–94.

15. For events in the Illinois "war zone," see Marc Cooper, "Harley-Riding, Picket-Walking Socialism Haunts Decatur," *Nation*, 8 Apr. 1996, 23–25.

16. On recent developments in the AFL-CIO, see Jeremy Brecher and Tim Costello, "A New Labor Movement in the Shell of the Old," *Z Magazine*, Apr. 1996, 45–49.

17. Dennis Hayes, *Behind the Silicon Curtain: The Seductions of Work in a Lonely*

Era (Boston: South End, 1989); Michael Hardesty and Nina Wurgaft, "Silicon Valley: A Tale of Two Classes," *Z Magazine*, Sept. 1992, 63–65; Navid Mohseni, "The Labor Process and Control of Labor in the U.S. Computer Industry," in *The Labor Process and Control of Labor: The Changing Nature of Work Relations in the Late Twentieth Century*, Berch Berberoglu (Westport, Conn.: Praeger, 1993), 59–77; Andrew Gorry, "Silicon Valley: A Divided Workforce," *CPU: Working in the Computer Industry* 3 (1993), on-line, Internet. *CPU* is available at no cost from listserv@cpsr.org. It can be found on the World Wide Web at http://www.mcs.com/~jdav/CPU/cpu.html.

18. Hardesty and Wurgaft, "Silicon Valley," 62.

19. Lenny Siegel, "New Chips in Old Skins: Work, Labor and Silicon Valley," *CPU: Working in the Computer Industry* 6 (1993), on-line, Internet; Lisa Hoyos and Mai Hoang, "Workers at the Centre: Silicon Valley Campaign for Justice," *CrossRoads* 43 (1994): 24–27; David Bacon, "Silicon Valley on Strike," *CPU: Working in the Computer Industry* 3 (1993), on-line, Internet.

20. Hoyos and Hoang, "Workers at the Centre."

21. Nathan Newman, "'Third Wave Unionism' Takes to the Net," on-line, Internet, Red Rock Eater News Service, 22 Aug. 1996.

22. See Paul Almeida, "The Network for Environmental and Economic Justice for the Southwest: Interview with Richard Moore," *Capitalism, Nature, Socialism* 5.1 (1994): 21–54, and Richard Moore and Louis Head, "Acknowledging the Past, Confronting the Present," in *Toxic Struggles: The Theory and Practice of Environmental Justice*, ed. Richard Hofrichter (Philadelphia: New Society, 1993).

23. David Bacon, "L.A. Labor—A New Militancy," *Nation*, 27 Feb. 1995, 273–76.

24. Mike Davis, "Armaggedon at the Emerald City: Local 226 vs MGM Grand." *Nation*, 11 July 1994: 46–49.

25. Pamela Chiang, "501 Blues," *Breakthrough* 18:2 (1994): 3–7.

26. The best single source for reporting the unfolding of these movements is the U.S. dissident trade union journal *Labor Notes*. Other interesting discussion can be found in Jeremy Brecher and Tim Costello, eds., *Building Bridges: The Emerging Grassroots Coalition of Labor and Community* (New York: Monthly Review Press, 1990); Collective Action Notes, "The U.S.A.: A Transitional Period—But to Where?" *Collective Action Notes* 9 (1996): 1–4; Peter Rachleef, "Seeds of a Labor Resurgency," *Nation*, 21 Feb. 1994: 226–29.

27. See Chiang, "501 Blues," and Andrew Banks, "Jobs with Justice: Florida's Fight against Worker Abuse," in Brecher and Costello, *Building Bridges*, 25–37.

28. Labor Resource Center, *Holding the Line in '89: Lessons of the NYNEX Strike* (Somerville, Mass.: Labor Resource Center, 1990); David Dyssegaard Kallick, "Toward a New Unionism," *Social Policy* 25.2 (1994): 2–6.

29. Brecher and Costello, *Building Bridges*.

30. See Moody, *Workers in a Lean World*.

31. Ibid., 277.

32. For developments in Italy, see Gregor Gall, "The Emergence of a Rank and File Movement: The Comitati di Base in the Italian Workers' Movement," *Capital and Class* 55 (1995): 9–20; and in France, Steve Jefferys, "France 1995: The Backwards March of Labour Halted?" *Capital and Class* 59 (1996): 7–22.

33. Karl Marx, *Capital: A Critique of Political Economy*, vol. 1 (New York: Vintage Books, 1977), 718.

34. See Michael Hardt and Antonio Negri, *Labor of Dionysus: A Critique of the State-Form* (Minneapolis: University of Minnesota Press, 1994).

35. See James O'Connor, *The Fiscal Crisis of the State* (New York: St. Martin's, 1973).

36. See Andrew Gamble, *The Free Market and the Strong State: The Politics of Thatcherism,* (Basingstoke, Eng.: Macmillan, 1988).

37. Michel Foucault, *Discipline and Punish: The Birth of the Prison* (Harmondsworth, Eng.: Penguin, 1979), 195–228.

38. On contemporary prison labor, see Massimo De Angelis, "The Autonomy of the Economy and Globalisation," *Vis-A-Vis,* Winter 1996, on-line, available from http:// www.lists.village.virginia.edu/spoons/aut_html/glob.html; and Daniel Burton-Rose, ed., with Dan Pens and Paul Wright, *The Celling of America: An Inside Look at the U.S. Prison Industry* (Monroe, Me.: Common Courage Press, 1998). For an account of the automated security systems now in use in the U.S. "prison-industrial complex," see Mike Davis, "Hell Factories in the Fields," *Nation,* 20 Feb. 1995, 229–34.

39. Mike Davis, *Beyond Blade Runner: Urban Control—The Ecology of Fear* (Westfield, N.J.: Open Magazine Pamphlet Series, 1992), 16.

40. See Mike Davis, "Los Angeles Was Just the Beginning," in *Open Fire: The Open Magazine Pamphlet Series Anthology,* ed. Greg Ruggiero and Stuart Sahulka (New York: New Press, 1993), 220–44.

41. Reprinted in Haki R. Madhubuti, ed., *Why L.A. Happened: Implications of the '92 Los Angeles Rebellion* (Chicago: Third World, 1993). For accounts of how the document emerged from the riot and the subsequent gang truce, see Alexander Cockburn, "Beat the Devil," and Mike Davis, "Urban America Sees Its Future in L.A.," both in *Nation,* 1 June 1992, 738–39.

42. The classic study of "poor people's movements" is, of course, Frances Fox Piven and Richard Cloward, *Poor People's Movements: Why They Succeed, How They Fail* (New York: Pantheon, 1977).

43. Alex Vitale and Keith McHenry, "Food Not Bombs," *Z Magazine,* Sept. 1994, 19–21.

44. Information on the "Andersen Conversion Project" is available from Toronto Action for Social Change, P.O. Box 73620, 509 St. Clair Ave. West, Toronto, ON M6C 1Co, Canada; email burch@web.net

45. On this point see Rosalind Polack Petchesky, *Abortion and Woman's Choice: The State, Sexuality and Reproductive Freedom* (Boston: Northeastern University Press, 1990), 241–52.

46. Maria Mies, "Why Do We Need All this? A Call against Genetic Engineering and Reproductive Technology," in *Made to Order: The Myth of Reproductive and Genetic Progress,* ed. Patricia Spallone and Deborah Lynn Steinberg (Oxford: Pergamon, 1987), 34–47, and Kathryn Russell, "A Value-Theoretic Approach to Childbirth and Reproductive Engineering," *Science and Society* 58.3 (1994): 287–314.

47. According to Andrew Kimbrell (*The Human Body Shop* [San Francisco: Harper, 1993], 101), such agreements "routinely require that the prospective mother submit to massive doses of fertility drugs, hormone injections, amniocentesis and an array of genetic probes and tests at the discretion of the client; require that the mother agrees to abort the fetus on demand, and is totally liable for all 'risks' associated with conception, pregnancy and childbirth." Payment is in the region of ten thousand dollars— one thousand dollars if the child is stillborn.

48. See Françoise Labone, "Looking for Mothers You Only Find Fetuses," in Spallone and Steinberg, *Made to Order*, 51–53.

49. On the relation of biotechnological investment to post-Fordist restructuring, see Edward Yoxen, "Life as a Productive Force: Capitalizing the Science and Technology of Molecular Biology," in *Science, Technology and the Labor Process: Marxist Studies*, vol. 1, ed. Les Levidow and Bob Young (Atlantic Highlands, N.J.: Humanities Press, 1981), 66–123; Herbert Gottweis, "Genetic Engineering, Democracy, and the Politics of Identity," *Social Text* 42 (1995): 127–52.

50. As in other areas of capitalist technological development, these innovations have to be understood not simply as means to increase productivity, but as tools to change social relations. For example, many of the developments in biotechnology have been central to the extension of large-scale capitalist techniques to farming—agribusiness. In his classic essay, "Technology as Political Weaponry" (in *Science, Politics and the Agricultural Revolution in Asia*, ed. Robert Anderson [Boulder, Colo.: Westview, 1981], 261–76), Harry Cleaver has described how the "Green Revolution" was used to break down forms of rural community resistant to capitalist modernization. The same process is now enacted on a multitude of fronts: through the establishment of patent rights over food sources cultivated in the wild by peasant and indigenous communities; the creation of herbicide-resistant plant strains tied to the products of particular chemical companies; the ability to bypass rural and Third World producers by artificial synthesis of naturally occurring substances; and the institution of methods—such as pharmacological augmentation of cows by the use of bovine growth hormone—which favor large-scale enterprises. For recent developments on this front, see Tom Athanasiou, "Greenwashing Agricultural Biotechnology," *Processed World* 28 (1991/92): 16–21; Cary Fowler and Pat Mooney, *Shattering: Food, Politics and the Loss of Genetic Diversity* (Tucson: University of Arizona Press, 1990); Vandana Shiva, *The Violence of the Green Revolution: Third World Agriculture, Ecology and Politics* (London: Zed Books, 1991); Jeremy Seabrook, "Biotechnology and Genetic Diversity," *Race and Class* 34.3 (1993): 15–30; Brian Tokar, "The False Promise of Biotechnology," *Z Magazine*, Feb. 1992, 27–32.

51. Gottweis, "Genetic Engineering," 137.

52. For a multiperspectival collection of readings on the Genome Project see Daniel Kevles and Leroy Hood, eds., *The Code of Codes: Scientific and Social Issues in the Human Genome Project* (Cambridge, Mass.: Harvard University Press, 1992).

53. See the discussion in Ruth Hubbard and Elijah Ward, *Exploding the Gene Myth* (Boston: Beacon, 1993).

54. Henry Greely, "Health Insurance, Employment Discrimination, and the Genetics Revolution," in Kevles and Hood, *The Code of Codes*, 264–80.

55. Seth Shulmann, "Preventing Genetic Discrimination," *Technology Review* (July 1995): 16–18.

56. The implications are most immediately apparent in the United States where many employers directly carry the costs of workers health insurance and thus have a powerful incentive not to hire workers who can be expected to get sick. Although the peculiar atavism of the U.S. health care system makes these issues surface very rapidly, the implications are far wider. For even where capital bears the costs of laborers' ill health only indirectly—through the welfare state programs—genetic screening offers the potential for lowering this expense.

57. See Kimbrell, *Human Body Shop*, and Giovanni Berlinguer, "The Body as Commodity and Value," *Capitalism, Nature, Socialism* 5.3 (1994): 35–49.

58. For speculations along this line by an elite "Eurocrat," see Jacques Attali, *Millennium: Winners and Losers in the Coming World Order* (New York: Random House, 1991).

59. Peter Linebaugh, *The London Hanged: Crime and Civil Society in the Eighteenth Century* (London: Penguin, 1991), 121–22.

60. See Marlene Fried, ed., *From Abortion to Reproductive Freedom: Transforming a Movement* (Boston: South End, 1990).

61. Rosalind Copelon, "From Privacy to Autonomy: The Conditions for Sexual and Reproductive Freedom," in Fried, *From Abortion to Reproductive Freedom*, 39. In the same anthology see also the essay by Angela Davis, "Racism, Birth Control, and Reproductive Rights," 15–26.

62. For a collection stating the FINRAGE position, see Spallone and Steinberg, *Made to Order*. FINRAGE takes the position that such technologies are inherently dominative, and aims at an outright ban on their development and new research agendas to discover different remedies for the problems that the repro-tech industries purport to "fix" technologically—for example, the investigation of social and environmental causes of infertility. Other feminist groups believe that although currently patriarchal and corporate control make these technologies inimical to women, it may be possible to bend their trajectory in positive directions. They therefore call not for the halting of development but for free and nondiscriminatory access—for example, making new reproductive possibilities available to lesbians and gays. There have also been concerns that the FINRAGE position fails to build links with the women participants in *in vitro* programs, who constitute not only the consumers but also the unwaged, experimental labor force of the repro-tech industry. There are real contradictions, and heated controversy, between these different positions. But both dissent strongly from the move toward reproductive commodification.

63. Sue Cox, "Strategies for the Present, Strategies for the Future: Feminist Resistance to New Reproductive Technologies," *Canadian Woman Studies* 13.2 (1993): 87.

64. On the Canadian events see Gwynne Basen, Margaret Eichler, and Abby Lippman, eds., *Misconceptions: The Social Construction of Choice and the New Reproductive and Genetic Technologies* (Quebec: Voyageur, 1993).

65. Debra Harry, "The Human Genome Diversity Project and Its Implications for Indigenous Peoples," *Information about Intellectual Property Rights No. 6* (Minneapolis, Minn.: Institute for Agriculture and Trade Policy, 1995). Available by email from iatp@iatp.org; Pat Mooney, *The Conservation and Development of Indigenous Knowledge in the Context of Intellectual Property Systems* (Ottawa: Rural Advancement Foundation International, 1993). The Ottawa-based Rural Advancement Foundation International has served as a clearing house and information source both for these efforts and for First and Third World farmers fighting agribusinesses' biotechnological enclosures.

66. Patrick Novotny, "Popular Epidemiology and the Struggle for Community Health: Alternative Perspectives from the Environmental Justice Movement," *Capitalism, Nature, Socialism* 5.2 (1994): 29–50. On other aspects of the crisis of health care under neoliberal regimes, see Donald Lowe, *The Body in Late Capitalist USA* (Durham, N.C.: Duke University Press, 1995), and Vicente Navarro, *Crisis, Health and*

Medicine: A Social Critique (New York: Tavistock, 1986) and *Dangerous to Your Health* (New York: Monthly Review Press, 1993).

67. See Peter Arno and Karyn Feiden, *Against the Odds: The Story of AIDS Drug Development, Politics and Profits* (New York: Harper Collins, 1992); Steven Epstein, "Democratic Science? AIDS Activism and the Contested Construction of Knowledge," *Socialist Review* 21.2 (1991): 35–61; Paula A. Treichler, "How to Have Theory in an Epidemic: The Evolution of AIDS Treatment Activism," in *Technoculture,* ed. Constance Penley and Andrew Ross (Minneapolis: University of Minnesota Press, 1991), 57–156.

68. Arno and Felden, *Against the Odds,* 137.

69. See George Carter, "ACT UP, the AIDS War, and Activism," in *Open Fire: The Open Magazine Pamphlet Series Anthology,* ed. Greg Ruggiero and Stuart Sahulka (New York: New Press, 1993), 123–50.

70. Epstein, "Democratic Science?" 37.

71. My analysis of this process draws on the following: David Noble, "Insider Trading: University Style," *Our Schools/Our Selves* 4.3 (1993): 45–52; Martin Kenney, *Biotechnology: The University Industrial Complex* (New Haven: Yale University Press, 1986); Janice Newson and Howard Buchbinder, *The University Means Business: Universities, Corporations and Academic Work* (Toronto: Garamond, 1988); Sheldon Krimsky, "The New Corporate Identity of the American University," *Alternatives* 14.2 (1987): 20–29.

72. On this point, see Noble, "Insider Trading."

73. See Tony Vellela, *New Voices: Student Activism in the 80s and 90s* (Boston: South End, 1988); Paul Loeb, *Generation at the Crossroads: Apathy and Action on the American Campus* (New Brunswick, N.J.: Rutgers University Press, 1994).

74. See Robert Ovetz, "Assailing the Ivory Tower: Student Struggles and the Entrepreneurialization of the University," *Our Generation* 24.1 (1993): 70–95.

75. James Laxer, speech at the Annual General meeting, Council of Canadians, Vancouver, 17 Oct. 1996.

76. On these developments, see Stanley Aronowitz, "The Last Good Job in America," in *Post-Work: The Wages of Cybernation,* ed. Stanley Aronowitz and Jonathan Cutler (New York: Routledge, 1998), 216 and 213, and also Cary Nelson, ed., *Will Teach for Food: Academic Labor in Crisis* (Minneapolis: University of Minnesota Press, 1998).

77. Marx, *Capital,* vol. 1, 638.

78. See Robert Gottlieb, *Forcing the Spring: The Transformation of the American Environmental Movement.* (Washington, D.C.: Island Press, 1993).

79. For an autonomist analysis of the class composition of the antinuclear movement, see p.m., "Strange Victories," in Midnight Notes Collective *Midnight Oil: Work, Energy, War, 1973–1992* (Brooklyn: Autonomedia, 1992), 193–215.

80. See Joseph Boland, "Ecological Modernisation," *Capitalism, Nature, Socialism* 52.2 (1994): 135–41, and Tim Luke, "Informationalism and Ecology," *Telos* 56 (1983): 59–73.

81. This shift follows a path that, as Harry Cleaver points out, Marx seems to have foreseen. In volume 1 of *Capital* "nature" appears as an object outside of and opposed to humans. But in its later volumes, Marx suggests that as capital increases the scope of its organization, nature is englobed by technology to a degree that its original fea-

tures become largely unidentifiable. Harry Cleaver, *Reading Capital Politically* (Brighton, Eng.: Harvester, 1979), 134.

82. Marx, *Capital*, vol. 1, 1037.

83. See Hayes, *Behind the Silicon Curtain*, and Robert Howard, *Brave New Workplace* (New York: Viking, 1985).

84. See Hofrichter, *Toxic Struggles*.

85. Ibid.

86. Ibid.

87. In addition to the example offered below, the Oil, Chemical, and Atomic Workers Union are fighting for a superfund to clean up hazardous waste sites; organizations such as the Network for Environmental and Economic Justice for the Southwest ally community and workplace fights against high-tech wastes; striking paperworkers in Jay, Maine, put control of plant effluents on their agenda; and Judith Bari's wing of Earth First has built links with forestry workers whose jobs are threatened by super-mechanized logging. For these and other instances, see Gottlieb, *Forcing the Spring*; Hofrichter, *Toxic Struggles*; and Almeida, "Network for Environmental and Economic Justice."

88. See Eric Mann, *Taking On General Motors: A Case Study of the UAW Campaign to Keep GM Van Nuys Open* (Los Angeles, Calif.: University of California at Los Angeles, Center for Labor Research, 1987).

89. See Eric Mann, "Labor-Community Coalitions as a Tactic for Labor Insurgency," in Brecher and Costello, *Building Bridges*, 113–34; idem, "Labor's Environmental Agenda in the New Corporate Climate," in Hofrichter, *Toxic Struggles*, 179–85; Mann, with the WATCHDOG Organizing Committee, *L.A.'s Lethal Air: New Strategies for Policy, Organizing and Action* (Los Angeles: Labor/Community Strategy Center, 1991); R. Bloch and R. Keil, "Planning for a Fragrant Future: Air Pollution Control, Restructuring and Popular Alternatives in Los Angeles," *Capitalism, Nature, Socialism* 2.1 (1991): 44–65.

90. Steven Gray, "Ontario's 'Green Work Alliance' Hopes Environmentally-Friendly Projects Can Reopen Plants," *Labor Notes* (Nov. 1992): 15–17; Roger Keil, "Green Work Alliances: The Political Economy of Social Ecology," *Studies in Political Economy* 44 (1994): 7–38.

91. Ken Tsuzuku, "Presentation to the 1991 Labor Notes Conference," in *A Conference on Labour and Team Concepts*, Proceedings of a Conference Co-Sponsored by Capilano College Labour Studies Programme and Vancouver and District Labour Council, Vancouver, Canada, 18–19 Oct. 1991.

92. As Tsuzuku notes, this strategy was initially a matter of financial necessity, but it led to broader perspectives: "we could not help but ask ourselves whether or not the system as ordered would really promote the interests of the workers of the client company, and if not[,] how the design concept could be improved" (ibid., 266).

93. Karl Marx, *Economic and Philosophical Manuscripts* (New York: International Publishers, 1964), 112.

94. Fredric Jameson, "Postmodernism and the Market," in *Socialist Register*, ed. R. Miliband, L. Panitch, and J. Saville (London: Merlin, 1990), 95–110.

95. The treatment of the sphere of consumption offered in this section is truncated in that it deals only with struggles surrounding capital's attempt to sell commodities

and not with its activities as a purchaser of the labor power and raw materials required for "productive consumption." This issue is, however, later picked up, in regard to raw materials, in the section on the "Reproduction of Nature," and in the next chapter, which discusses capital's global cheap-labor strategy. Although Marx distinguished the extraction of surplus value in the workplace from its realization in the market, he also noted that the faster capital circulates, the more often in a given period it can flow through the production process and be augmented by the addition of surplus value. Increasing the speed with which commodities are bought and sold can thus have the same consequence as increasing the productivity of labor: more profits. See Marx, *Grundrisse*, 539.

96. For just one of many possible examples see Marx, *Grundrisse*, 516–44.

97. See the discussion of the work of Herbert Schiller in chap. 3.

98. Ben Bagdikian, *The Media Monopoly* (Boston: Beacon, 1990). See also Noam Chomsky and Edward S. Herman, *Manufacturing Consent: The Political Economy of the Mass Media* (New York: Pantheon, 1988).

99. Karl Marx and Friedrich Engels, *The German Ideology* (London: Lawrence and Wishart, 1938), 41.

100. The classic study of this process is Stuart Ewen, *Captains of Consciousness* (New York: McGraw-Hill, 1976).

101. This section draws on Robert McChesney, "The Global Struggle for Democratic Communication," *Monthly Review* 48.3 (1996): 1–20; Nicholas Garnham, *Capitalism and Communication: Global Culture and the Economics of Information* (London: Sage, 1990); Vincent Mosco, *The Pay-Per Society: Computers and Communication in the Information Age: Essays in Critical Theory and Public Policy* (Toronto: Garamond, 1989).

102. On this point see Giuseppe Cocco and Carlo Vercellone, "Les Paradigmes sociaux du post-Fordisme," *Futur Antérieur* 4 (1990): 71–94.

103. According to Michael Dawson and John Bellamy Foster ("Virtual Capitalism: The Political Economy of the Information Society," *Monthly Review* 48.3 [1995]: 47), in 1992 U.S. business spent one trillion dollars—one-sixth of the gross domestic product—on marketing.

104. See Eileen Meehan, "Technical Capability vs. Corporate Imperatives: Towards a Political Economy of Cable Television and Information Diversity," in *The Political Economy of Information*, ed. Vincent Mosco and Janet Wasko (Madison: University of Wisconsin Press, 1988).

105. Kevin Wilson, *Technologies of Control: The New Interactive Media for the Home* (Madison: University of Wisconsin Press, 1988), 36. See also Oscar Gandy, *The Panoptic Sort: Towards a Political Economy of Information* (Boulder, Colo.: Westview, 1993), and David Lyon, *The Electronic Eye: The Rise of Surveillance Society* (Cambridge: Polity, 1994).

106. Dallas Smythe, "Communications: Blindspot of Western Marxism," *Canadian Journal of Political and Social Theory* 1.3 (1977): 6. See also his *Dependency Road: Communications, Capitalism, Consciousness and Canada* (Norwood, N.J.: Ablex, 1981). This line of thought has subsequently been developed by Sut Jhally, *The Codes of Advertising: Fetishism and the Political Economy of Meaning in the Consumer*

Society (New York: St. Martin's, 1987). In a personal conversation shortly before his death Smythe agreed that his perspective converged with the autonomist's "social factory" analysis.

107. Smythe, "Communications," 4.

108. Ibid.

109. Such "active audience" analysis has been particularly developed from "cultural studies" perspectives such as that of John Fiske, *Understanding Popular Culture* (London: Unwin Hyman, 1989).

110. For discussion of these terms see John Downing, *Radical Media: The Political Experience of Alternative Communication* (Boston: South End, 1984).

111. The seminal essay on alternative media in this period is Hans Magnus Enzensberger, "Constituents of a Theory of the Media," in his *The Consciousness Industry* (New York: Seabury Press), 95–128. For a contemporary survey see Jesse Drew, "Media Activism and Radical Democracy," in Brook and Boal, *Resisting the Virtual Life*, 71–84, and David Trend, "Rethinking Media Activism: Why the Left Is Losing the Culture War," *Socialist Review* 93.2 (1993): 5–34.

112. See Bruce Girard, ed., *A Passion for Radio* (Montreal: Black Rose, 1992); Greg Boozell, "The Revolution Will Be Microwaved: The FCC, Microwatt Radio, and Telecommunication Networks," *Afterimage* 22.1 (1994): 12–14; Wes Thomas, "Hyperwebs: Pirate Radio," *Mondo 2000* 11 (1993): 26–39.

113. See the excellent discussions in Nancy Thede and Alain Ambrosi, eds., *Video the Changing World* (Montreal: Black Rose, 1991), and Dee Dee Halleck, "Watch Out, Dick Tracy! Popular Video in the Wake of the Exxon Valdez," in Penley and Ross, *Technoculture*, 211–29.

114. See Patricia Aufderheide, "Underground Cable: A Survey of Public Access Programming," *Afterimage* 22:1 (1994): 5–7; Douglas Kahn, "Satellite Skirmishes: An Interview with Paper Tiger West's Jesse Drew," *Afterimage* 20.10 (1993): 9–11; Douglas Kellner, *Television and the Crisis of Democracy* (Boulder, Colo.: Westview, 1990); Martin Lucas and Martha Wallner, "Resistance by Satellite: The Gulf Crisis and Deep Dish Satellite TV Network," in *Channels of Resistance*, ed. Tony Dowmunt (London: British Film Institute, 1993), 176–94; Dot Tuer, "All in the Family: An Examination of Community Access Cable in Canada," *Fuse* 17.3 (1994): 23–29.

115. Rafael Roncaglio, "Notes on the Alternative," in Thede and Ambrosi, *Video the Changing World*, 207.

116. For a fascinating analysis of media politics surrounding the L.A. rebellion, on which I draw heavily here, see John Fiske, *Media Matters: Everyday Culture and Political Change* (Minneapolis: University of Minnesota Press, 1994).

117. As Clarence Lusane ("Rap, Race, and Rebellion," *Z Magazine*, Sept. 1992, 36) comments, "On the one hand, rap is the voice of alienated, frustrated, and rebellious black youth who recognize their marginality and vulnerability in post-industrial America. On the other hand, rap is the packaging and marketing of discontent by some of the best ad agencies and largest record producers in the world. It's this duality that . . . made rap and rappers an explosive issue in the 1992 elections."

118. On these points see Fiske, *Media Matters*, and also Playthell Benjamin, "Seeing Is Not Believing," *Guardian*, 9 May 1992, 34.

119. Reprinted in Madhubuti, *Why L.A. Happened*. On the role of radio in the revolt, see Ron Sakolsky, "Zoom Black Magic Liberation Radio: The Birth of the Micro-Radio Movement in the USA," in Girard, *A Passion For Radio*, 106–13.

120. Wilmette Brown, *No Justice, No Peace: The 1992 Los Angeles Rebellion from a Black/Woman's Perspective* (London: Wages for Housework Campaign, 1993), 5–8.

121. Jason Frank, "Television That Works: Labor Video in the 1990s," *Socialist Review* 93.2 (1993): 37–78.

122. Davis, "Armaggedon at the Emerald City," 46–49.

123. Steven Zeltzer estimates that there are more than forty labor TV shows on public access cable in the United States alone. See his paper "Labor Media Communication: Voices in the Global Economy," available on the World Wide Web at http://www.igc.org/lvpsf.

124. Information about the Labor Video project can be found at its World Wide Web site, http://www.igc.org/lvpsf.

125. Information on the campaign to free Mumia Abu Jamal can be found on the World Wide Web page of Refuse and Resist at http://www.walrus.com/~resist/altindex.html. *Adbusters: Journal of the Mental Environment* is published by the Media Foundation, Vancouver, Canada; email adbusters@adbusters.org.

126. See William H. Davidow, *The Virtual Corporation: Structuring and Revitalizing the Corporation for the Twenty-first Century* (New York: Harper, 1992).

127. Marx, *Capital*, vol. 1, 1056.

128. Peter Childers and Paul Delany, "Wired World, Virtual Campus: Universities and the Political Economy of Cyberspace," *Works and Days* 12.1/2 (1994): 62. See also Michael Hauben, "The Social Forces behind the Development of Usenet News," *Amateur Computerist* 5.1/2 (1993): 13–21, and Bruce Sterling, "A Short History of the Internet," *Magazine of Fantasy and Science Fiction*, Feb. 1993 [accessed 12 Nov. 1993], on-line, Internet, ACTIV-L@MIZZOU1.BITNET.

129. Andrew Ross, "Hacking Away at the Counterculture," in Penley and Ross, *Technoculture*, 107–34.

130. See Katie Hafner and John Markoff, *Cyberpunk: Outlaws and Hackers on the Computer Frontier* (New York: Simon and Schuster, 1991); Bryan Clough, *Approaching Zero: Data Crime and the Computer Underworld* (London: Faber and Faber, 1992); Bruce Sterling, *The Hacker Crackdown: Law and Disorder on the Electronic Frontier* (New York: Bantam, 1992).

131. See Howard Rheingold, *The Virtual Community* (Reading, Mass.: Addison-Wesley, 1993).

132. For examples of this sort of optimistic position see Adam Jones, "Wired World: Communications Technology, Governance and the Democratic Uprising," in *The Global Political Economy of Communication: Hegemony, Telecommunications, and the Information Economy*, ed. Edward Comor (New York: St. Martin's, 1994), 145–64.

133. On this point see Peter Golding, "World Wide Wedge: Division and Contradiction in the Global Information Infrastructure," *Monthly Review* 48.3 (1996): 78–79.

134. See, for example, Laura Miller, "Women and Children First: Gender and the Settling of the Electronic Frontier," in Brook and Boal, *Resisting the Virtual Life*, 49–59.

135. See, for example, Ken Hirschop, "Democracy and New Technologies," *Monthly Review* 48.3 (1996): 86–98. For a somewhat different line, based largely on Frankfurt

School critique of the reifying powers of technology, see Julian Stallabrass, "Empowering Technology: The Exploration of Cyberspace," *New Left Review* 211 (1995): 3–32.

136. On the possibilities and problems for women on the Internet, see Ellen Balka, "Womantalk Goes On-line: The Use of Computer Networks in the Context of Feminist Social Change" (Ph.D. diss., Simon Fraser University, Vancouver, Canada, 1991); Leslie Regan Shade, "Gender Issues in Computer Networking," paper presented at Community Networking: The International Free-Net Conference, Carleton University, Ottawa, Canada, 17–19 Aug. 1993; Hoai-An Truong, with Gail Williams, Judi Clark, Anna Couey, and others of Bay Area Women in Telecommunications, "Gender Issues in Online Communications," on-line, Internet, ACTIV-L@MIZZOU1.BITNET, 13 Jan. 1994; and Dale Spender, *Nattering on the Net: Women, Power and Cyberspace* (Melbourne: Spinifex, 1995).

137. See Eric Lee, *The Labour Movement and the Internet: The New Internationalism* (London: Pluto Press, 1997).

138. Ibid., 48.

139. See Montieth Illingworth, "Workers on the Net, Unite!: Labor Goes Online to Organize, Communicate, and Strike," *Information Week*, 22 Aug. 1994, on-line, Internet, ACTIV-L@MIZZOU1.BITNET, 3 Sept. 1994.

140. See Peter Waterman, *International Labour Communication by Computer: The Fifth International?* Working Paper Series 129 (The Hague: Institute of Social Studies, 1992).

141. For example, Solinet exploded with contending views about the appropriate response to a social democratic provincial government in Ontario that launched a major assault on public service workers.

142. Siegel, "New Chips in Old Skins."

143. For further details on the *San Francisco Free Press*, and a similar use of computer networks by striking newspaper workers in Detroit, see Lee, *The Labour Movement and the Internet*, 79–84.

144. The Hotel and Restaurant Employees International used the Internet in its campaign to organize a chain of luxury hotels known as the Western Lodging Group. According to Nathan Newman, when mass firings of workers took place at the Lafayette Park Hotel, Oakland, California, the publicizing of this news on the Internet generated hundreds of letters, calls, and emails to management. As the campaign evolved, the union targeted corporate customers of the hotel who regularly use it to house employees or visiting clients. One of these was a software corporation called PeopleSoft. The HREI highlighted negative facts from this company's own financial reports and posted them to a series of computer-oriented newsgroups. PeopleSoft claims that within a week the value of its stock dropped by sixty-three million dollars because of reactions by investors. Soon after, it announced it was moving customers and other visitors to a different hotel. See Nathan Newman, "'Third Wave Unionism' Takes to the Net," on-line, Internet, Red Rock Eater News Service, 22 Aug. 1996.

145. Jim Davis, personal email to the author, 11 June 1994.

146. Ibid.

147. On the "information highway" see Howard Besser, "From Internet to Information Superhighway," in Brook and Boal, *Resisting the Virtual Life*, 59–71; Andrew Shapiro, "Street Corners in Cyberspace," *Nation*, 3 July 1995, 10–13; Herbert J. Schiller,

"The Information Superhighway: Paving Over the Public," *Z Magazine,* Mar. 1994, 46–51; Grant Kestner, "Access Denied: Information Policy and the Limits of Liberalism," *Afterimage* 21.6 (1994): 5–10. For Canadian perpectives, see David McIntosh, "Cyborgs in Denial: Technology and Identity in the Net," *Fuse* 18.3 (1994): 14–21, and Heather Menzies, *Whose Brave New World? The Information Highway and the New Economy* (Toronto: Between the Lines, 1996). On the repression of hackers, see Sterling, *Hacker Crackdown,* and Jason Wehling, "'Netwars' and Activist Power on the Internet," on-line, Internet, ACTIV-L@MIZZOU1.BITNET [accessed 25 Mar. 1995].

148. Schiller, "The Information Superhighway," 46.

149. See the discussion of these terms in the previous chapter.

150. For the claim that cyber-struggles displace street-level activism see Mark Poster, *The Mode of Information: Poststructuralism and Social Context* (Chicago: University of Chicago Press, 1990), 154.

151. European Counter Network, UK, "INFO: European Counter Network Online," on-line, Internet, ACTIV-L@MIZZOU1.BITNET, 25 Dec. 1994.

152. Dorothy Kidd and Nick Witheford, "Counterplanning from Cyberspace and Videoland: or, Luddites on Monday and Friday, Cyberpunks the Rest of the Week," paper presented at Monopolies of Knowledge: A Conference Honoring the Work of Harold Innis, Vancouver, Canada, 12 Nov. 1994, 23.

153. See Waterman, *International Labour Communication by Computer,* and chap. 6 of this book.

154. See Edward Stewart, *The Paris Commune 1871* (London: Eyre and Spottiswoode, 1971).

155. I owe the phrase "communication commons" to Dorothy Kidd, "Talking the Walk: The Communication Commons amidst the Media Enclosures" (Ph.D. diss., Simon Fraser University, Vancouver, Canada, 1998).

6 PLANETS

1. Karl Marx, *Capital: A Critique of Political Economy,* vol. 1 (New York: Vintage Books, 1977), 929.

2. Karl Marx, *Grundrisse* (Harmondsworth, Eng.: Penguin, 1973), 524–25.

3. Karl Marx and Friedrich Engels, *The Communist Manifesto* (New York: Washington Square Press, 1969), 64.

4. Ibid.

5. Ibid., 73.

6. See Manuel Castells, *The Informational City: Information Technology, Economic Restructuring and the Urban-Regional Process* (Oxford: Blackwell, 1989); David Harvey, *The Condition of Postmodernity: An Enquiry into the Origins of Cultural Change* (Oxford: Blackwell, 1989); Joyce Kolko, *Restructuring the World Economy* (New York: Pantheon, 1988); Robert Ross and Kent Trachte, *Global Capitalism: The New Leviathan* (New York: New York University Press, 1990); Gary Teeple, *Globalization and the Decline of Social Reform* (Toronto: Garamond, 1995).

7. This approach is strongly influenced by Harry Cleaver's observation that most Marxist analyses of the expanding planetary scope of capitalism—e.g., the Hobson-Bukharin-Leninist theory of imperialism, dependency theory, and world-systems

theory—focus on the totalizing, worldwide imposition of commodity relationships to the neglect of the resistances and alternatives that challenge this process. Cleaver ("Secular Crisis in Capitalism: The Insurpassability of Class Antagonism," paper presented at the Rethinking Marxism Conference, University of Massachusetts, Amherst, 1992) says, "Because of the top-down orientation of these projects, nowhere has there been an attempt to grasp the logic of capitalist development in terms of the autonomous self-activity of the people struggling against it."

8. See Zerowork Collective, "Introduction," *Zerowork: Political Materials* 1 (1975): 1–7, and Harry Cleaver, *Reading Capital Politically* (Brighton, Eng.: Harvester, 1979).

9. Cleaver, *Reading Capital,* 43.

10. In addition, the appearance in the Second World of dissident movements experimenting in alternative forms of socialism and self-management, such as the Prague Spring, undermined the stability not only of the Stalinist regime, but of the cold war polarization on which so much capitalist control rested, and breathed new life into the European Left.

11. To focus on the role of this wave of international unrest in precipitating the economic crisis is not to discount other factors, for example, the intensified international competition resulting from the postwar recovery of European and Japanese industry. It is, however, to place such issues of intercapitalist rivalry in the context of the greater issue that faced capital as a whole, namely, the control of socialized labor. If Japanese capital constituted a challenge to North American capital, this is precisely because the former, having, with U.S. help, inflicted a significant defeat on its own working class in the immediate postwar period, was in a position to squeeze more cooperation and creativity out of "its" workers for less money, while social militancy confronted U.S. business with rigid or rising wages and social costs. See Joe Moore, *Japanese Workers and the Struggles for Power, 1945–1947* (Madison: University of Wisconsin Press, 1983).

12. On the place of the "oil shock" in restructuring see Midnight Notes Collective, *Midnight Oil.*

13. Mario Montano, "Notes on the International Crisis," *Zerowork* 1 (1975): 52–53; article reprinted in Midnight Notes, *Midnight Oil,* 115–43.

14. Ibid.

15. Ibid.

16. See Folker Fröbel, Jürgen Heinrichs, and Otto Kreye, *The New International Division of Labor: Structural Unemployment in Industrialised Countries and Industrialisation in Developing Countries,* trans. Pete Burgess (New York: Cambridge University Press, 1988).

17. Montano, "Notes on the International Crisis," 31.

18. On these points, see, in addition to the sources already cited, Arthur MacEwan, "What's 'New' About the 'New International Economy'?" *Socialist Review* 21.3/4 (1991): 111–31.

19. George Caffentzis and Silvia Federici, "Modern Land Wars and the Myth of the High-Tech Economy," in *The World Transformed: Gender, Labour and International Solidarity in the Era of Free Trade, Structural Adjustment and GATT,* ed. Cindy Duffy and Craig Benjamin (Guelph, Ontario: RhiZone, 1994), 144; See also Maria Mies, *Patriarchy and Accumulation on a World Scale: Women in the International Division*

of Labour (London: Zed Books, 1986); Maria Mies, Veronika Bennholdt-Thomsen, and Claudia von Werlhof, *Women: The Last Colony* (London: Zed Books, 1988); and various essays in Mariarosa Dalla Costa and Giovanna F. Dalla Costa, eds., *Paying the Price: Women and the Politics of International Economic Strategy* (London: Zed Books, 1995).

20. Heather Menzies (*Fast Forward and Out of Control* [Toronto: Macmillan, 1989], 96) observes that through its global information network, the giant U.S. firm Bechtel Corporation can

> take advantage of differential labor costs by employing lower salaried architects in India to draft construction plans, which become instantaneously available via satellite to supervisors in one corner of the world and project managers in another. Bechtel can use up-to-the-minute financial information to get the best financing rates from New York banks and insurance from a London company. It can then manage the construction of the project in the middle of Saudi Arabia by using Korean workers, Indian architects, American managers and European material managers. Computer communications makes it all possible.

21. See Kim Moody, "When High Wage Jobs Are Gone, Who Will Buy What We Make?" *Labor Notes* (June 1994): 8.

22. Ben Bagdikian, "Cornering Hearts and Minds: The Lords of the Global Village," *Nation*, 12 June 1989, 805–20. For discussion of the degree of U.S. domination in global media see David Morely and Kevin Robins, *Spaces of Identity: Global Media, Electronic Landscapes and Cultural Boundaries* (London: Routledge, 1995), and Benjamin Barber, *Jihad vs McWorld: How the Planet Is Both Falling Apart and Coming Together—And What This Means for Democracy* (New York: Times, 1995).

23. On this form of globalization see Pico Iyer, *Video Night in Kathmandu: And Other Reports from the Not So Far East* (New York: Knopf, 1988).

24. On these tendencies see Armand Mattelart, *Advertising International: The Privatization of Public Space* (New York: Routledge, 1991).

25. Theodore Levitt, "The Globalization of Markets," *Harvard Business Review* 6.1 (1983): 92–102.

26. Quoted in Arif Dirlik, "Post-Socialism/Flexible Production: Marxism in Contemporary Radicalism," *Polygraph* 6/7 (1993): 156–57.

27. An egregious example is the radio series by Gwnn Dyer, "Millennium," CBC Radio, Jan. 1996.

28. See James O'Connor, *The Fiscal Crisis of the State* (New York: St. Martin's, 1973). On this point see also Harry Cleaver, "The Subversion of Money-as-Command within the Current Crisis," paper presented at Conference on Money and the State, FLACSO [Facultad Latino-Americano de Ciencias Sociales], Mexico City, Mexico, 14–17 July 1992.

29. MacEwan, "What's 'New'?" 123. See also Howard Wachtel, *The Money Mandarins: The Making of a Supranational Economic Order* (New York: Pantheon, 1986).

30. Richard J. Barnet and John Cavanagh, *Global Dreams: Imperial Corporations and the New World Order* (New York: Simon and Schuster, 1994), 399.

31. Ibid.

32. Christian Marazzi, "Money in the World Crisis: The New Basis of Capitalist Power," *Zerowork* 2 (1977): 107.

33. Cleaver, "The Subversion of Money-as-Command," 11.

34. On the debt crisis see Cleaver, "Close the IMF, Abolish Debt and End Development," *Capital and Class* 39 (1990): 17–50, and Dalla Costa and Dalla Costa, *Paying the Price.*

35. As George Caffentzis observes ("Rambo on the Barbary Shore," in Midnight Notes Collective, *Midnight Oil*, 299–300), this dual military strategy has mirrored economic development: "It is premised on the Vietnam era revolt against mass military service between 1965–73, just as recent economic strategy premises the revolt of the mass factory worker in the late 1960 and early 1970s. . . . [T]he military's 'solution'—a combination of buying automated death machines and hiring out the 'dirty jobs' to low wage mercenaries abroad—is identical to the economic 'solution'—automation and computerization of domestic production and the exportation of 'dirty work' to the 'dirt wages' of the 'free trade zones' of the Philippines, Singapore, South Korea, Mexico and so on."

36. Hamid Mowlana, "Roots of War: The Long Road to Intervention," in *Triumph of the Image: The Media's War in the Persian Gulf—A Global Perspective*, ed. Hamid Mowlana, George Gerbner, and Herbert Schiller (Boulder, Colo.: Westview, 1992), 35. This collection contains many other excellent papers of the issues discussed in this section. See also Haim Bresheeth and Nira Yuval-Davis, eds., *The Gulf War and the New World Order* (London: Zed Books, 1992), and John MacArthur, *Second Front: Censorship and Propaganda in the Gulf War* (New York: Hill and Wang, 1992).

37. See Robert Reich, *The Work of Nations: Preparing Ourselves for Twenty-first Century Capitalism* (New York: Knopf, 1991).

38. Ibid., 111.

39. Ibid., 312.

40. Jeremy Brecher and Tim Costello, *Global Village or Global Pillage: Economic Reconstruction from the Bottom Up* (Boston: South End, 1994), 12.

41. On this point see Butch Lee and Red Rover, *Night-Vision: Illuminating War and Class on the Neo-Colonial Terrain* (New York: Vagabond Press), 1993.

42. On these tendencies in the computer industry see Eric Auchard, "Discount Programming: The Global Labour Market," *CPU: Working in the Computer Industry* 1 (1993): 13–24, on-line, Internet.

43. This is the route followed by Nike as its footwear plants flee across the world. See John Cavanagh and Robin Broad, "Global Reach: Workers Fight the Multinationals," *Nation*, 18 Mar. 1996, 21–24.

44. See the important analysis in Midnight Notes Collective, "The New Enclosures," in *Midnight Oil*, 317–33. On tendencies to global immiseration, Michel Chossudovsky, "The Globalisation of Poverty and the New World Economic Order," Working Paper #9114E, Department of Economics, Faculty of Social Sciences, University of Ottawa, 1991, and Riccardo Petrella, "World City States of the Future." *New Perspectives Quarterly* 8.4 (1991): 59–63.

45. See, for example, Cynthia Hamilton, "Urban Insurrection and the Global Crisis of Industrial Society," in Duffy and Benjamin, *The World Transformed*, 169–79.

46. On this point see Midnight Notes, "The New Enclosures."

47. See Robin Cohen, *The New Helots: Migrants in the International Division of Labor* (Brookfield, Vt.: Gower, 1987).

48. See Caffentzis and Federici, "Modern Land Wars," 140.

49. Mariarosa Dalla Costa, "Development and Reproduction," *Common Sense* 17 (1995): 29.

50. Romano Alquati, "The Network of Struggles in Italy," unpublished paper, Red Notes Archive, 1974.

51. Ibid.

52. Ibid.

53. Frantz Fanon, "This Is the Voice of Algeria," in *A Dying Colonialism* (New York: Monthly Review Press, 1965), 69–98.

54. Peter Golding, "World Wide Wedge: Division and Contradiction in the Global Information Infrastructure," *Monthly Review* 48.3 (1996): 82.

55. Ibid., 81.

56. Lee and Rover, *Night-Vision*, 129.

57. On the Kayapo see Ella Shohat and Robert Stam, *Unthinking Eurocentricism: Multiculturalism and the Media* (London: Routledge, 1994).

58. "Seeds of Struggle," talk presented by Brewster Kneen at Simon Fraser University, Vancouver, Canada, 4 Nov. 1996. Kneen, editor of *Ram's Horn* magazine and author of *Trading Up: How Cargill, the World's Largest Grain Company Is Changing Canadian Agriculture* (Toronto: NC Press, 1990), was invited to India to assist in the "seed *satyagraha*," whose demonstrations culminated in the destruction of Cargill's India offices. My account of the circulation of struggles drives from personal conversations with him.

59. See Bruce Girard, ed., *A Passion for Radio* (Montreal: Black Rose, 1992).

60. See Nancy Thede and Alain Ambrosi, eds., *Video the Changing World* (Montreal: Black Rose, 1991).

61. See Howard Frederick, "Electronic Democracy," *Edges* 5.1 (1992): 13–18.

62. See MacEwan, "What's 'New'?"

63. These examples are drawn from Brecher and Costello, *Global Village or Global Pillage*, and Kim Moody, *An Injury to One: The Decline of American Labour* (London: Verso, 1988). Information on the Merseyside dock strike can be found at http://www.labournet.org.uk/docks2/other/dockhome.htm.

64. Moody, *An Injury to One*, 297–301.

65. See Peter Waterman, *International Labour Communication by Computer: The Fifth International?* Working Paper Series 129 (The Hague: Institute of Social Studies, 1992). See also his "Communicating Labour Internationalism: A Review of Relevant Literature and Resources," *Communications: European Journal of Communications* 15.1/2 (1990): 85–103, and "Reconceptualising the Democratisation of International Communication," *International Social Science Journal* 123 (1990): 78–91.

66. Waterman, *International Labour Communication by Computer*, 38.

67. Ibid., 35.

68. Quoted in Waterman, "From Moscow with Electronics: A Communication Internationalism for an Information Capitalism," *Democratic Communique* 11.2/3 (1994): 11.

69. Ibid., 15.

70. Waterman, *International Labour Communication by Computer*, 67, quoting Fred Stangelaar, "An Outline of Basic Principles of Alternative Communication," paper presented at Workshop on International Communication by Computer, Institute of Social Studies, The Hague, 27 Oct. 1985.

71. The periodical *Boycott Quarterly* is devoted entirely to this strategy.

72. Cavanagh and Broad, "Global Reach," 22.

73. *The Guardian*, 22 Feb. 1996. The Web site McSpotlight can be found at http://www.mcspotlight.org.

74. The Clean Clothes site can be found at http://www.cleanclothes.org/.

75. Moreover, as Cavanagh and Broad note in "Global Reach," corporations often evade boycotts by implementing tokenist "codes of conduct" or greenwashing campaigns. Boycotts tend to be effective against products with strong brand loyalties, but less so in industries where this is not so important, or where the consumers are other companies. And there are scores of boycott attempts that fail because, in a context where they can expect little or no attention from mainstream media, they lack the resources to command public attention.

76. In the United States and Canada, the threat to incomes, social programs, and environmental conditions posed by direct exposure to the low-wage Mexican economy was obvious. In Mexico, although the government was able to muster significant support for the agreement through lavish promises of development and modernization, there was opposition from peasants and small farmers threatened by the influx of agribusiness, workers in telecommunications and other public-sector industries confronting privatization, and those who feared the generalization of "maquila" conditions. See Cindy Duffy and Craig Benjamin, "Women and the World Transformed," in Duffy and Benjamin, *The World Transformed*, 83.

77. Joseph Brenner, "Internationalist Labor Communication by Computer Network: The United States, Mexico and NAFTA," paper presented at School of International Service, American University, Washington D.C., 1994; Howard Frederick, "North American NGO Networking against NAFTA: The Use of Computer Communications in Cross-Border Coalition Building," Seventeenth International Congress of the Latin American Studies Association, Los Angeles, 24–27 Sept. 1994.

78. See also Mujer a Mujer Collective, "Communicating Electronically: Computer Networking and Feminist Organising," *RFR/DRF* 20.1/2: (1991): 10.

79. See Praful Bidwai, "Making India Work—For the Rich," *Multinational Monitor* 16.7/8 (1995): 9–13, and Michel Chossudovsky, "India under IMF Rule," *The Ecologist* 22.6 (1992): 270–74.

80. Brecher and Costello, *Global Village or Global Pillage*, 86.

81. See Salim Lakha, "Resisting Globalization: The Alternative Discourse in India," *Arena Journal* 4 (1994/95): 41–50.

82. On the Narmada dam struggle, see the film *Narmada: A Valley Rises*, dir. Ali Kazimi, Colour Canada, 1994; on the anti-GATT uprisings, see note 58 and Vandana Shiva, "Seeds of Struggle," in Duffy and Benjamin, *The World Transformed*, 57–70; on biopiracy activism see Yuli Ismartono and Teena Gill, "Asian Farmers Struggle against Transnationals," Third World Network, on-line, Internet, ACTIV-L@MIZZOU1.BITNET [accessed 17 Jan. 1996]; Martin Kohr, "Global Fight against 'Bio-

Piracy," Third World Network, on-line, Internet, ACTIV-L@MIZZOU1.BITNET [accessed 6 Nov. 1995]. The Third World Network can be contacted at twn@igc.apc.org. Also, visit the World Wide Web site of Rural Advancement Foundation International at http://www.rafi.ca/; on child labor activism see Cavanagh and Broad, "Global Reach."

83. Harry Cleaver, "The Chiapas Uprising," *Studies in Political Economy* 44 (1994): 155.

84. Ibid., 156.

85. Harry Cleaver, "Zapatistas in Cyberspace," paper presented at Simon Fraser University, Vancouver, Canada, 15 Nov. 1994. The text whose translation was coordinated via email is *Zapatistas: Document of the New Mexican Revolution* (New York: Autonomedia, 1994). A communiqué from Subcomandante Marcos of 17 Mar. 1995 (quoted by Jason Wehling, "'Netwars' and Activists Power on the Internet," on-line, Internet, ACTIV-L@MIZZOU1.BITNET [accessed 25 Mar. 1995] refers to the importance of international support for the revolt in the following terms:

> we learned that there were marches and songs and movies and other things that were not war in Chiapas, which is the part of Mexico where we live and die. And we learned that these things happened, and that "NO TO WAR!" was said in Spain and in France and in Italy and in Germany and in Russia and in England and in Japan and in Korea and in Canada and in the United States and in Argentina and in Uruguay and in Chile and in Venezuela and in Brazil and in other parts where it wasn't said but it was thought. And so we saw that there are good people in many parts of the world. . . . When they are old, then they can talk with the children and young people of their country that, "I struggled for Mexico at the end of the 20th century, and from over here I was there with them . . . and I did not know their faces but I did know their hearts and it was the same as ours."

86. John Arquilla and David Ronfeldt, "Cyberwar Is Coming!" *Comparative Strategy* 12.2 (1993): 141–65. Citations are from an electronic copy accessed at http://www.rand.org/publications/RRR/RRR.fall95.cyber/cyberwar.html. All quotations in this paragraph of my text are from this source.

87. Quoted in Joel Simon, "Netwar Could Make Mexico Ungovernable," Pacific News Service, on-line, Internet, ACTIV-L@MIZZOU1.BITNET [accessed 20 Mar. 1995]. Other quotations from Ronfeldt in this paragraph of my text are from this source.

88. See *Manufacturing Consent,* film directed by Mark Achbar and Peter Wintonick, coproduced by Necessary Illusions and the National Film Board of Canada, 1992. On networked information about the Timorese resistance movement, see Charles Scheiner, "Electronic Resources on East Timor," on-line, Internet, ACTIV-L@MIZZOU1.BITNET [accessed 25 Oct. 1996], available from timor_info@igc.apc.org.

89. See "US Labor Dispute Raises East Timor" and "Ramos Horta to Speak in Charleston," both on-line, Internet, ACTIV-L@MIZZOU1.BITNET [accessed 25 Apr. 1996]. The latter describes how Timorese resistance leader Ramos Horta and Allan Nairn, one of the journalists who reported the Dilli massacre, spoke to members of the United Paperworkers International Union on strike against Trailmobile Corporation "before a backdrop of banners demanding Indonesia end its occupation of East Timor and calling for justice for locked-out Trailmobile workers."

90. "Trailer Workers Meet Timor Resistance Leader," on-line, Internet, ACTIV-L@MIZZOU1.BITNET [accessed 15 Aug. 1996].

91. Joshua Hammer, "Nigeria Crude," *Harper's Magazine*, June 1996, 58–71.

92. See, for example, the film *Delta Force*, CARNA Films, 1995.

93. Raghu Krishnan, "December 1995: The First Revolt against Globalization," *Monthly Review* 48.1 (1996): 1–22.

94. "Net Strike—Grève en Réseau," on-line, Internet, counter@francenet.fr, 19 Dec. 1995 (my trans.). Valuable archives of computer messages at the time of the strikes can be found at http://www.ainfos.ca/A-Infos95/france.html.

95. One dissenter argued that the Internet was not an appropriate site for such sabotage, that the "Net Strike" would simply exasperate webmasters of the particular sites, but not be understood as a form of social protest. Moreover, if generalized, this type of sabotage would set in motion a destructive logic: "anti-fascists paralyze the sites of fascists, fascists paralyze the sites of anti-fascists, Christians paralyze the sites of Muslims, Muslims paralyze the sites of Hindus, Hindus paralyze the sites of Christians, Christians paralyze the sites of gays and lesbians, Macintoshiennes paralyze sites dedicated to Windows." This would be damaging because "the Net is the best tool of counter power (because it is the best means of diffusion of information and knowledge is power) which we have been given. It is our greatest treasure, for us militants, too—don't count on me to engage in an action that would damage it." Another said, "To see the net as a means of direct action seems to me an error for two reasons: one is its inefficiency, the other is that it can very well be turned against sites and means of diffusion such as this one here." On Boxing Day 1995, Strano issued an email bulletin, "Echoes of the Net Strike," presenting a preliminary assessment of the action. Strano observed that while it was not possible to determine the precise number of participants, the count of hits on French government sites suggested that there had been "several thousand of strikers." While there had been rapid access to the sites before the strike hour, it had declined rapidly thereafter, so that within fifteen minutes several sites—that of the Ministry of Education first—were jammed. Strano Network also reported many on-line messages of support for the action—including one from U.S. cyberpunk celebrity Bruce Sterling—and several news items about the "Net Strike" in the press and radio. The organizer of the "Net Strike" concluded that the action showed a "widespread desire to take actions against the French government," "the potential effectiveness of the on-line strike as an instrument of action, and the willingness of cybernauts to use it, and "the extreme speed and spontaneity of organisation of such a movement."

96. See Samir Amin, *Accumulation on a World Scale: A Critique of the Theory of Underdevelopment* (New York: Monthly Review Press, 1974).

97. Brecher and Costello, *Global Village or Global Pillage*, 96–97. For a collection of articles—very mixed in both perspective and quality—that shares this general orientation see Jeremy Brecher, John Brown Childs, and Jill Cutler, eds., *Global Visions: Beyond the New World Order* (Boston: South End, 1993).

98. Brecher and Costello, *Global Village or Global Pillage*, 96–97.

99. Ibid., 174.

100. Waterman, *International Labour Communication by Computer*, 47. See also Dirlik, "Post-Socialism."

7   POSTMODERNISTS

1. See, for example, Jean François Lyotard's acknowledgement of his debt to American sociology, in the opening pages of *The Postmodern Condition: A Report on Knowledge* (Minneapolis: University of Minnesota Press, 1984). Early uses of the term "postmodern" from this quarter can be found in Peter Drucker, *Landmarks of Tomorrow* (New York: Harper and Row, 1957), and Amitai Etzioni, *The Active Society: A Theory of Societal and Political Processes* (New York: Free Press, 1968).

2. For useful studies of this relationship, see Krishan Kumar, *From Post-Industrial to Post-Modern Society: New Theories of the Contemporary World* (Oxford: Blackwell, 1995); David Lyon, *Postmodernity* (Minneapolis: University of Minnesota Press, 1994); Margaret Rose, *The Post-Modern and the Post-Industrial: A Critical Analysis* (Cambridge: Cambridge University Press, 1991); Frank Webster, *Theories of the Information Society* (London: Routledge, 1995). The term "postmodern" is of course a bewildering category: distinctions can be made between *postmodernism*, an artistic movement, *poststructuralism*, a philosophic (or antiphilosophic) tendency, and concepts of *postmodernity* as a particular social formation. I use the term "postmodern theory" to designate the work of those thinkers who believe that a distinctively postmodern moment can be recognized in any or all of these fields, particularly the last.

3. For an interesting account of this context, see Mark Poster, *Existential Marxism in Postwar France: From Sartre to Althusser* (Princeton: Princeton University Press, 1975).

4. For a very economical statement of this position, see Gianni Vattimo, *The Transparent Society* (Baltimore: Johns Hopkins University Press, 1992), 1–11. On the relation between poststructuralism and informatics, see Mark Poster, *The Mode of Information: Poststructuralism and Social Context* (Chicago: University of Chicago Press, 1990).

5. Lyotard, *The Postmodern Condition*, xxxiv.

6. Jean Baudrillard, "Interview: Game with Vestiges," *On the Beach* 6 (1984): 19–25, quoted in Steven Best and Douglas Kellner, *Postmodern Theory: Critical Interrogations* (London: Macmillan, 1992).

7. Jean François Lyotard, *Political Writings* (Minneapolis: University of Minnesota Press, 1993), 115.

8. See Mark Poster, *Foucault, Marxism and History: Mode of Production versus Mode of Information* (Cambridge: Polity, 1984).

9. Marshall Berman, *All That Is Solid Melts into Air: The Experience of Modernity* (New York: Simon and Schuster, 1982), 33, 348.

10. Perry Anderson, *In the Tracks of Historical Materialism* (London: Verso, 1983), 44–45.

11. Stephen Eric Bronner, *Socialism Unbound* (New York: Routledge, 1990), 171.

12. Andrew Britton, "The Myth of Postmodernism: The Bourgeois Intelligentsia in the Age of Reagan," *Cineaction* 13/14 (1988): 17.

13. Alex Callincos, "Postmodernism, Post-Structuralism, Post-Marxism?" *Theory, Culture and Society* 2.3 (1985): 85–101.

14. Todd Gitlin, "Images Wild," *Tikkun* 4.4 (1989): 112.

15. Raymond Williams, *Marxism and Literature* (Oxford: Oxford University Press, 1977), 128.

16. Fredric Jameson, "Foreword" to Lyotard, *The Postmodern Condition*, xii.

17. Fredric Jameson, "Postmodernism: or the Cultural Logic of Late Capitalism," *New Left Review* 146 (1984): 55–92.

18. For analysis of Mandel's work, see chap. 3.

19. For the debate around Jameson's work see the essays in *Postmodernism/Jameson/Critique*, ed. Douglas Kellner (Washington D.C.: Maisonneuve Press, 1989).

20. See David Harvey, *The Condition of Postmodernity: An Enquiry into the Origins of Cultural Change* (Oxford: Blackwell, 1989).

21. Ibid., 302.

22. Ibid.

23. Ibid., 112.

24. See Jacques Derrida, *Specters of Marx: The State of the Debt, the Work of Mourning, and the New International* (London: Routledge, 1994).

25. Ibid., 168. See also Adrian Wilding, review of *Specters of Marx*, by Jacques Derrida, *Common Sense* 17 (1995): 92–95.

26. At different points in his *oeuvre* Jameson oscillates between suggesting an eventual return to more "normal" conditions of class struggle, development of vaguely described "cognitive mapping" practices (1984; 1988) and "homeopathic" adoptions of postmodernism (1987). See his "Periodizing the 60's," *The Ideologies of Theory: Essays, 1971–1986*, vol. 2 (Minneapolis: University of Minnesota Press, 1988), 178–210; "Cognitive Mapping," in *Marxism and the Interpretation of Culture*, ed. Cary Nelson and Lawrence Grossberg (Urbana: University of Illinois Press, 1988), 347–58; "Reading without Interpretation: Post-Modernism and the Video-Text," in *The Linguistics of Writing: Arguments Between Language and Literature*, ed. Nigel Fabb, Derk Attridge, Alan Durant, and Colin McCabe (Manchester, Eng.: Manchester University Press, 1987), 199–224.

27. Harvey, *The Condition of Postmodernity*, 353–56.

28. Antonio Negri, *Marx beyond Marx: Lessons on the Grundrisse* (New York: Bergin and Garvey, 1984), xvi.

29. Antonio Negri, "Interpretation of the Class Situation Today: Methodological Aspects," in *Open Marxism*, vol. 2, *Theory and Practice*, ed. Werner Bonefeld, Richard Gunn, and Kosmas Psychopedis (London: Pluto, 1992), 85.

30. Negri, *The Politics of Subversion: A Manifesto for the Twenty-first Century* (Cambridge: Polity, 1989), 87.

31. Karl Marx, *Capital: A Critique of Political Economy*, vol. 1 (New York: Vintage Books, 1977), 1054. Marx wrote, "When the worker co-operates in a planned way with others, he strips off the fetters of his individuality, and develops the capabilities of his species" (447). For his extended reflections on the topic of cooperation see *Capital*, vol. 1, chap. 13, "Co-Operation," 439–54.

32. Negri, *The Politics of Subversion*, 52.

33. Ibid., 116.

34. Marx writes:

It was an immense step forward for Adam Smith to throw out every limiting specification of wealth creating activity—not only manufacturing, or commercial, or agricultural labor, but one as well as the others, labor in general. With the abstract generality of wealth-creating activity we now have the universality of the object defined as wealth, the product as such or again labor as such, but labor as past objectified labor. How difficult and great this transition was may be seen from how Adam Smith himself from time to time still falls back into the Physiocratic system. Now it might seem that all that had been achieved thereby was to discover the abstract expression for the simplest and most ancient relation in which human beings—in whatever form of society—play the role of producers. This is correct in one respect. Nor in another. . . . Indifference towards specific labors correspond to a society in which individuals can with ease transfer from one labor to another, and where the specific kind is a matter of chance for them, hence of indifference. Not only the category "labor," but labor in reality has here become the means of creating wealth in general, and has ceased to be organically linked with individuals in any specific form. Such a state of affairs is at its most developed in the modern form of existence of bourgeois society—in the United States. Here, then, for the first time, the point of departure of the category "labor," "labor as such," labor pure and simple, becomes true in practice. . . . Thus the simplest abstraction which modern economics places at the head of its discussions and which expresses an immeasurably ancient relation valid in all forms of society, nevertheless achieves practical truth as an abstraction only as a category of the most modern society.

Marx, *Grundrisse* (Harmondsworth, Eng.: Penguin, 1973), 104–5.

35. Negri, *The Politics of Subversion*, 201.

36. Ibid., 200.

37. Ibid., 202.

38. Ibid., 200.

39. Ibid., 203.

40. Ibid., 206.

41. Ibid., 203.

42. Ibid., 202.

43. See Jürgen Habermas, *The Theory of Communicative Action*, vol. 1, *Reason and the Rationalization of Society* (Boston: Beacon, 1984), and vol. 2, *A Critique of Functionalist Reason* (Boston: Beacon, 1987). Negri writes that "in the productive community of advanced capitalism we find ourselves confronted by a primary phenomenon which, following Habermas, we will call 'communicative action.' It is on the basis of the interaction of communicative acts that the horizon of reality comes to be constituted. . . . Above all, communicative action gives rise to the extraordinary possibility of activating dead socialised labour. Communication is the Direct Current of these relationships." *The Politics of Subversion*, 117.

44. Habermas in an interview with Peter Dews upholds this segregation as follows:

Marxists . . . have to ask themselves whether socialism today, under present conditions, can still really mean a *total* democratic restructuration from top to bottom, and visa versa, of the economic system: that is a transformation of the capitalist economy according to models of self-management and council-based administration. I myself do not believe so. . . . I wonder—this is an empirical question which cannot be answered abstractly, but only through experimental practice—if we should not preserve part of the today's complexity within the economic system, limiting the discursive formation of the collective will precisely to the decisive and central structures of political power: that is, apart from the labor process as such. . . . We must start from the fact that social systems as complex as highly developed capitalist societies would founder in chaos under any attempt to transform their fundamental structures overnight. . . . Such a path would . . . accomplish a prudent and long-term process of transformation. The task is a very difficult one, for which an extraordinarily intelligent party is necessary.

Quoted in Peter Dews, ed., *Autonomy and Solidarity: Interview with Jürgen Habermas* (London: Verso, 1985). Critics sympathetic to autonomist Marxism, such as Michael Ryan, have strongly attacked this scaling-down of leftist ambitions for thoroughgoing social transformation as "managerial social democracy"—see Ryan's "The Joker's Not Wild: Critical Theory and Social Policing," in his *Politics and Culture: Working Hypotheses for a Post-Revolutionary Society* (London: Macmillan, 1989), 27–45.

45. For a valuable account of Baudrillard's trajectory see Douglas Kellner, *Jean Baudrillard: From Marxism to Postmodernism and Beyond* (Stanford, Calif.: Stanford University Press, 1989).

46. Jean Baudrillard, *Simulations* (New York: Semiotext(e), 1983), 3.

47. Baudrillard, "The Implosion of Meaning in the Media and the Implosion of the Social in the Masses," in *The Myths of Information: Technology and Post-industrial Culture,* ed. Kathleen Woodward (Madison, Wis.: Coda, 1980), 137–50. At some points in his writings Baudrillard suggests that this indifference, the extreme inertia of the "silent majorities," might constitute the only possible resistance to a regime that incessantly solicits the participation of the subjectivities it has itself created. But in his later works, even this possibility evaporates, in increasingly "fatal" scenarios. All natural functions have become artificial, the senses are technologized, automation has liquidated labor, and images supplanted things. Wired at all points to Walkmans, cellular phones, and media, we float through a vertiginous world devoid of truth or foundation, which increasingly tends toward the dimensions of a technologically created hallucination, a virtual reality in which it becomes impossible to tell the difference between actual and imaginary, someone and something.

48. See Jean Baudrillard, *The Gulf War Did Not Take Place* (Bloomington: Indiana University Press, 1995).

49. Ibid.

50. Ibid.

51. Ibid.

52. Ibid.

53. Ibid.

54. Negri, *The Politics of Subversion,* 203.

55. Indeed, Baudrillard's idea that in such a condition oppositional impulses can express themselves only negatively, in terms of extreme passivity, indifference, and nonparticipation, was also mooted by the autonomist Mario Tronti in his discussion of labor during periods of defeat—see his *Ouvriers et capital* (Paris: Christian Bourgeois, 1977).

56. Michael Hardt and Antonio Negri, *Labor of Dionysus: A Critique of the State-Form* (Minneapolis: University of Minnesota Press, 1994), 268.

57. Ibid., 271.

58. In regard to Baudrillard, Negri's critique of the French *nouveaux philosophes* is relevant. He observes that while their vision of totalized capitalist power displays "hatred for the despotic powers that dead labour tries increasingly to exercise over living labour," the problem with their position is that "this pessimism aborts into a philosophy which simply reflects the destructured power of capital, inasmuch as it uses the categories within an absoluteness which is neither dialectical nor revolutionary. It is not dialectical because it looks at power in unqualified terms, 'without adjectives'; it is not revolutionary because, consequently, it cannot develop a logic of separation." "Domination and Sabotage," in *Working Class Autonomy and the Crisis*, ed. Red Notes Collective (London: Red Notes, 1979), 166.

59. Antonio Negri, *Revolution Retrieved: Selected Writings on Marx, Keynes, Capitalist Crisis and New Social Subjects* (London: Red Notes, 1988), 192.

60. See Christopher Norris, *Uncritical Theory: Postmodernism, Intellectuals and the Gulf War* (London: Lawrence and Wishart, 1992).

61. For accounts of this activity see Douglas Kahn, "Satellite Skirmishes: An Interview with Paper Tiger West's Jesse Drew," *Afterimage* 20.10 (1993): 9–11, and Martin Lucas and Martha Wallner, "Resistance by Satellite: The Gulf Crisis and Deep Dish Satellite TV Network," in *Channels of Resistance*, ed. Tony Dowmunt (London: British Film Institute, 1993), 176–94.

62. See Robert Hackett, *Engulfed: Peace Protest and America's Press during the Gulf War* (New York: New York University, Center for War and Peace and the New Media, 1993).

63. See Donna Haraway, "A Manifesto for Cyborgs: Science, Technology, and Socialist Feminism in the 1980's," *Socialist Review* 80 (1985): 65–107.

64. Ibid., 65, 66.

65. Ibid., 68.

66. Ibid.

67. Ibid., 91.

68. Ibid.

69. Ibid., 101.

70. Negri, *Politics of Subversion*, 93.

71. Ibid., 85–86.

72. Hardt and Negri, *Labor of Dionysus*, 10.

73. Andrew Ross, "Hacking Away at the Counterculture," in *Technoculture*, ed. Constance Penley and Andrew Ross (Minneapolis: University of Minnesota Press, 1991), 107–34. I supplement Ross's account with information from the Amateur Computerist, on-line, http://www.columbia.edu/~hauben/acn.index.html.

74. See, for example, Arthur Kroker and Michael Weinstein, *Data Trash: The Theory of the Virtual Class* (Montreal: New World Perspectives, 1994).

75. See, for example, the sardonic dismissal of these thinkers by Harvey, *The Condition of Postmodernity*, 352, or the more measured critique in Best and Kellner, *Postmodern Theory*. For Deleuze's affirmation of Marxism see his *Negotiations* (New: York: Columbia University Press, 1995), and for Guattari's, his "Institutional Practices and Politics," in *The Guattari Reader*, ed. Gary Genosko (Oxford: Blackwell, 1996), 123.

76. Guattari, "The Postmodern Impasse," Genosko, *Guattari Reader,* 110. See also Guattari's "Postmodernism and Ethical Abdication," in the same collection, 114–17.

77. I am particularly indebted to Kenneth Surin, "Reinventing a Physiology of Collective Liberation: Going 'Beyond Marx' in the Marxism(s) of Toni Negri, Felix Guattari, and Gilles Deleuze," paper presented at the Rethinking Marxism Conference, University of Massachusetts, Amherst, 1992; Michael Hardt, "The Art of Organization: Foundations of a Political Ontology in Gilles Deleuze and Antonio Negri" (Ph.D. diss., University of Washington, 1990), and his chapter on Deleuze and Guattari in Best and Kellner, *Postmodern Theory,* 76–110.

78. Gilles Deleuze and Felix Guattari, *Anti-Oedipus: Capitalism and Schizophrenia* (New York: Viking, 1983), 116. On "molecular" and "molar" formations, see 183.

79. Ibid., 139.

80. On "territorialization" and "deterritorialization," ibid., 222–40.

81. On "nomadism" see Deleuze and Guattari, *A Thousand Plateaus: Capitalism and Schizophrenia* (London: Athlone, 1987), 380–85.

82. Ibid., 204.

83. Deleuze and Guattari, *Anti-Oedipus,* 222–40.

84. Deleuze and Guattari, *A Thousand Plateaus,* 454–73.

85. Ibid., 465.

86. Ibid., 454.

87. Felix Guattari, *Molecular Revolution: Psychiatry and Politics* (Harmondsworth, Eng.: Penguin, 1984), 260.

88. Ibid., 263.

89. Ibid. For Deleuze and Guattari "minorities" are defined not by numbers but by distinction from the concept of majority as "the national worker, qualified, male and over thirty-five." Minority struggles are characterized by connectability. In all the struggles, around votes, abortions, jobs, Third World, "there is also always a sign to indicate that these struggles are the index of another, coexistent combat." *A Thousand Plateaus,* 471.

90. Guattari, *Molecular Revolution,* 263.

91. Ibid.

92. Deleuze and Guattari, *A Thousand Plateaus,* 351–423.

93. Ibid., 3–25.

94. Ibid., 6–7.

95. Guattari, *Molecular Revolution,* 110.

96. Sergio Bologna, "The Tribe of Moles," 51, and Sylvere Lotringer and Christian Marazzi, "The Return of Politics," 8, both in *Italy: Autonomia—Post-Political Politics,* ed. Sylvere Lotringer and Christian Marazzi (New York: Semiotext(e), 1980).

97. Deleuze and Guattari, *A Thousand Plateaus,* 473.

98. Ibid.

99. Felix Guattari, "The Three Ecologies," *New Formations* 8 (1989): 146.

100. Ibid.

101. Ibid.

102. Harry Cleaver, "The Chiapas Uprising," *Studies in Political Economy* 44 (1994): 141–57.

103. My analysis in this section draws on the following: Conor Foley, "Virtual Protest," *New Statesman and Society*, 18 Nov. 1994, 47–49; Neil Goodwin and Julia Guest, "By-Pass Operation," *New Statesman and Society*, 19 Jan. 1996, 14–15; Tim Maylon, "Killing the Bill," *New Statesman and Society*, 8 July 1994, 12–13; Camilla Berens, "Folk Law," *New Statesman and Society*, 5 May 1995, 34–36; Aufheben Collective, "Auto-Struggles: The Developing War against the Road Monster," *Aufheben* 3 (1994): 3–23; and on my own research among activists in Britain.

104. Foley, "Virtual Protest," 47.

105. Deleuze and Guattari, *Anti-Oedipus*, 105.

106. Deleuze and Guattari, *A Thousand Plateaus*, 224.

107. Deleuze and Guattari, *Anti-Oedipus*, 258.

108. Richard Butler, quoted in Hannah Nordhaus, "Underground by Modem," *Terminal City*, 11 Aug. 1993, 7.

109. On these developments see Crawford Killian, "Nazis on the Net," *Georgia Straight*, 11–18 Apr. 1996, 13–17.

110. Killian, "Nazis on the Net," 15.

111. Deleuze and Guattari, *A Thousand Plateaus*, 215.

112. Ibid., 214.

113. See Best and Kellner, *Postmodern Theory*.

114. Negri, *Politics of Subversion*, 145–46.

115. Felix Guattari and Toni Negri, *Communists Like Us* (New York: Autonomedia, 1990).

116. Ibid., 7, 8.

117. Ibid., 128.

118. Ibid., 103.

119. Ibid., 123.

120. Ibid., 16.

121. Ibid., 146.

122. Ibid., 17.

123. Ibid., 13.

124. Ibid., 42.

125. See Ernesto Laclau and Chantal Mouffe, *Hegemony and Socialist Strategy: Towards a Radical Democratic Politics* (London: Verso, 1985). In making this contrast between Negri and Guattari and Laclau and Mouffe, I am considerably indebted to Richard Hutchinson, "Machines of Desire: Class, Identity and the Potential of the New Social Movements," paper presented at New Directions in Critical Theory Conference, University of Arizona, Tucson, 17 Apr. 1993.

126. Guattari and Negri, *Communists Like Us*, 125.

127. This general deficit is particularly marked by contrast with one brief, surprising, and indeed anomalous section in the final chapter of Laclau and Mouffe's *Hegemony and Socialist Strategy* (161–75). For in these pages, the authors offer an all-too-cursory outline of the historical and material conditions that, they say, provide the grounds for the emergence of post-Marxist theory and radical democratic politics. It is—strange to say—an analysis of changes within the mode of production. Since the end of the Second World War, Laclau and Mouffe argue, there has been in the advanced industrial nations a significant shift in the organization of capitalism, involving trans-

formations in the labor process, state structure, and popular culture. Commodification and bureaucratization have reached into previously untouched areas of social life: at the same time, there has been a growth in the complexity and density of civil society. Taken together, these tendencies have resulted in a multiplicity of new—and not necessarily class based—points of social antagonism. It is these conditions that make necessary and possible the emergence of new forms of social struggle—and of theories such as those of Laclau and Mouffe, which attempt to account for these fresh forms of praxis. Many readers of the book have pointed out that this section is inconsistent with the main thrust of *Hegemony and Socialist Strategy*. For while the bulk of the book is devoted to repudiating Marxism's insistence on the correspondence between economics and politics, its conclusion suddenly explains the necessity of a new politics by shifts in the pattern of capital accumulation. On this point see Michael Rustin, "Absolute Voluntarism: Critique of a Post-Marxist Concept of Hegemony," *New German Critique* 43 (1988): 146–71; A. Belden Fields, "In Defense of Political Economy and Systemic Analysis: A Critique of Prevailing Theoretical Approaches to the New Social Movements," in Grossberg and Nelson, *Marxism and the Interpretation of Culture*, 141–56; and Michelle Barratt, *The Politics of Truth: From Marx to Foucault* (Cambridge: Polity, 1991).

128. See chap. 3 for discussion of the British "New Times" initiative, which was heavily influenced by Laclau and Mouffe's post-Marxism.

129. Harry Cleaver, "Secular Crisis in Capitalism: The Insurpassability of Class Antagonism," paper presented at the Rethinking Marxism Conference, University of Massachusetts, Amherst, 1992.

130. Ibid.

131. Antonio Negri, *Marx beyond Marx: Lessons on the Grundrisse* (New York: Bergin and Garvey, 1984), 150.

132. See, for example, Harry Cleaver, "Socialism," in *The Development Dictionary: A Guide to Knowledge as Power*, ed. Wolfgang Sachs (London: Zed Books, 1992), 233–49.

8   ALTERNATIVES

1. Frederick Engels, *Socialism: Utopian and Scientific* (Peking: Foreign Languages Press, 1975), 88.

2. Karl Marx and Friedrich Engels, *The German Ideology* (London: Lawrence and Wishart, 1938), 26. There is of course an enormous literature on the relation of Marxism to utopian thought: two of the sources we have found most stimulating are Krishan Kumar, *Utopia and Anti-Utopia in Modern Times* (Oxford: Blackwell, 1987), 48–65, and E. P. Thompson, *William Morris: Romantic to Revolutionary* (London: Merlin, 1977).

3. Pierre Bordieu, *Liberation*, 14 Dec. 1995, quoted in Massimo De Angelis, "The Autonomy of the Economy and Globalisation," *Vis-A-Vis*, Winter 1996, on-line, available from http://www.lists.village.virginia.edu/spoons/aut_html/glob.html.

4. De Angelis, "The Autonomy of the Economy."

5. Ibid.

6. Ibid.

7. Massimo De Angelis, email, "Discussing neo-liberalism and utopia," on-line, to aut-op-sy discussion group, aut-op-sy@lists.village.virginia.edu [accessed 4 Apr. 1996].

8. Ibid.

9. The most accessible English-language discussion of "self-valorization" is Harry Cleaver, "The Inversion of Class Perspective in Marxian Theory: From Valorisation to Self-Valorisation," in *Open Marxism*, vol. 2, *Theory and Practice*, ed. Werner Bonefeld, Richard Gunn, and Kosmas Psychopedis (London: Pluto, 1992), 106–44. See also "An Interview with Harry Cleaver," *Vis-A-Vis* 1 (1993), on-line, available from http://www.eco.utexas.edu/Homepages/faculty/Cleaver/index2.html.

10. Among works that have influenced my account are Tessa Morris-Suzuki's sketch of "information democracy" in *Beyond Computopia: Information, Automation and Democracy in Japan* (London: Kegan Paul International, 1988); Rudolf Bahro, *The Alternative in Eastern Europe* (London: New Left, 1978); and Raymond Williams, *Towards 2000* (London: Chatto and Windus, 1983). Two other valuable sources have recently come to my attention: a long essay by Brian Milani, "Beyond Globalisation: The Struggle to Redefine Wealth," which is available on the World Wide Web at http://www.web.net/bmilani/MAI.htm, and Thad Williamson's survey of contemporary anticapitalist visions, *What Comes Next? Proposals for a Different Society* (Washington, D.C.: National Center for Economic Security Alternatives, 1998).

11. Karl Marx, *Grundrisse* (Harmondsworth, Eng.: Penguin, 1973), 699–700.

12. For a collection of essays from diverse perspectives but within this broad orientation, see Wolfgang Sachs, ed., *The Development Dictionary: A Guide to Knowledge as Power* (London: Zed Books, 1992).

13. For discussion of this type of problem see Maurizio Viano and Vincenzo Binnetti, "What Is to Be Done? Marxism and the Academy," in *Marxism beyond Marxism*, ed. Saree Makdisi, Cesare Casarino, and Rebecca E. Karl (London: Routledge, 1996), 243–54.

14. See, for example, Harry Cleaver, "Socialism," in Sachs, *The Development Dictionary*, 233–49.

15. See Cornelius Castoriadis, *Political and Social Writings. Volume 3, 1961–1979: Recommencing the Revolution: From Socialism to the Autonomous Society* (Minneapolis: University of Minnesota Press, 1993).

16. See Christopher Hill, *The World Turned Upside Down: Radical Ideas during the English Revolution* (New York: Viking, 1973).

17. Karl Marx, *Capital: A Critique of Political Economy*, vol. 3 (New York: Vintage Books, 1981), 958.

18. See Norbert Weiner, "A Letter to Walther Reuther, UAW President," 13 Aug. 1949, reprinted in David Noble, *Progress without People: New Technology, Unemployment, and the Message of Resistance* (Toronto: Between the Lines, 1995), 161–63.

19. David Noble, "The Truth About the Information Highway," *CPU: Working in the Computer Industry* 13, on-line, Internet, ACTIV-L@MIZZOU1.BITNET, 15 Feb. 1995.

20. On this point see Jim Davis and Michael Stack, "The Digital Advantage," in *Cutting Edge: Technology, Information, Capitalism and Social Revolution*, ed. Jim Davis, Thomas Hirschl, and Michael Stack (London: Verso, 1997), 121–44.

21. See, for example, Ethan B. Kaplan, "Workers and the World Economy," *Foreign*

Affairs 75.3 (1996): 16–63, and Barrie Sherman and Phil Judkins, *Licensed to Work* (London: Cassell, 1995).

22. Stanley Aronowitz and William DiFazio, *The Jobless Future: Sci-Tech and the Dogma of Work* (Minneapolis: University of Minnesota Press, 1994); Jeremy Rifkin, *The End of Work: The Decline of the Global Labor Force and the Dawn of the Post-Market Era* (New York: Putnam, 1995); Barrie Sherman and Phil Judkins, *Licensed to Work* (London: Cassell, 1995); Stanley Aronowitz and Jonathan Cutler, eds., *Post-Work: The Wages of Cybernation* (New York: Routledge, 1998). For an important collection of essays about the crisis of automation, see Davis, Hirschl, and Stack, *Cutting Edge.*

23. Paolo Virno, "Notes on the General Intellect," in Makdisi, Casarino, and Karl, *Marxism beyond Marxism,* 267.

24. Ibid.

25. Karl Marx, *Capital,* vol. 1 (New York: Vintage Books, 1977), 781–802. For further discussion on the issue of technological unemployment, see chap. 8 of the present book.

26. Adam Przeworski, "Less Is More: In France the Future of Unemployment Lies in Leisure," *Dollars and Sense,* July/Aug. 1995, 12, 15.

27. See Juliet Schor, *The Overworked American: The Unexpected Decline of Leisure* (New York: Harper, 1991), and Benjamin Hunnicutt, *Work without End: Abandoning Shorter Hours for the Right to Work* (Philadelphia: Temple University Press, 1988).

28. For example, a shortened work week was proposed by the Labor Notes group—see the pamphlet by Kim Moody and Simone Sagovac, *Time Out: The Case for a Shorter Work Week* (Detroit: Labor Notes, 1995).

29. Franco Berardi, *Le Ciel est enfin tombé sur la terre* (Paris: Seuil, 1978), 27 (my trans.).

30. For exposition and discussion of these arguments, see the collection *Arguing for Basic Income: Ethical Foundations for a Radical Reform,* ed. Philippe Van Parijs (London: Verso, 1992).

31. Steve Wright, "Confronting the Crisis of Fordism: Italian Debates around Guaranteed Income," unpublished ms., 1995.

32. Zerowork Collective, "Introduction," *Zerowork: Political Materials* 1 (1975): 3.

33. Gorz, as editor of *Les Tempes Moderne,* ran a special issue on the Italian New left, and approvingly cites Negri in several of his works.

34. See Andre Gorz, *Paths to Paradise: On the Liberation from Work* (London: Pluto, 1985). For other statements of Gorz's position, see his *Critique of Economic Reason* (London: Verso, 1989) and *Capitalism, Socialism, Ecology* (London: Verso, 1994).

35. Gorz, *Paths to Paradise,* 41.

36. Ibid.

37. Ibid.

38. See Andre Gorz, *Farewell to the Working Class* (London: Pluto, 1982).

39. See Wright, "Confronting the Crisis of Fordism." For a scathing critique of Gorz by some North American autonomists, see Midnight Notes Collective, "The Working Class Waves Bye-Bye," *Midnight Notes* 7 (1984): 12–18; David Byrne, "Just Haad on a Minute There: A Rejection of Andre Gorz's 'Farewell to the Working Class,'" *Capital and Class* 24 (1985): 74–98, and R. Hyman, "Andre Gorz and His Disappear-

ing Proletariat," in *Socialist Register*, ed. Ralph Miliband and John Saville (London: Merlin, 1983), 272–95.

40. Milton Friedman, "The Case for the Negative Income Tax: A View from the Right," in *Issues in American Public Policy*, ed. J. H. Bunzel (Englewood Cliffs, N.J.: Prentice Hall, 1968), 111–20.

41. Jean Swanson, "GAI: Guaranteed Disaster," *Canadian Dimension*, Dec./Jan. 1994/95, 24.

42. De Angelis, "'Crisi dell'occupazione,' strategie del capitale e ipotesi strategiche per l'autonomia," *Vis-A-Vis* 2 (1994), quoted in Wright, "Confronting the Crisis of Fordism."

43. Ibid.

44. See Swanson, "GAI: Guaranteed Disaster," and her debate with Eric Shragge in the same issue of *Canadian Dimension*.

45. Wright, "Confronting the Crisis of Fordism."

46. In addition to works already cited see Philippe Van Parijs, *Marxism Recycled* (Cambridge: Cambridge University Press, 1993) and *Real Freedom for All* (Oxford: Clarendon, 1995); Diane Elson, "Market Socialism or Socializing the Market?" *New Left Review* 172 (1987): 3–44; David Purdy, "Citizenship, Basic Income and the State," *New Left Review* 208 (1994): 30–48; Sally Lerner, "How Will North America Work in the Twenty-first Century?" in Davis, Hirschl, and Stack, *Cutting Edge*, 177–94.

47. Mariarosa Dalla Costa and Selma James, *The Power of Women and the Subversion of the Community* (Bristol, Eng.: Falling Wall Press, 1972).

48. See Marilyn Waring, *If Women Counted: A New Feminist Economics* (San Francisco: Harper and Row, 1988).

49. See Mike Cooley, *Architect or Bee? The Human Price of Technology* (London: Hogarth, 1987).

50. See Shosana Zuboff, *In the Age of the Smart Machine: The Future of Work and Power* (New York: Basic, 1988).

51. Van Parijs, *Marxism Recycled*, 236.

52. For discussion of this point, see Przeworski, "Less Is More," 15.

53. On these developments see Vincent Mosco, *The Pay-Per Society: Computers and Communication in the Information Age: Essays in Critical Theory and Public Policy* (Toronto: Garamond, 1989); Nicholas Garnham, *Capitalism and Communication: Global Culture and the Economics of Information* (London: Sage, 1990); W. D. Rowland Jr. and M. Tracey, "Worldwide Challenges to Public Service Broadcasting," *Journal of Communication* 40.2 (1990): 8–27; Marc Raboy, Ivan Bernier, Florian Sauvageau, Dave Atkinson, "Cultural Development and the Open Economy: A Democratic Issue and a Challenge to Public Policy," *Canadian Journal of Communication* 19 (1994): 291–315; and Edward Comor, ed., *The Global Political Economy of Communications: Hegemony, Telecommunications and the Information Economy* (New York: St. Martin's, 1994).

54. For a handy sketch of this spectrum of activity see John Chesterman and Andy Lipman, *The Electronic Pirates* (London: Comedia, 1988).

55. John Keane, *Media and Democracy* (Oxford: Blackwell, 1990), 159.

56. As Nicholas Garnham observes in *Capitalism and Communication*, there is a contradiction at the heart of the communication commodity, arising from media

business's need to simultaneously maximize and restrict distribution. On the one hand, they want to sell as *much* as they can. Therefore they increase the speed and efficiency of communication methods to reach more people, more of the time, in more and more various ways. But on the other hand, media corporations want to *sell* as much they can; they are interested not just in getting out messages, but in getting back money. So they confront a paradoxical requirement to expand and restrict communication at the same time—simultaneously creating plenty and imposing scarcity.

57. Marx, *Grundrisse*, 548.

58. Ibid., 548–49.

59. See Mark Chen, "Pandora's Mailbox," *Z Magazine*, Dec. 1994, 39–41.

60. See John P. Barlow, "The Economy of Ideas: A Framework for Rethinking Patents and Copyrights in the Digital Age," *Wired* 2.3 (1994): 85–129, and Anne Branscombe, *Who Owns Information? From Privacy to Public Access* (New York: Harper Collins, 1994).

61. See Dorothy Kidd, "Taking the Walk: The Communication Commons amidst the Media Enclosures" (Ph.D. diss., Simon Fraser University, Vancouver, Canada, 1998).

62. Keane, *Media and Democracy*, 159.

63. Ibid., 161.

64. See Douglas Kellner, *Television and the Crisis of Democracy* (Boulder, Colo.: Westview, 1990).

65. Computer Professionals for Social Responsibility, "A Public Interest Vision of the National Information Infrastructure," on-line, Internet, ACTIV-L@MIZZOU1.BITNET, 11 Mar. 1994.

66. Ibid.

67. Ibid.

68. Ibid.

69. Computer Professionals for Social Responsibility, Berkeley Chapter Peace and Justice Working Group, "A Computer and Information Technologies Platform," on-line, Internet, ACTIV-L@MIZZOU1.BITNET, 19 Sept. 1992.

70. This suggestion is also made by Morris-Suzuki in her *Beyond Computopia*. This call for the elimination of property rights will seem problematic for many artists, researchers, and cultural workers. For people in these areas the potential for piracy inherent in electronic technologies threatens their livelihood. They would like to see intellectual property rights in this area revised to strengthen the rights of the immediate producers against the large enterprises—publishing houses, music businesses, software giants—who purchase and dispose of their work. And they would be aghast at the prospect of forgoing royalties and other revenues on which their activities depend. Within the current, capitalist context, legal protection for these producers is appropriate and necessary. The dismantling of intellectual property rights should start from an assault on the legal fortifications of the corporations who are the beneficiaries of the current system, not those of individuals and small organizations. Its full realization could only be part of the overall constitution of a commonwealth aimed to break fundamentally with market logic, as part of an entire "package" that includes measures such as a guaranteed income, which would fundamentally undercut the need for people to worry about selling their intellectual or physical labor power.

71. Marx, *Grundrisse*, 533.

72. Ibid., 526.

73. Ibid., 531.

74. Ibid.

75. Ibid., 161.

76. Ibid., 160.

77. Ibid., 161.

78. Gilles Deleuze and Felix Guattari, *Anti-Oedipus: Capitalism and Schizophrenia* (New York: Viking, 1983), 252.

79. See, for example, the London Edinburgh Weekend Return Group, *In and Against the State* (London: Pluto, 1980).

80. I use this term of Foucault's to designate the operations of public administration, without necessarily identifying such activities with the functions of the centralized state. See Michel Foucault, "Governmentality," in *The Foucault Effect*, ed. Graham Burchell, Colin Gordon, Peter Miller (London: Harvester, 1991), 87–104.

81. See Kenneth Surin, "Marxism(s) and the Withering Away of the State," *Social Text* 27 (1990): 35–54.

82. I agree with Joachim Hirsch, who, in "The New Leviathan and the Struggle for Democratic Rights" (*Telos* 48 [1981]: 79–89), argues that what is necessary is "reduction of the role of the state" (87–88) in the sense of "a collective reappropriation of responsibilities, self-management, autonomous interest organisation, debureaucratisation and decentralisation" (88); this program is valid even though it will necessarily conflict with attempts "to exploit anti-state and anti-bureaucratic resentments for quasi-populist, reactionary mobilisation and social-political austerity strategies."

83. Steven Wright describes how some Italian autonomists now argue a guaranteed income based upon the free distribution of selected social services could provide a starting point from which to build for the further extension of a self-managed sector where need takes priority over profit and provide a "motor of a process of reappropriation of the welfare state's institutions and services, based upon the expansion of self-managed social labour and cooperation." He also shows how the idea of reappropriation of the welfare state "from below" has been expounded by Marco Revelli, a member of the more traditional Left with autonomist affinities, who believes that "the welfare state has finally run its course." Since it is now both "useless to the bosses [and] alienating to workers," a political strategy based upon an unconditional defense of the welfare state would be "suicidal." Its current crisis can, according to Revelli, lead only to two outcomes: either toward a social "free-for-all" such as can be found in the United States, where each must fend as best they can, or toward "a more mature sociality" based upon mutual aid. Many examples of the latter, Ravelli points out, already exist in Italy: above all, the thousands of cooperatives and mutual societies which provide health care and other social benefits to their members. The Left's aim, Revelli suggests, should be to expand this area of welfare from below, and "to reconstruct those autonomies that the inevitably bureaucratic apparatuses of the parties, unions and state have dispersed." (Wright, "Confronting the Crisis of Fordism," 4.) For an example of such thinking from an Anglo-Saxon perspective, see Hilary Wainwright, *Arguments for a New Left: Answering the Free Market Right* (Oxford: Blackwell, 1994).

84. See Nicole Cox and Silvia Federici, *Counterplanning from the Kitchen—Wages*

for Housework: A Perspective on Capital and the Left (Bristol, Eng.: Falling Wall Press, 1975); Craig Benjamin and Terisa Turner, "Counterplanning from the Commons: Labour, Capital and the 'New Social Movements,'" *Labour, Capital and Society* 25.2 (1992): 218–48; and Dorothy Kidd and Nick Witheford, "Counterplanning from Cyberspace and Videoland: or, Luddites on Monday and Friday, Cyberpunks the Rest of the Week," paper presented at Monopolies of Knowledge: A Conference Honoring the Work of Harold Innis, Vancouver, Canada, 12 Nov. 1994.

85. Douglas Schuler, *New Community Networks: Wired for Change* (Reading, Mass.: Addison Wesley, 1996); Benjamin Goldman, "The Environment and Community Right to Know: Information for Participation," *Computers in Human Services* 8.1 (1991): 19–40.

86. RAFI can be reached at http://www.rafi.ca.

87. For extensive discussions of this activity see Schuler, *New Community Networks*; Eric Lee, *The Labor Movement and the Internet* (London: Pluto, 1997); and Jay Weston, "Old Freedoms and New Technologies: The Evolution of Community Networking," paper presented at Free Speech and Privacy in the Information Age Symposium, University of Waterloo, Canada, 26 Nov. 1994, available on-line, Internet, Red Rock Eater news group.

88. The classic statement of this view remains F. A. Hayek, "The Use of Knowledge in Society," *American Economic Review* 35 (1945): 519–30. For a fascinating discussion and rebuttal of Hayek, see Wainwright, *Arguments for a New Left.*

89. For discussion of these issues see the debate between Alec Nove, the champion of "market socialism," and Ernest Mandel, an advocate of state planning: Alec Nove, *The Economics of Feasible Socialism* (London: Allen and Unwin, 1985), and "Markets and Socialism," *New Left Review* 161 (1987): 98–104; Ernest Mandel, "In Defence of Socialist Planning," *New Left Review* 159 (1986): 5–39, and "The Myth of Market Socialism," *New Left Review* 169 (1988): 108–21. See also Boris Frankel, *Beyond the State? Dominant Theories and Socialist Strategies* (London: Macmillan, 1983) and "The Historical Obsolescence of Market Socialism—A Reply to Alec Nove," *Radical Philosophy* 39 (1985): 28–33; Hans Breitenbach, Tom Burden, and David Coates, *Features of a Viable Socialism* (New York: Harvester, 1990); Pat Devine, *Democracy and Economic Planning* (Cambridge: Polity, 1988).

90. See Hayek, "Use of Knowledge."

91. See Elson, "Market Socialism or Socializing the Market?"

92. Elson writes,

There has been a tendency among Marxists (beginning with Marx) to interpret conscious control in terms of gathering all relevant information at one decision-making point and of taking decisions with full knowledge of all inter-connections and ramifications. That is an impossible, and an undesirable, goal. Conscious control is better interpreted as open access to all available information concerning the product and its price, so that any decision-maker has access to the same information as any other. . . . Such a system of coordination does not require simultaneous processing of large amounts of information, of the kind necessary for effective central planning (which, even with the latest computer technology is argued to be unfeasible). Rather it requires the gathering and processing, at discrete intervals, in separate bundles, of information already generated by enterprises for their own use, such as unit costs and levels of inventories, and process and product specifications. The barrier to this is not technical: current levels of micro-processor technology can certainly handle this kind of information processing very rapidly. . . . The barrier is not technical: it is social and political.

Ibid., 43.

93. Ibid.

94. Ibid., 44.

95. See Michael Albert and Robin Hahnel, *The Political Economy of Participatory Economics* (Princeton: Princeton University Press, 1991), and *Looking Forward: Participatory Economics for the Twenty-first Century* (Boston: South End, 1991).

96. Marx, *Grundrisse*, 706.

97. Marx's other writings contain elements of the frank scientific triumphalism common to his age (elements later amplified in scientific socialism) and also insights into the metabolic interconnection of humanity and nature that prefigure contemporary ecological thought. For discussions of the "green" Marx see the journal *Capitalism, Nature, Socialism*, edited by James O'Connor, and David Pepper, *Eco-Socialism: From Deep Ecology to Social Justice* (London: Routledge, 1993).

98. See Theodor Adorno and Max Horkheimer, *Dialectic of Enlightenment* (1947; New York: Herder and Herder, 1972).

99. Maria Mies, "Why Do We Need All This? A Call against Genetic Engineering and Reproductive Technology," in *Made to Order: The Myth of Reproductive and Genetic Progress*, ed. Patricia Spallone and Deborah Lynn Steinberg (Oxford: Pergamon, 1987), 46. Despite the title of the essay, Mies addresses issues across the span of high-technology development, including what she terms "inundation with 'technical means of communication.'" See also her *Patriarchy and Accumulation on a World Scale: Women in the International Division of Labour* (London: Zed Books, 1986).

100. Mies, "Why Do We Need All This?" 46.

101. On this point see Felix Guattari, "The Three Ecologies," *New Formations* 8 (1989): 146–47:

Increasingly in the future, the maintenance of natural equilibrium will be dependent upon human intervention; the time will come, for example, when massive programs will have to be set in train to regulate the relationship between oxygen, ozone, and carbon dioxide in the earth's atmosphere. . . . What is required for the future is more than the mere defense of nature. If the Amazonian "lung" is to be regenerated, the Sahara desert made fertile again, we need, immediately, to go on the offensive. Even the human creation of new plant and animal species looms unavoidably on the horizon; the urgent task we face is, then, to fashion an ethics appropriate to a scenario that is both terrifying and fascinating, and, more importantly, a politics appropriate to the general destiny of humanity.

102. Herbert Marcuse, *One-Dimensional Man: Studies in the Ideology of Advanced Industrial Society* (Boston: Beacon, 1964), 156–69.

103. Jürgen Habermas, *Toward a Rational Society: Student Protest, Science and Politics* (Boston: Beacon, 1970).

104. For discussions of the debate that tend toward this viewpoint, see Steven Vogel, "New Science, New Nature: The Habermas-Marcuse Debate Revisited," *Research in*

Philosophy and Technology 11 (1991): 157–78, and Andrew Feenberg, *Critical Theory of Technology* (Oxford: Oxford University Press, 1991).

105. Andrew Feenberg, "Marcuse and the Critique of Technology," in his *Alternative Modernity: The Technical Turn in Philosophy and Social Theory* (Berkeley: University of California Press, 1995), 19–40.

106. On these tendencies see Richard Levins, "Toward the Renewal of Science," *Rethinking Marxism* 3.3/4 (1990): 102–25; David Dickson, *The New Politics of Science* (Chicago: University of Chicago Press, 1988); Cooley, *Architect or Bee?*; Sandra Harding, ed., *The "Racial" Economy of Science: Toward a Democratic Future* (Bloomington: Indiana University Press, 1993): Richard Hofrichter, ed., *Toxic Struggles: The Theory and Practice of Environmental Justice* (Philadelphia: New Society, 1993).

107. See Evelyn Fox Keller, *Reflections on Gender and Science* (New Haven: Yale University Press, 1985).

108. Tom Athanasiou, "Greenwashing Agricultural Biotechnology," *Processed World* 28 (1991/92): 21.

109. Feenberg, *Critical Theory of Technology*; Wainwright, *Arguments for a New Left*; Richard Sclove, *Democracy and Technology* (New York: Guilford, 1995); Michael Goldhaber, *Reinventing Technology: Policies for Democratic Values* (New York: Routledge and Kegan Paul, 1986).

110. Schuler, *New Community Networks*, 383.

9   INTELLECTS

1. Karl Marx, *Grundrisse* (Harmondsworth, Eng.: Penguin, 1973), 699–743.
2. Ibid., 705.
3. Ibid., 694, 705, 706, 709.
4. Ibid., 706.
5. Ibid., 692.
6. Ibid., 694.
7. Ibid., 700.
8. Some of the writings of this group can be found in the collection edited by Paolo Virno and Michael Hardt, *Radical Thought in Italy: A Potential Politics* (Minneapolis: University of Minnesota Press, 1996).
9. Paolo Virno, "Notes on the General Intellect," in *Marxism beyond Marxism*, ed. Saree Makdisi, Cesare Casarino, and Rebecca E. Karl (London: Routledge, 1996), 268.
10. Ibid.
11. Ibid., 270.
12. Ibid.
13. Antonio Negri, "Constituent Republic," *Common Sense* 16 (1994): 89; also in Virno and Hardt, *Radical Thought in Italy*, 213–24.
14. Maurizio Lazzarato and Toni Negri, "Travail immaterial and subjectivité," *Futur Antérieur* 6 (1994): 86.
15. Virno, "Notes on the General Intellect," 270.
16. Jean-Marie Vincent, "Les Automatismes sociaux et le 'general intellect,'" *Futur Antérieur* 16 (1993): 121 (my trans.).
17. Ibid.

18. Ibid., 123.

19. See Maurizio Lazzarato, "General Intellect: Towards an Inquiry into Immaterial Labour," *Immaterial Labour, Mass Intellectuality, New Constitution, Post Fordism and All That* (London: Red Notes, 1994), 1–14. See also Lazzarato and Negri, "Travail immaterial and subjectivité," 86–99, and Lazzarato, "Immaterial Labor," in Virno and Hardt, *Radical Thought in Italy,* 133–50.

20. Lazzarato, "General Intellect," 4.

21. Lazzarato and Negri, "Travail immaterial and subjectivité," 86.

22. Harland Prechel describes such a situation in his "Transformations in Hierarchy and Control of the Labor Process in the Post-Fordist Era: The Case of the U.S. Steel Industry," in *The Labor Process and Control of Labor: The Changing Nature of Work Relations in the Late Twentieth Century,* ed. Berch Berberoglu (Westport, Conn.: Praeger, 1993), 44–58. In the steel plants he examines, workers are required to communicate, interact, and participate, but only within certain predetermined parameters embodied in the various forms of "premise" or "algorithmic control" associated with informatic production systems. In "premise control," top management calculates cost-efficient ways to conduct each step in the operation: these are then transmitted as rules through the computer system. Parameters within which choices can be made are already embedded in the programs directing production. Responsibility thus takes the form of adjustments within a preset process, rather than control over it. Command resides at the level of the total system, so that the autonomy bestowed on the parts is strictly limited. As Prechel puts it in his study of the post-Fordist U.S. steel industry, management can thus "centralize command, while decentralizing responsibility for the decision," pushing responsibility down the organizational hierarchy, yet maintaining control at the top. It is this sort of "premise control" that allows some companies to devolve responsibilities from middle-level and line management to the shop floor, while still maintaining ultimate authority within upper-level management. This creates a paradoxical position: while the worker may indeed be "skilled," this skill is divorced from any individual or group "control" over the production process. On this point see also Steven Vallas, *Power in the Workplace: The Politics of Production at AT&T* (New York: State University of New York Press, 1993).

23. Lazzarato, "General Intellect," 6.

24. Ibid., 5–6.

25. Ibid., 6–7.

26. Franco Barchiesi, on-line, Internet, aut-op-sys [accessed 5 Apr. 1996].

27. Maurizio Lazzarato, "'Pas de sous pas de totos!': La grève des ouvriers Peugeot," *Futur Antérieur* 1 (1990): 63–76; "Les caprices du flux—les mutations technologiques du point de vue de ceux qui les vivent," *Futur Antérieur* 4 (1992): 156–64.

28. Quoted in Massimo De Angelis, "The Autonomy of the Economy and Globalisation," *Vis-A-Vis,* Winter 1996, on-line. This article is available from http://www.lists.village.virginia.edu/spoons/aut_html/glob.html.

29. Peter Downs, "Striking against Overtime: The Example of Flint," *Against the Current* 54 (1995): 7–8; Jane Slaughter, "Addicted to Overtime," *The Progressive,* Apr. 1995, 31–33.

30. Kim Moody, *Workers in a Lean World: Unions in the International Economy* (London: Verso, 1997), 278.

31. See Thomas Greven, "Can We Convert Defense Jobs to Peacetime Uses?" *Labor Notes* (Nov. 1992): 7; Carl Boggs, "Economic Conversion as a Radical Strategy: Where Social Movements and Labor Meet," in *Building Bridges: The Emerging Grassroots Coalition of Labor and Community*, ed. Jeremy Brecher and Tim Costello (New York: Monthly Review Press, 1990), 302–10.

32. The concept of "socially useful production" is usually associated with the famous shop-stewards' movement at the Lucas plant, a British aerospace firm, in the 1970s. See Hilary Wainwright, *Arguments for a New Left: Answering the Free Market Right* (Oxford: Blackwell, 1994).

33. "Autonomous production" is a term used by Japanese workers who for eight years occupied and ran a high-tech Toshiba plant. See the discussion in chap. 5, and also Ken Tsuzuku, "Presentation to the 1991 Labor Notes Conference," in *A Conference on Labour and Team Concepts*, Proceedings of a Conference Co-Sponsored by Capilano College Labour Studies Programme and Vancouver and District Labour Council, Vancouver, Canada, 18–19 Oct. 1991, 261–70.

34. Vincent, "Les automatismes sociaux," 127.

35. Antonio Negri, "Infinité de la communication/finitude du désir," *Futur Antérieur* 11 (1992/93): 5–8 (my trans.).

36. Ibid., 7.

37. Maurizio Lazzarato and Antonio Negri, *Le Bassin de Travail Immateriel (B.T.I.) dans la métropole Parisienne: Définition, recherches, perspectives* (Paris: Tekne-Logos, 1993).

38. Maurizio Lazzarato, "La 'Panthère' et la communication," *Futur Antérieur* 2 (1990): 54–67 (my trans.).

39. For a valuable discussion of the social subjectivities emergent in cyberspace, see Harry Cleaver, "Marxian Categories, the Crisis of Capital and the Constitution of Social Subjectivity Today," *Common Sense* 14 (1993): 32–57.

40. Chris Carlsson, "The Shape of Truth to Come," in *Resisting the Virtual Life*, ed. James Brook and Iain Boal (San Francisco: City Lights, 1995), 242.

41. Ibid.

42. "How the Net Killed the MAI," *The Globe and Mail*, 29 Apr. 1998, 1.

43. Ibid.

44. Ibid.

45. Ibid.

46. Negri, "Constituent Republic," 88, 93–94.

47. Virno, "Notes on the General Intellect," and also "Virtuosity and Revolution: The Political Theory of Exodus," in Virno and Hardt, *Radical Thought in Italy*, 189–213.

48. This discussion is, so far, largely unpublished, at least in English. My account of it draws on discussions that have taken place on the aut-op-sys email group, aut-op-sy@lists.village.virginia.edu, and among the "Infra-Reds" collective in Vancouver.

49. Virno and Hardt, *Radical Thought in Italy*, 280.

50. Ibid.

51. Ibid.

52. Michael Hardt and Antonio Negri, *Labor of Dionysus: A Critique of the State Form* (Minneapolis: University of Minnesota Press, 1994), 280.

53. Hardt and Negri, *Radical Thought in Italy*, 280.

54. Ed Emery, "No Politics without Inquiry: A Proposal for a Class Composition Inquiry 1996–7," *Common Sense* 18 (1995): 1–11.

55. A comrade close to *Futur Antérieur* but living in the East End of London remarked to me that unemployed black youths in the neighborhood were, through involvement in rap, reggae, and other music, immersed in highly technological and "immaterial" networks of production, taping, mixing, sampling, and pirating, in a field that constituted one of the most dynamic sectors of the contemporary cultural industry, and were doing so sometimes in a politicized and oppositional way.

56. Negri and Lazzarato, "Travail immatérial and subjectivité," 88.

57. David Noble, "Digital Diploma Mills," available from http://www.firstmonday.dk/issues/issue3_1/noble/index.html.

58. Ibid.

59. Ibid.

60. Ibid.

61. See Negri and Lazzarato, "Travail immatérial and subjectivité."

62. The change Michel Foucault describes as the shift from the "universal" to the "specific" intellectual catches something of this transition. See his *Power/Knowledge* (New York: Pantheon, 1980).

63. Roberto Rodriguez, "Information Highway: Latino Student Protesters Create Nationwide Link Up," *Black Issues in Higher Education*, 16 June 1994, on-line, Internet, ACTIV-L@MIZZOU1.BITNET [accessed 5 Aug. 1994].

64. "Students Fight the Contract," *Progressive*, 16 May 1995.

65. Dorothy Kidd and Nick Witheford, "Counterplanning from Cyberspace and Videoland: or, Luddites on Monday and Friday, Cyberpunks the Rest of the Week," paper presented at Monopolies of Knowledge: A Conference Honoring the Work of Harold Innis, Vancouver, Canada, 12 Nov. 1994.

66. Raymond Williams, "The Politics of Nuclear Disarmament," in *Exterminism and Cold War*, ed. New Left Review (London: Verso, 1982), 85.

67. Karl Marx and Friedrich Engels, *The Communist Manifesto* (New York: Washington Square, 1969), 34.

BIBLIOGRAPHY

Adler, Paul. "Automation, Skill and the Future of Capitalism." *Berkeley Journal of Sociology* 13 (1988): 3–36.

———. "Marx, Machines, and Skill." *Technology and Culture* 3.4 (1990): 780–812.

Adorno, Theodor, and Max Horkheimer. *Dialectic of Enlightenment.* New York: Herder and Herder, 1972.

Adrian X, Robert. "Infobahn Blues." In *Digital Delirium,* edited by Arthur and Marilouise Kroker, 84–88. New York: St. Martin's Press, 1997.

Aglietta, Michel. *A Theory of Capitalist Regulation: The US Experience.* London: New Left Books, 1979.

Albert, Michael, and Robin Hahnel. *Looking Forward: Participatory Economics for the Twenty-first Century.* Boston: South End, 1991.

———. *The Political Economy of Participatory Economics.* Princeton: Princeton University Press, 1991.

Almeida, Paul. "The Network for Environmental and Economic Justice for the Southwest: Interview with Richard Moore." *Capitalism, Nature, Socialism* 5.1 (1994): 21–54.

Alquati, Romano. "The Network of Struggles in Italy." Unpublished paper, 1974. Red Notes Archive, London.

Amin, Ash, ed. *Post-Fordism: A Reader.* Oxford: Blackwell, 1994.

Amin, Samir. *Accumulation on a World Scale: A Critique of the Theory of Underdevelopment.* New York: Monthly Review Press, 1974.

———. *Empire of Chaos.* New York: Monthly Review Press, 1992.

Anderson, Perry. *Considerations on Western Marxism.* London: New Left Books, 1976.

———. *In the Tracks of Historical Materialism.* London: Verso, 1983.

Arno, Peter, and Karyn Feiden. *Against the Odds: The Story of AIDS Drug Development, Politics and Profits.* New York: Harper Collins, 1992.

Aronowitz, Stanley, and Jonathan Cutler, eds. *Post-Work: The Wages of Cybernation.* New York: Routledge, 1998.

Aronowitz, Stanley, and William DiFazio. *The Jobless Future: Sci-Tech and the Dogma of Work.* Minneapolis: University of Minnesota Press, 1994.

Aronowitz, Stanley, Barbara Marhnsons, and Michael Merser, eds. *Technoscience and Cyberculture.* New York: Routledge, 1996.

Arquilla, John, and David Ronfeldt. "Cyberwar Is Coming!" *Comparative Strategy* 12.2 (1993): 141–65. Citations are from an electronic copy.

Athanasiou, Tom. "Greenwashing Agricultural Biotechnology." *Processed World* 28 (1991/92): 16–21.

Attali, Jacques. *Millennium: Winners and Losers in the Coming World Order.* New York: Random House, 1991.

Auchard, Eric. "Discount Programming: The Global Labour Market." *CPU: Working in the Computer Industry* 1 (1993). On-line, Internet.

Aufderheide, Patricia. "Underground Cable: A Survey of Public Access Programming." *Afterimage* 22.1 (1994): 5–7.

Aufheben Collective. "Auto-Struggles: The Developing War against the Road Monster." *Aufheben* 3 (1994): 3–23.

———. "Decadence: The Theory of Decline or the Decline of Theory—Part II." *Aufheben* 3 (1994): 24–34.

———. Review of *Midnight Oil,* by Midnight Notes Collective. *Aufheben* 3 (1994): 35–41.

Babbage, Charles. *On the Economy of Machinery and Manufactures.* London, 1835.

Bacon, David. "L.A. Labor—A New Militancy." *Nation,* 27 Feb. 1995, 273–76.

———. "Silicon Valley on Strike." *CPU: Working in the Computer Industry* 3 (1993). On-line, Internet.

Bagdikian, Ben H. "Cornering Hearts and Minds: The Lords of the Global Village." *Nation,* 12 June 1989, 805–20.

———. *The Media Monopoly.* Boston: Beacon, 1990.

Bahro, Rudolf. *The Alternative in Eastern Europe.* London: New Left Books, 1978.

Baldi, Guido. "Negri beyond Marx." *Midnight Notes* 8 (1985): 32–36.

Balestrini, Nanni. *The Unseen.* London: Verso, 1989.

Balka, Ellen. "Womantalk Goes On-line: The Use of Computer Networks in the Context of Feminist Social Change." Ph.D. diss., Simon Fraser University, Vancouver, Canada, 1991.

Banks, Andrew. "Jobs with Justice: Florida's Fight against Worker Abuse." In *Building Bridges: The Emerging Grassroots Coalition of Labor and Community,* edited by Jeremy Brecher and Tim Costello, 25–37. New York: Monthly Review Press, 1990.

Barber, Benjamin. *Jihad vs McWorld: How the Planet Is Both Falling Apart and Coming Together—And What This Means for Democracy.* New York: Times, 1995.

Barlow, John. "The Economy of Ideas: A Framework for Rethinking Patents and Copyrights in the Digital Age." *Wired* 2.3 (1994): 85–129.

Barnet, Richard J., and John Cavanagh. *Global Dreams: Imperial Corporations and the New World Order.* New York: Simon and Schuster, 1994.

Barratt, Michelle. *The Politics of Truth: From Marx to Foucault.* Cambridge: Polity, 1991.

Bartleby the Scrivener. "Marx beyond Midnight." *Midnight Notes* 8 (1985): 32–35.

Basen, Gwynne, Margaret Eichler, and Abby Lippman, eds. *Misconceptions: The Social Construction of Choice and the New Reproductive and Genetic Technologies.* Quebec: Voyageur, 1993.

Baudrillard, Jean. *The Gulf War Did Not Take Place.* Bloomington: Indiana University Press, 1995.

———. "The Implosion of Meaning in the Media and the Implosion of the Social in the Masses." In *The Myths of Information: Technology and Postindustrial Culture,* edited by Kathleen Woodward, 137–50. Madison, Wis.: Coda, 1980.

———. "Interview: Game with Vestiges." *On the Beach* 6 (1984): 19–25.

———. *Simulations.* New York: Semiotext(e), 1983.

Bell, Daniel. *The Coming of Post-Industrial Society.* New York: Basic, 1973.

———. *The Cultural Contradictions of Capitalism.* New York: Basic, 1976.

———. *The End of Ideology.* New York: Free Press, 1961.

———. "First Love and Early Sorrow." *Times Higher Education Supplement,* 16 Jan. 1981, 9–11.

———. "The Social Framework of the Information Society." In *The Computer Age: A Twenty Year View,* edited by Michael L. Dertouzous and Joel Moses, 163–212. Cambridge, Mass.: MIT Press, 1979. Essay reprinted in *The Complete Guide to the New Technology and Its Impact on Society,* ed. Tom Forester, 500–549. Cambridge, Mass.: MIT Press, 1980.

Bell, Peter F., and Harry Cleaver. "Marx's Crisis Theory as a Theory of Class Struggle." *Research in Political Economy* 5 (1982): 189–261.

Beniger, James. *The Control Revolution: Technological and Economic Origins of the Information Society.* Cambridge, Mass.: Harvard University Press, 1986.

Benjamin, Craig, and Terisa Turner. "Counterplanning from the Commons: Labour, Capital and the 'New Social Movements.'" *Labour, Capital and Society* 25.2 (1992): 218–48.

Benjamin, Playthell. "Seeing Is Not Believing." *Guardian,* 9 May 1992, 34.

Benston, Margaret Lowe. "For Women, the Chips Are Down." In *The Technological Woman: Interfacing with Tomorrow,* edited by Jan Zimmerman, 44–54. New York: Praeger, 1983.

Berardi, Franco. *Le Ciel est enfin tombé sur la terre.* Paris: Seuil, 1978.

Berberoglu, Berch, ed. *The Labor Process and Control of Labor: The Changing Nature of Work Relations in the Late Twentieth Century.* Westport, Conn.: Praeger, 1993.

Berens, Camilla. "Folk Law." *New Statesman and Society,* 5 May 1995, 34–36.

Berg, Maxine. *The Machinery Question and the Making of Political Economy, 1815–1848.* Cambridge: Cambridge University Press, 1980.

Berlinguer, Giovanni. "The Body as Commodity and Value." *Capitalism, Nature, Socialism* 5.3 (1994): 35–49.

Berman, Marshall. *All That Is Solid Melts into Air: The Experience of Modernity.* New York: Simon and Schuster, 1982.

Bernal, J. D. *The Social Function of Science.* Cambridge, Mass.: MIT Press, 1939.

Besser, Howard. "From Internet to Information Superhighway." In *Resisting the Virtual Life: The Culture and Politics of Information,* edited by James Brook and Iain A. Boal, 59–71. San Francisco: City Lights, 1995.

Best, Steven, and Douglas Kellner. *Postmodern Theory: Critical Interrogations.* London: Macmillan, 1992.

Bidwai, Praful. "Making India Work—For the Rich." *Multinational Monitor* 16.7/8 (1995): 9–13.

Bierra, Mariella, and Marco Revelli. "Absentéisme et conflictualité: L'Usine reniée. Crise de la centralité de l'usine et nouveaux comportements ouvriers." In *Usines et ouvriers: Figures du nouvel ordre productif,* edited by Jean Paul de Gaudemar, 105–36. Paris: François Maspero, 1980.

Billington, James. *Fire in the Minds of Men: Origins of the Revolutionary Faith.* New York: Basic, 1980.

Blauner, Robert. *Alienation and Freedom: The Factory Worker and His World.* Chicago: University of Chicago Press, 1964.

Bloch, R., and R. Keil. "Planning for a Fragrant Future: Air Pollution Control, Restructuring and Popular Alternatives in Los Angeles." *Capitalism, Nature, Socialism* 2.1 (1991): 44–65.

Block, Fred, and Larry Hirschhorn. "New Productive Forces and the Contradictions of Contemporary Capitalism: A Post-Industrial Perspective." *Theory and Society* 7.3 (1979): 363–95.

Boggs, Carl. "Economic Conversion as a Radical Strategy: Where Social Movements and Labour Meet." In *Building Bridges: The Emerging Grassroots Coalition of Labor and Community,* edited by Jeremy Brecher and Tim Costello, 302–10. New York: Monthly Review Press, 1990.

Boland, Joseph. "Ecological Modernization." *Capitalism, Nature, Socialism* 5.2 (1994): 135–41.

Bologna, Sergio. "Class Composition and the Theory of the Party at the Origin of the Workers Councils Movement." In *The Labour Process and Class Strategies,* edited by Conference of Socialist Economists, 68–91. London: Conference of Socialist Economists, 1976.

———. "The Tribe of Moles." In *Italy: Autonomia—Post-Political Politics,* edited by Sylvere Lotringer and Christian Marazzi, 36–61. New York: Semiotext(e), 1980.

Bonefeld, Werner. "Human Practice and Perversion: Beyond Autonomy and Structure." *Common Sense* 15 (1994): 43–42.

Bonefeld, Werner, Richard Gunn, and Kosmas Psychopedis, eds. *Open Marxism.* Vol. 1, *Dialectics and History.* London: Pluto, 1992.

———. *Open Marxism.* Vol. 2, *Theory and Practice.* London: Pluto, 1992.

Bonefeld, Werner, and John Holloway, eds. *Post-Fordism and Social Form: A Marxist Debate on the Post-Fordist State.* London: Macmillan, 1991.

Boozell, Greg. "The Revolution Will Be Microwaved: The FCC, Microwatt Radio, and Telecommunication Networks." *Afterimage* 22.1 (1994): 12–14.

Bottomore, Tom, et al., eds. *A Dictionary of Marxist Thought.* Cambridge, Mass.: Harvard University Press, 1983.

Branscombe, Anne. *Who Owns Information? From Privacy to Public Access.* New York: Harper Collins, 1994.

Brants, Kees. "The Social Construction of the Information Revolution." *European Journal Of Communication* 4 (1989): 79–87.

Braverman, Harry. *Labor and Monopoly Capital: The Degradation of Work in the Twentieth Century.* New York: Monthly Review Press, 1974.

Brecher, Jeremy, John Brown Childs, and Jill Cutler, eds. *Global Visions: Beyond the New World Order.* Boston: South End, 1993.

Brecher, Jeremy, and Tim Costello. *Global Village or Global Pillage: Economic Reconstruction From the Bottom Up.* Boston: South End, 1994.

———. "A New Labor Movement in the Shell of the Old." *Z Magazine,* Apr. 1996, 45–49

Brecher, Jeremy, and Tim Costello, eds. *Building Bridges: The Emerging Grassroots Coalition of Labor and Community.* New York: Monthly Review Press, 1990.

Breitenbach, Hans, Tom Burden, and David Coates. *Features of a Viable Socialism.* New York: Harvester, 1990.

Brenner, Joseph E. "Internationalist Labor Communication by Computer Network: The United States, Mexico and NAFTA." Paper presented at School of International Service, American University, Washington, D.C., 1994.

Bresheeth, Haim, and Nira Yuval-Davis, eds. *The Gulf War and the New World Order.* London: Zed Books, 1991.

Brinton, Maurice. *The Bolsheviks and Workers' Control 1917 to 1921: The State and Counter Revolution.* Montreal: Black Rose, 1975.

Britton, Andrew. "The Myth of Postmodernism: The Bourgeois Intelligentsia in the Age of Reagan." *Cineaction* 13/14 (1988): 3–17.

Bronner, Stephen Eric. *Socialism Unbound.* New York: Routledge, 1990.

Brook, James, and Iain A. Boal, eds. *Resisting the Virtual Life: The Culture and Politics of Information.* San Francisco: City Lights, 1995.

Brown, Wilmette. *No Justice, No Peace: The 1992 Los Angeles Rebellion from a Black/ Woman's Perspective.* London: Wages for Housework Campaign, 1993.

Brzezinski, Zbignew. *Between Two Ages: America's Role in the Technotronic Era.* New York: Viking, 1970.

———. *The Grand Failure: The Birth and Death of Communism in the Twentieth Century.* New York: Macmillan, 1988.

———. *Power and Principle: Memoirs of the National Security Adviser.* New York: Farrar, Strauss, Giroux, 1983.

Bukharin, Nikolai. *Historical Materialism: A System of Sociology.* New York: International Publishers, 1925.

Burnham, James. *The Managerial Revolution.* 1941. Reprint. Bloomington: Indiana University Press, 1966.

Burton-Rose, Daniel, ed., with Dan Pens and Paul Wright. *The Celling of America: An Inside Look at the U.S. Prison Industry.* Monroe, Me.: Common Courage Press, 1998.

Byrne, David. "Just Haad on a Minute There: A Rejection of Andre Gorz's 'Farewell to the Working Class.'" *Capital and Class* 24 (1985): 74–98.

Caffentzis, George. "The Work/Energy Crisis and the Apocalypse." *Midnight Notes* 3 (1980). Reprinted in Midnight Notes Collective, *Midnight Oil: Work, Energy, War, 1973–1992,* 215–17. New York: Autonomedia, 1992.

Caffentzis, George, and Silvia Federici. "Modern Land Wars and the Myth of the High-Tech Economy." In *The World Transformed: Gender, Labour and International*

Solidarity in the Era of Free Trade, Structural Adjustment and GATT, edited by Cindy Duffy and Craig Benjamin, 131–45. Guelph, Ontario: RhiZone, 1994.

Callincos, Alex. "Postmodernism, Post-Structuralism, Post-Marxism?" *Theory, Culture and Society* 2.3 (1985): 85–101.

Cantor, Charles. "The Challenges to Technology and Informatics." In *The Code of Codes: Scientific and Social Issues in the Human Genome Project*, edited by Daniel J. Kevles and Leroy Hood, 98–111. Cambridge, Mass.: Harvard University Press, 1992.

Carlsson, Chris. "The Shape of Truth to Come." In *Resisting the Virtual Life: The Culture and Politics of Information*, edited by James Brook and Iain Boal, 235–44. San Francisco: City Lights, 1995.

Carter, George. "ACT UP, the AIDS War, and Activism." In *Open Fire: The Open Magazine Pamphlet Series Anthology*, edited by Greg Ruggiero and Stuart Sahulka, 123–50. New York: New Press, 1993.

Castells, Manuel. *The Informational City: Information Technology, Economic Restructuring and the Urban-Regional Process.* Oxford: Blackwell, 1989.

Castoriadis, Cornelius. *Political and Social Writings.* Vol. 1, *1946–1965: From the Critique of Bureaucracy to the Positive Content of Socialism.* Translated by David Ames Curtis. Minneapolis: University of Minnesota Press, 1988.

———. *Political and Social Writings.* Vol. 2, *1955–1960: From the Workers' Struggle against Bureaucracy to Revolution in the Age of Modern Capitalism.* Translated by David Ames Curtis. Minneapolis: University of Minnesota Press, 1988.

———. *Political and Social Writings.* Vol. 3, *1961–1979: Recommencing the Revolution: From Socialism to the Autonomous Society.* Translated by David Ames Curtis. Minneapolis: University of Minnesota Press, 1993.

Cavanagh, John, and Robin Broad. "Global Reach: Workers Fight the Multinationals." *Nation,* 18 Mar. 1996, 21–24.

Charo, R. Alta. "A Political History of RU-486." In *Biomedical Politics*, edited by Kathi E. Hama, 43–97. Washington, D.C.: National Academy Press.

Chen, Mark. "Pandora's Mailbox." *Z Magazine,* Dec. 1994, 39–41.

Chesterman, John, and Andy Lipman. *The Electronic Pirates.* London: Comedia, 1988.

Chiang, Pamela. "501 Blues." *Breakthrough* 18.2 (1994): 3–7.

Childers, Peter, and Paul Delany. "Wired World, Virtual Campus: Universities and the Political Economy of Cyberspace." *Works and Days* 12.1/2 (1994): 61–73.

Chomsky, Noam, and Edward S. Herman. *Manufacturing Consent: The Political Economy of the Mass Media.* New York: Pantheon, 1988.

Chossudovsky, Michel. "The Globalisation of Poverty and the New World Economic Order." Working Paper #9114E, Department of Economics, Faculty of Social Sciences, University of Ottawa, 1991.

———. "India under IMF Rule." *The Ecologist* 22.6 (1992): 270–74.

Clark, Collin. *The Conditions of Economic Progress.* London: Macmillan, 1940.

Clarke, John. *New Times and Old Enemies: Essays on Cultural Studies and America.* London: Harper Collins, 1991.

Cleaver, Harry. "The Chiapas Uprising." *Studies in Political Economy* 44 (1994): 141–57.

———. "Close the IMF, Abolish Debt and End Development." *Capital and Class* 39 (1990): 17–50.

———. "An Interview with Harry Cleaver." *Vis-A-Vis* 1 (1993). On-line, World Wide Web, http://www.eco.utexas.edu.80/Homepages/faculty/Cleaver/index2.html.

———. "The Inversion of Class Perspective in Marxian Theory: From Valorisation to Self-Valorisation." In *Open Marxism*, vol. 2, *Theory and Practice*, edited by Werner Bonefeld, Richard Gunn, and Kosmas Psychopedis, 106–44. London: Pluto, 1992.

———. "Karl Marx: Economist or Revolutionary?" In *Marx, Schumpter and Keynes: A Centenary Celebration of Dissent*, edited by Suzanne W. Helburn and David F. Bramhall, 121–48. New York: M. E. Sharpe, 1986.

———. "Malaria, the Politics of Public Health and the International Crisis." *The Review of Radical Political Economics* 9.1 (1977): 81–103.

———. "Marxian Categories, the Crisis of Capital and the Constitution of Social Subjectivity Today." *Common Sense* 14 (1993): 32–57.

———. *Reading Capital Politically*. Brighton, Eng.: Harvester, 1979.

———. "Secular Crisis in Capitalism: The Insurpassability of Class Antagonism." Paper presented at the Rethinking Marxism Conference, University of Massachusetts, Amherst, 1992.

———. "Socialism." In *The Development Dictionary: A Guide to Knowledge as Power*, edited by Wolfgang Sachs, 233–49. London: Zed Books, 1992.

———. "The Subversion of Money-as-Command within the Current Crisis." Paper presented at the Conference on Money and the State, Mexico City, Mexico, 14–17 July 1992.

———. "Technology as Political Weaponry." In *Science, Politics and the Agricultural Revolution in Asia*, edited by Robert Anderson, 261–76. Boulder, Colo.: Westview, 1981.

———. "Zapatistas in Cyberspace." Paper presented at Simon Fraser University, Vancouver, Canada, 15 Nov. 1994.

Clough, Bryan. *Approaching Zero: Data Crime and the Computer Underworld*. London: Faber and Faber, 1992.

Cocco, Giuseppe, and Maurizio Lazzarato. "Au-delà du welfare state." *Futur Antérieur* 15 (1993): 57–69.

Cocco, Giuseppe, and Carlo Vercellone. "Les Paradigmes sociaux du post-Fordisme." *Futur Antérieur* 4 (1990): 71–94.

Cockburn, Cynthia. *Brothers: Male Dominance and Technological Change*. London: Pluto, 1991.

———. *Machinery of Dominance: Women, Men and Technical Know How*. London: Pluto, 1985.

Cohen, Gerald. *Karl Marx's Theory of History: A Defence*. Oxford: Clarendon, 1978.

Cohen, Jean. *Class and Civil Society: The Limits of Marxian Critical Theory*. Amherst: University of Massachusetts Press, 1982.

Cohen, Robin. *The New Helots: Migrants in the International Division of Labour*. Brookfield, Vt.: Gower, 1987.

Collectif A/Traverso. *Radio Alice, Radio Libre*. Paris: J. P. Delarge, 1977.

Collective Action Notes. "The U.S.A.: A Transitional Period—But to Where?" *Collective Action Notes* 9 (1996): 1–4.

Collettivo Strategie. "The 'Technetronic Society' According to Brezezinski." In *Compulsive Technology*, edited by Tony Solominides and Les Levidow, 126–38. London: Free Association Books, 1985.

Comor, Edward, ed. *The Global Political Economy of Communications: Hegemony, Telecommunications and the Information Economy.* New York: St. Martin's, 1994.

Computer Professionals for Social Responsibility, Berkeley Chapter, Peace and Justice Working Group. "A Computer and Information Technologies Platform." On-line, Internet, ACTIV-L@MIZZOU1.BITNET [accessed 19 Sept. 1992].

———. "A Public Interest Vision of the National Information Infrastructure." On-line, Internet, ACTIV-L@MIZZOU1.BITNET [accessed 11 Mar. 1994].

Cooley, Mike. *Architect or Bee? The Human Price of Technology.* London: Hogarth, 1987.

Cooper, Julian M. "The Scientific and Technical Revolution in Soviet Theory." In *Technology and Communist Culture: The Socio-Cultural Impact of Technology Under Socialism,* edited by Frederic J. Fleron Jr., 146–79. New York: Praeger, 1977.

Cooper, Marc. "Harley-Riding, Picket-Walking Socialism Haunts Decatur." *Nation,* April 8, 1996, 23–25.

Copelon, Rosalind. "From Privacy to Autonomy: The Conditions for Sexual and Reproductive Freedom." In *From Abortion to Reproductive Freedom: Transforming a Movement,* edited by Marlene Fried, 27–44. Boston: South End, 1990.

Coriat, Benjamin. *L'Atelier et le robot.* Paris: Christian Bourgeois, 1990.

Corn, David. "CyberNewt." *Nation,* 6 Feb. 1995, 154–55.

Cowan, Ruth Schwartz. "Genetic Technology and Reproductive Choice: An Ethics for Autonomy." In *The Code of Codes: Scientific and Social Issues in the Human Genome Project,* edited by Daniel J. Kevles and Leroy Hood, 244–63. Cambridge, Mass.: Harvard University Press, 1992.

Cox, Nicole, and Silvia Federici. *Counterplanning from the Kitchen—Wages for Housework: A Perspective on Capital and the Left.* Bristol, Eng.: Falling Wall Press, 1975.

Cox, Sue. "Strategies for the Present, Strategies for the Future: Feminist Resistance to New Reproductive Technologies." *Canadian Woman Studies* 13.2 (1993): 86–89.

Crouch, Colin, and Alessandro Pizzorno. *The Resurgence of Class Conflict in Western Europe since 1968.* Vols. 1 and 2. London: Macmillan, 1978.

Dalla Costa, Mariarosa. "Development and Reproduction." *Common Sense* 17 (1995): 11–33.

Dalla Costa, Mariarosa, and Giovanna F. Dalla Costa, eds. *Paying the Price: Women and the Politics of International Economic Strategy.* London: Zed Books, 1995.

Dalla Costa, Mariarosa, and Selma James. *The Power of Women and the Subversion of the Community.* Bristol, Eng.: Falling Wall Press, 1972.

Davidow, William H. *The Virtual Corporation: Structuring and Revitalizing the Corporation for the Twenty-first Century.* New York: Harper, 1992.

Davis, Angela. "Racism, Birth Control, and Reproductive Rights." In *From Abortion to Reproductive Freedom: Transforming a Movement,* edited by Marlene Fried, 15–26. Boston: South End, 1990.

Davis, Jim, and Michael Stack. "The Digital Advantage." In *Cutting Edge: Technology, Information, Capitalism and Social Revolution,* edited by Jim Davis, Thomas Hirschl, and Michael Stack, 121–44. London: Verso, 1997.

———. "Knowledge in Production." *Race and Class* 34.3 (1992): 1–34.

Davis, Jim, Thomas Hirschl, and Michael Stack, eds. *Cutting Edge: Technology, Information, Capitalism and Social Revolution.* London: Verso, 1997.

Davis, Mike. "Armaggedon at the Emerald City: Local 226 vs MGM Grand." *Nation,* 11 July 1994, 46–49.

———. *Beyond Blade Runner: Urban Control—The Ecology of Fear.* Westfield, N.J.: Open Magazine Pamphlet Series, 1992.

———. *City of Quartz: Excavating the Future in L.A.* London: Verso, 1990.

———. "Hell Factories in the Fields." *Nation,* 20 Feb. 1995, 229–34.

———. "Los Angeles Was Just the Beginning." In *Open Fire: The Open Magazine Pamphlet Series Anthology,* edited by Greg Ruggiero and Stuart Sahulka, 220–44. New York: New Press, 1993.

Dawson, Michael, and John Bellamy Foster. "Virtual Capitalism: The Political Economy of the Information Society." *Monthly Review* 48.3 (1996): 40–58.

De Angelis, Massimo. "The Autonomy of the Economy and Globalization." *Vis-A-Vis,* Winter 1996. On-line, World Wide Web, http://www.lists.village.virginia.edu/spoons/aut_html/glob.html.

De Gaudemar, Jean Paul, ed. *Usines et ouvriers: Figures du nouvel ordre productif.* Paris: François Maspero, 1980.

De la Haye, Yves. *Marx and Engels on the Means of Communication: The Movement of Commodities, People, Information and Capital.* New York: International General, 1979.

De Landa, Manuel. 1996. "Markets and Antimarkets in the World Economy." In *Technoscience and Cyberculture,* edited by Stanley Aronowitz, Barbara Marhnsons, and Michael Merser, 181–94. New York: Routledge, 1996.

———. *War in the Age of Intelligent Machines.* New York: Zone, 1991.

Debord, Guy. *Society of the Spectacle.* Detroit: Black and Red, 1977.

Deleuze, Gilles. *Negotiations.* New York: Columbia University Press, 1995.

Deleuze, Gilles, and Felix Guattari. *Anti-Oedipus: Capitalism and Schizophrenia.* New York: Viking, 1983.

———. *A Thousand Plateaus: Capitalism and Schizophrenia.* London: Athlone, 1987.

Derrida, Jacques. *Specters of Marx: The State of the Debt, the Work of Mourning, and the New International.* London: Routledge, 1994.

Devine, Pat. *Democracy and Economic Planning.* Cambridge: Polity, 1988.

Dews, Peter, ed. *Autonomy and Solidarity: Interviews with Jürgen Habermas.* London: Verso, 1985.

Dickson, David. *The New Politics of Science.* Chicago: University of Chicago Press, 1988.

Dirlik, Arif. "Post-Socialism/Flexible Production: Marxism in Contemporary Radicalism." *Polygraph* 6/7 (1993): 134–69.

Disch, Thomas M. "Newt's Futurist Brain Trust." *Nation,* 27 Feb. 1995, 266–70.

Dizard, W. P. *The Coming Information Age.* New York: Longman, 1982.

Dowmunt, Tony, ed. *Channels of Resistance: Global Television and Local Empowerment.* London: British Film Institute, 1993.

Downing, John. *Radical Media: The Political Experience of Alternative Communication.* Boston: South End, 1984.

Downs, Peter. "Striking against Overtime: The Example of Flint." *Against the Current* 54 (1995): 7–8.

Drew, Jesse. "Media Activism and Radical Democracy." In *Resisting the Virtual Life:*

The Culture and Politics of Information, edited by James Brook and Iain A. Boal,
71–85. San Francisco: City Lights, 1995.

Drucker, Peter. *The Age of Discontinuity.* New York: Harper and Row, 1968.

———. *Landmarks of Tomorrow.* New York: Harper and Row, 1957.

Duffy, Cindy, and Craig Benjamin, eds. *The World Transformed: Gender, Labour and
International Solidarity in the Era of Free Trade, Structural Adjustment and GATT.*
Guelph, Ontario: RhiZone, 1994.

Early, Steve, and Larry Cohen. "Jobs with Justice: Building a Broad-Based Movement
for Workers' Rights." *Social Policy* 25.2 (1994): 7–18.

Edelstein, Alex S., John E. Bass, and Sheldon M. Hasel, eds. *Information Societies:
Comparing the Japanese and American Experiences.* Seattle: University of Washington Press, 1978.

Edmond, Wendy, and Suzie Fleming, eds. *All Work and No Pay: Women, Housework
and the Wages Due.* London: Power of Women Collective and Falling Wall Press,
1975.

Elson, Diane. "Market Socialism or Socializing the Market?" *New Left Review* 172
(1987): 3–44.

Emery, Ed. "No Politics without Inquiry: A Proposal for a Class Composition Inquiry
1996–7." *Common Sense* 18 (1995): 1–11.

Engels, Frederick. *Socialism: Utopian and Scientific.* Peking: Foreign Languages Press,
1975.

Enzensberger, Hans Magnus. *The Consciousness Industry.* New York: Seabury Press,
1974.

Epstein, Steven. "Democratic Science? AIDS Activism and the Contested Construction of Knowledge." *Socialist Review* 21.2 (1991): 35–61.

Etzioni, Amitai. *The Active Society: A Theory of Societal and Political Processes.* New
York: Free Press, 1968.

European Counter Network, UK. "INFO: European Counter Network Online." On-
line, Internet, ACTIV-L@MIZZOU1.BITNET [accessed 25 Dec. 1994].

Ewen, Stuart. *Captains of Consciousness.* New York: McGraw-Hill, 1976.

———. Review of *The Mind Managers,* by Herbert Schiller. *Telos* 17 (1973): 185–87.

Fanon, Frantz. "This Is the Voice of Algeria." In Fanon, *A Dying Colonialism,* 69–98.
Translated by Haaken Chevalier. New York: Monthly Review Press, 1965.

Federici, Silvia. *Wages against Housework.* London: Power of Women Collective and
Falling Wall Press, 1975.

Feenberg, Andrew. *Alternative Modernity: The Technical Turn in Philosophy and
Social Theory.* Berkeley: University of California Press, 1995.

———. *Critical Theory of Technology.* Oxford: Oxford University Press, 1991.

Feenberg, Andrew, and Alistair Hannay, eds. *Technology and the Politics of Knowledge.* Bloomington: Indiana University Press, 1995.

Feikert, Dave. "Britain's Miners and New Technology." In *Issues in Radical Science,*
Radical Science Series 17, edited by the Radical Science Collective, 22–30. London:
Free Association Books, 1985.

Fields, A. Belden. "In Defense of Political Economy and Systemic Analysis: A Critique
of Prevailing Theoretical Approaches to the New Social Movements." In *Marxism
and the Interpretation of Culture,* ed. Cary Nelson and Lawrence Grossberg, 141–
56. Urbana: University of Illinois Press, 1988.

Finlay, Marike. *Powermatics: A Discursive Critique of New Communications Technology.* London: Routledge and Kegan Paul, 1987.

Fiske, John. *Media Matters: Everyday Culture and Political Change.* Minneapolis: University of Minnesota Press, 1994.

———. *Understanding Popular Culture.* London: Unwin Hyman, 1989.

Fleron, Frederic J., Jr., ed. *Technology and Communist Culture: The Socio-Cultural Impact of Technology under Socialism.* New York: Praeger, 1977.

Florida, Richard. "The New Industrial Revolution." *Futures* July/August (1991): 559–76.

Foley, Conor. "Virtual Protest." *New Statesman and Society,* 18 Nov. 1994, 47–49.

Fortunati, Leopoldina. *The Arcane of Reproduction: Housework, Prostitution, Labor and Capital.* New York: Autonomedia, 1995.

Foucault, Michel. *Discipline and Punish: The Birth of the Prison.* Harmondsworth, Eng.: Penguin, 1979.

———. "Governmentality." In *The Foucault Effect,* edited by Graham Burchell, Colin Gordon, and Peter Miller, 87–104. London: Harvester, 1991.

———. *Power/Knowledge: Selected Interviews and Other Writings 1972–1977.* New York: Pantheon, 1980.

Fowler, Cary, and Pat Mooney. *Shattering: Food, Politic and the Loss of Genetic Diversity.* Tucson: University of Arizona Press, 1990.

Frank, Jason, et al. "Television That Works: Labor Video in the 1990s." *Socialist Review* 93/2 (1993): 37–78.

Frankel, Boris. *Beyond the State? Dominant Theories and Socialist Strategies.* London: Macmillan, 1983.

———. "The Historical Obsolescence of Market Socialism—A Reply to Alec Nove." *Radical Philosophy* 39 (1985): 28–33.

———. *The Post-Industrial Utopians.* Cambridge: Polity, 1987.

Frederick, Howard. "Electronic Democracy." *Edges* 5.1 (1992): 13–28.

———. *Global Communication and International Relations.* Belmont, Calif.: Wadsworth, 1993.

———. "North American NGO Networking against NAFTA: The Use of Computer Communications in Cross-Border Coalition Building." Paper presented at Seventeenth International Congress of the Latin American Studies Association, Los Angeles, 24–27 Sept. 1994.

Freeman, Christopher. *Technology, Policy and Economic Performance.* London: Frances Pinter, 1987.

Fried, Marlene, ed. *From Abortion to Reproductive Freedom: Transforming a Movement.* Boston: South End, 1990.

Friedman, Andy. "Responsible Autonomy versus Direct Control Over the Labour Process." *Capital and Class* 1 (1977): 43–58.

Friedman, Milton. "The Case for the Negative Income Tax: A View from the Right." In *Issues in American Public Policy,* edited by J. H. Bunzel, 111–20. Englewood Cliffs, N.J.: Prentice Hall, 1968.

Fröbel, Folker, Jürgen Heinrichs, and Otto Kreye. *The New International Division of Labor: Structural Unemployment in Industrialized Countries and Industrialization in Developing Countries.* Translated by Pete Burgess. New York: Cambridge University Press, 1988.

Fuentes, Annette, and Barbara Ehrenreich. *Women in the Global Factory.* Boston: South End, 1983.

Fukuyama, Francis. *The End of History and the Last Man.* New York: Macmillan, 1992.

Gall, Gregor. "The Emergence of a Rank and File Movement: The Comitati di Base in the Italian Workers' Movement." *Capital and Class* 55 (1995): 9–20.

Gamble, Andrew. *The Free Market and the Strong State: The Politics of Thatcherism.* Basingstoke, Eng.: Macmillan, 1988.

Gandy, Oscar. *The Panoptic Sort: Towards a Political Economy of Information.* Boulder, Colo.: Westview, 1993.

Garnham, Nicholas. *Capitalism and Communication: Global Culture and the Economics of Information.* London: Sage, 1990.

Gates, Bill. *The Road Ahead.* New York: Norton, 1995.

Gibson, William, and Bruce Sterling. *The Difference Engine.* New York: Bantam, 1991.

Girard, Bruce, ed. *A Passion for Radio.* Montreal: Black Rose, 1992.

Gitlin, Todd. "Images Wild." *Tikkun* 4.4 (1989): 110–12.

Goldhaber, Michael. *Reinventing Technology: Policies for Democratic Values.* New York: Routledge and Kegan Paul, 1986.

Golding, Peter. "World Wide Wedge: Division and Contradiction in the Global Information Infrastructure." *Monthly Review* 48.3 (1996): 70–86.

Goldman, Benjamin. "The Environment and Community Right to Know: Information for Participation." *Computers in Human Services* 8.1 (1991): 19–40.

Goodwin, Neil, and Julia Guest. "By-Pass Operation." *New Statesman and Society,* 19 Jan. 1996, 14–15.

Gore, Albert. "Information Superhighways: The Next Information Revolution." *The Futurist,* Jan.–Feb. (1991), 21–23.

———. "The National Information Infrastructure: Information Conduits, Providers, Appliances and Consumers." *Vital Speeches of the Day,* Feb. 1994, 229–33.

Gorry, Andrew. "Silicon Valley: A Divided Workforce." *CPU: Working in the Computer Industry* 3. On-line, Internet, 1993. Available from listserv@cpsr.org at http://www.mcs.com/~jdav/CPU/cpu.html.

Gorz, Andre. *Capitalism, Socialism, Ecology.* London: Verso, 1994.

———. *Critique of Economic Reason.* London: Verso, 1989.

———. *Farewell to the Working Class.* London: Pluto, 1982.

———. *Paths to Paradise: On the Liberation from Work.* London: Pluto, 1985.

Gottlieb, Robert. *Forcing the Spring: The Transformation of the American Environmental Movement.* Washington, D.C.: Island, 1993.

Gottweis, Herbert. "Genetic Engineering, Democracy, and the Politics of Identity." *Social Text* 42 (1995): 127–52.

Graham, Julie. "Fordism/Post-Fordism, Marxism/Post-Marxism." *Rethinking Marxism* 4.1 (1991): 39–58.

Gramsci, Antonio. *Selections from the Prison Notebooks.* Edited by Q. Hoare and G. Nowell-Smith. New York: International Publishers, 1971.

Gray, Steven. "Ontario's 'Green Work Alliance' Hopes Environmentally-Friendly Projects Can Reopen Plants." *Labor Notes* (Nov. 1992): 15–17.

Greely, Henry T. "Health Insurance, Employment Discrimination, and the Genetics Revolution." In *The Code of Codes: Scientific and Social Issues in the Human*

Genome Project, edited by Daniel J. Kevles and Leroy Hood, 264–80. Cambridge, Mass.: Harvard University Press, 1992.

Greven, Thomas. "Can We Convert Defense Jobs to Peacetime Uses?" *Labor Notes* (Nov. 1992): 7, 15.

Grossman, Rachael. "Women's Place in the Integrated Circuit." *Radical America* 14.1 (1980): 29–50.

Guattari, Felix. *The Guattari Reader.* Edited by Gary Genosko. Oxford: Blackwell, 1996.

———. *Molecular Revolution: Psychiatry and Politics.* Harmondsworth, Eng.: Penguin, 1984.

———. "The Three Ecologies." *New Formations* 8 (1989): 131–47.

Guattari, Felix, and Toni Negri. *Communists Like Us: New Spaces of Liberty, New Lines of Alliance.* New York: Semiotext(e), 1990.

Habermas, Jürgen. *The Theory of Communicative Action.* Vol. 1, *Reason and the Rationalization of Society.* Boston: Beacon, 1984.

———. *The Theory of Communicative Action.* Vol. 2, *A Critique of Functionalist Reason.* Boston: Beacon, 1987.

———. *Toward a Rational Society: Student Protest, Science and Politics.* Boston: Beacon, 1970.

Hacker, Sally. *"Doing It the Hard Way": Investigations of Gender and Technology.* Boston: Unwin Hyman, 1990.

Hackett, Robert. *Engulfed: Peace Protest and America's Press During the Gulf War.* New York: New York University, Center for War and Peace and the New Media, 1993.

Hafner, Katie, and John Markoff. *Cyberpunk: Outlaws and Hackers on the Computer Frontier.* New York: Simon and Schuster, 1991.

Halal, William. *The New Capitalism: Democratic Free Enterprise in Post-Industrial Society.* New York: Wiley, 1986.

Hall, Peter, and Paschal Preston. *The Carrier Wave: New Information Technology and the Geography of Innovation, 1846–2003.* London: Unwin Hyman, 1988.

Hall, Stuart. "The Meaning of New Times." In *New Times: The Changing Face of Politics in the 1990s,* edited by Stuart Hall and Martin Jacques, 116–36. London: Lawrence and Wishart, 1989.

Hall, Stuart, and Martin Jacques. *New Times: The Changing Face of Politics in the 1990s.* London: Lawrence and Wishart, 1989.

Halleck, Dee Dee. "Watch Out, Dick Tracy! Popular Video in the Wake of the Exxon Valdez." In *Technoculture,* edited by Constance Penley and Andrew Ross, 211–29. Minneapolis: University of Minnesota Press, 1991.

Hamilton, Cynthia. 1994. "Urban Insurrection and the Global Crisis of Industrial Society." In *The World Transformed: Gender, Labour and International Solidarity in the Era of Free Trade, Structural Adjustment and GATT,* edited by Cindy Duffy and Craig Benjamin, 169–79. Guelph, Ontario: RhiZone, 1994.

Hammer, Joshua. "Nigeria Crude." *Harper's Magazine,* June 1996, 58–71.

Haraway, Donna. "A Manifesto for Cyborgs: Science, Technology, and Socialist Feminism in the 1980's." *Socialist Review* 80 (1985): 65–107.

Hardesty, Michael, and Nina Wurgaft. "Silicon Valley: A Tale of Two Classes." *Z Magazine,* Sept. 1992, 63–65.

Harding, Sandra, ed. *The "Racial" Economy of Science: Toward a Democratic Future.* Bloomington: Indiana University Press, 1993.

Hardt, Michael. "The Art of Organization: Foundations of a Political Ontology in Gilles Deleuze and Antonio Negri." Ph.D. diss., University of Washington, 1990.

Hardt, Michael, and Antonio Negri. *The Labor of Dionysus: A Critique of the State-Form.* Minneapolis: University of Minnesota Press, 1994.

Harris, Laurence. "Forces and Relations of Production." In *A Dictionary of Marxist Thought,* edited by Tom Bottomore, et al., 178–80. Cambridge, Mass.: Harvard University Press, 1983.

Harry, Debra. "The Human Genome Diversity Project and Its Implications for Indigenous Peoples." *Information about Intellectual Property Rights No. 6.* Minneapolis, Minn.: Institute for Agriculture and Trade Policy, 1995. Available by email from iatp@iatp.org.

Harvey, David. *The Condition of Postmodernity: An Enquiry into the Origins of Cultural Change.* Oxford: Blackwell, 1989.

Hauben, Michael. "The Social Forces behind the Development of Usenet News." *Amateur Computerist* 5.1/2 (1993): 13–21.

Hauben, Ronda. "Computers for the People: A History—or How the Hackers Gave Birth to the Personal Computer." Parts 1–3, 5. *Amateur Computerist* 3.4 (1991): 10–12; 4.2/3 (1992): 14–17; 4.4 (1992): 10–12; 5.1/2 (1993): 31–33.

Hayek, F. A. "The Use of Knowledge in Society." *American Economic Review* 35 (1945): 519–30.

Hayes, Dennis. *Behind the Silicon Curtain: The Seductions of Work in a Lonely Era.* Boston: South End, 1989.

Hebdige, Dick. "After the Masses." In *New Times: The Changing Face of Politics in the 1990s,* edited by Stuart Hall and Martin Jacques, 76–93. London: Lawrence and Wishart, 1989.

Henwood, Doug. "Info Fetishism." In *Resisting the Virtual Life: The Culture and Politics of Information,* edited by James Brook and Iain A. Boal, 163–72. San Francisco: City Lights, 1995.

Hertzberg, Hendrik. "Marxism: The Sequel." *New Yorker,* 13 Feb. 1995, 6–7.

Hill, Christopher. *The World Turned Upside Down: Radical Ideas during the English Revolution.* New York: Viking, 1973.

Hiltz, Starr Roxanne, and Maurice Turoff. *The Network Nation: Human Communication via Computer.* Reading, Mass.: Addison-Wesley, 1978.

Hirsch, Joachim. "The New Leviathan and the Struggle for Democratic Rights." *Telos* 48 (1981): 79–89.

Hirschop, Ken. "Democracy and New Technologies." *Monthly Review* 48.3 (1996): 86–98.

Hirschhorn, Larry. *Beyond Mechanization: Work and Technology in a Post-Industrial Age.* Cambridge, Mass.: MIT Press, 1984.

———. "The Post-Industrial Labor Process." *New Political Science* 2.3 (1981): 11–32.

Hofrichter, Richard, ed. *Toxic Struggles: The Theory and Practice of Environmental Justice.* Philadelphia: New Society, 1993.

Howard, Dick. *The Unknown Dimension: European Marxism since Lenin.* New York: Basic, 1972.

Howard, Robert. *Brave New Workplace.* New York: Viking, 1985.

Hoyos, Lisa, and Mai Hoang. "Workers at the Centre: Silicon Valley Campaign for Justice." *CrossRoads* 43 (1994): 24–27.

Hubbard, Ruth, and Elijah Ward. *Exploding the Gene Myth.* Boston: Beacon, 1993.

Hunnicutt, Benjamin. *Work without End: Abandoning Shorter Hours for the Right to Work.* Philadelphia: Temple University Press, 1988.

Hutchinson, Richard. "Machines of Desire: Class, Identity and the Potential of the New Social Movements." Paper presented at New Directions in Critical Theory Conference, University of Arizona, Tucson, 17 Apr. 1993.

Huws, Ursula. "Terminal Isolation: The Atomisation of Work and Leisure in the Wired Society." In *Issues in Radical Science,* Radical Science Series 16, edited by the Radical Science Collective, 9–26. London: Free Association Books, 1985.

Hyman, R. "Andre Gorz and His Disappearing Proletariat." In *Socialist Register,* edited by Ralph Miliband and John Saville, 272–95. London: Merlin, 1983.

Illingworth, Montieth M. "Workers on the Net, Unite!: Labor Goes Online to Organize, Communicate, and Strike." *Information Week,* 22 Aug. 1994. On-line, Internet, ACTIV-L@MIZZOU1.BITNET [accessed 3 Sept. 1994].

Ismartono, Yuli, and Teena Gill. "Asian Farmers Struggle against Transnationals." Third World Network, On-line, Internet, ACTIV-L@MIZZOU1.BITNET [accessed 17 Jan. 1996].

Iyer, Pico. *Video Night in Kathmandu: And Other Reports from the Not So Far East.* New York: Knopf, 1988.

Jacoby, Russell. *Dialectic of Defeat: Contours of Western Marxism.* Cambridge: Cambridge University Press, 1981.

———. "The Politics of Crisis Theory: Toward a Critique of Automatic Marxism II." *Telos* 23 (1975): 3–52.

———. "Towards a Critique of Automatic Marxism: The Politics of Philosophy From Lukács to the Frankfurt School." *Telos* 10 (1971): 119–46.

James, C. L. R. *At the Rendezvous of Victory.* London: Alison and Busby, 1984.

———. *The Black Jacobins: Toussaint L'Ouveture and the San Domingo Revolution.* New York: Vintage Books, 1989.

———. *The C. L. R. James Reader.* Edited by A. Grimshaw. Oxford: Blackwell, 1992.

———. *The Future in the Present.* London: Alison and Busby, 1977.

———. *Spheres of Existence.* London: Alison and Busby, 1980.

James, Selma. "Marx and Feminism." *Third World Book Review* 1.6 (1986): 1–6.

———. *Sex, Race and Class.* Bristol, Eng.: Falling Wall Press, 1975.

———. "Women's Unwaged Work—The Heart of the Informal Sector." *Women: A Cultural Review* 2.3 (1991): 267–71.

Jameson, Fredric. "Actually Existing Marxism." In *Marxism beyond Marxism,* edited by Saree Makdisi, Cesare Casarino, and Rebecca E. Karl, 14–54. London: Routledge, 1996.

———. "Cognitive Mapping." In *Marxism and the Interpretation of Culture,* ed. Cary Nelson and Lawrence Grossberg, 347–58. Urbana: University of Illinois Press, 1988.

———. "Foreword." In *The Postmodern Condition: A Report on Knowledge,* by Jean François Lyotard, vii–xxii. Minneapolis: University of Minnesota Press, 1984.

———. "Periodizing the 60's." In Jameson, *The Ideologies of Theory: Essays, 1971– 1986,* vol. 2, 178–210. Minneapolis: University of Minnesota Press, 1988.

————. "Postmodernism: or the Cultural Logic of Late Capitalism." *New Left Review* 146 (1984): 55–92.

————. "Postmodernism and the Market." In *Socialist Register*, edited by R. Miliband, L. Panitch, and J. Saville, 95–110. London: Merlin, 1990.

————. "Reading without Interpretation: Post-Modernism and the Video-Text." In *The Linguistics of Writing: Arguments Between Language and Literature*, edited by Nigel Fabb, Derk Attridge, Alan Durant, and Colin McCabe, 199–224. Manchester, Eng.: Manchester University Press, 1987.

Jefferys, Steve. "France 1995: The Backwards March of Labour Halted?" *Capital and Class* 59 (1996): 7–22.

Jhally, Sut. *The Codes of Advertising: Fetishism and the Political Economy of Meaning in the Consumer Society.* New York: St. Martin's Press, 1987.

Jones, Adam. 1994. "Wired World: Communications Technology, Governance and the Democratic Uprising." In *The Global Political Economy of Communications: Hegemony, Telecommunications and the Information Economy*, edited by Edward Comor, 145–64. New York: St. Martin's, 1994.

Jones, Barry. *Sleepers Wake!* Melbourne: Oxford University Press, 1982.

Kahn, Douglas. "Satellite Skirmishes: An Interview with Paper Tiger West's Jesse Drew." *Afterimage* 20.10 (1993): 9–11.

Kahn, Herman, and Anthony J. Weiner. *The Year 2000: A Framework for Speculation on the Next Thirty-three Years.* New York: Macmillan, 1967.

Kallick, David Dyssegaard. "Toward a New Unionism." *Social Policy* 25.2 (1994): 2–6.

Kaplan, Ethan. "Workers and the World Economy." *Foreign Affairs* 75.3 (1996): 16–63.

Katsiaficas, George. *The Subversion of Politics: European Autonomous Social Movements and the Decolonization of Everyday Life.* Atlantic Highlands, N.J.: Humanities Press, 1997.

Katz, Cindi, and Neil Smith. "L.A. Intifada: Interview with Mike Davis." *Social Text* 33 (1992): 19–33.

Keane, John. *Media and Democracy.* Oxford: Blackwell, 1990.

Keen, Brewster. *Trading Up: How Cargill, the World's Largest Grain Company, Is Changing Canadian Agriculture.* Toronto: NC Press, 1991.

Keil, Roger. "Green Work Alliances: The Political Economy of Social Ecology." *Studies in Political Economy* 44 (1994): 7–38.

Keller, Evelyn Fox. *Reflections on Gender and Science.* New Haven: Yale University Press, 1985.

Kellner, Douglas. *Jean Baudrillard: From Marxism to Postmodernism and Beyond.* Stanford, Calif.: Stanford University Press, 1989.

————. *Television and the Crisis of Democracy.* Boulder, Colo.: Westview, 1990.

————, ed. *Postmodernism/Jameson/Critique.* Washington D.C.: Maisonneuve Press, 1989.

Kennedy, Paul. *Preparing for the Twenty-first Century.* New York: Random House, 1993.

Kenney, Martin. *Biotechnology: The University Industrial Complex.* New Haven: Yale University Press, 1986.

Kenney, Martin, and Richard Florida. "Beyond Mass Production: Production and the Labor Process in Japan." *Politics and Society* 16.1(1988): 121–58.

Kestner, Grant. "Access Denied: Information Policy and the Limits of Liberalism." *Afterimage* 21.6 (1994): 5–10.

Kevles, Daniel J., and Leroy Hood, eds. *The Code of Codes: Scientific and Social Issues in the Human Genome Project*. Cambridge, Mass.: Harvard University Press, 1992.

Kidd, Dorothy. "Talking the Walk: The Communication Commons Amidst the Media Enclosures." Ph. D. diss., Simon Fraser University, Vancouver, Canada, 1998.

Kidd, Dorothy, and Nick Witheford. "Counterplanning from Cyberspace and Videoland: or Luddites on Monday and Friday, Cyberpunks the Rest of the Week." Paper presented at Monopolies of Knowledge: A Conference Honoring the Work of Harold Innis, Vancouver, Canada, 12 Nov. 1994.

Killian, Crawford. "Nazis on the Net." *Georgia Straight*, 11–18 Apr. 1996, 13–17.

Kimbrell, Andrew. *The Human Body Shop*. San Francisco: Harper, 1993.

Knabb, Ken, ed. *Situationist International Anthology*. Berkeley, Calif.: Bureau of Public Secrets, 1981.

Kohr, Martin. "Global Fight against 'Bio-Piracy.'" Third World Network, On-line, Internet, ACTIV-L@MIZZOU1.BITNET [accessed 6 Nov. 1995].

Kolko, Joyce. *Restructuring the World Economy*. New York: Pantheon, 1988.

Krimsky, Sheldon. "The New Corporate Identity of the American University." *Alternatives* 14.2 (1987): 20–29.

Krishnan, Raghu. "December 1995: "The First Revolt against Globalization." *Monthly Review* 48.1 (1996): 1–22.

Kroker, Arthur, and Michael Weinstein. *Data Trash: The Theory of the Virtual Class*. Montreal: New World Perspectives, 1994.

Kumar, Krishan. *From Post-Industrial to Post-Modern Society: New Theories of the Contemporary World*. Oxford: Blackwell, 1995.

———. "Futurology: The View from Eastern Europe." *Futures* 4.1 (1972): 90–95.

———. *Prophecy and Progress*. Harmondsworth, Eng.: Penguin, 1978.

———. *Utopia and Anti-Utopia in Modern Times*. Oxford: Blackwell, 1987.

Labone, Françoise. "Looking for Mothers You Only Find Fetuses." In *Made to Order: The Myth of Reproductive and Genetic Progress*, edited by Patricia Spallone and Deborah Lynn Steinberg, 48–57. Oxford: Pergamon, 1987.

Labor Notes. "Time Out: The Case for a Shorter Work Week." Detroit: Labor Notes, 1995.

Labor Resource Center. *Holding the Line in '89: Lessons of the NYNEX Strike: How Telephone Workers Can Fight Even More Effectively Next Time*. Somerville, Mass.: Labor Resource Center, 1990.

Laclau, Ernesto, and Chantal Mouffe. *Hegemony and Socialist Strategy: Towards a Radical Democratic Politics*. London: Verso, 1985.

Lakha, Salim. "Resisting Globalization: The Alternative Discourse in India." *Arena Journal* 4 (1994/95): 41–50.

Lazzarato, Maurizio. "Les Caprices du flux—les mutations technologiques du point de vue de ceux qui les vivent." *Futur Antérieur* 4 (1994): 156–64.

———. "General Intellect: Towards an Inquiry into Immaterial Labour." In *Immate-*

rial Labour, Mass Intellectuality, New Constitution, Post Fordism and All That, 1–
14. London: Red Notes, 1994.

———. "La 'Panthère' et la communication." *Futur Antérieur* 2 (1990): 54–67.

———. "'Pas de sous pas de totos!': La grève des ouvriers Peugeot." *Futur Antérieur* 1
(1990): 63–76.

Lazzarato, Maurizio, and Antonio Negri. *Le Bassin de Travail Immateriel (B.T.I.) dans
la métropole Parisienne: Définition, recherches, perspectives.* Paris: Tekne-Logos,
1993.

Lazzarato, Maurizio, and Toni Negri. "Travail immaterial and subjectivité." *Futur
Antérieur* 6 (1991): 86–89.

Lazzarato, Maurizio, A. Negri, and G. C. Santilli. *La Confection dans le quartier du
sentier: Restructuration des formes d'emploi et expansion dans un secteur en crise.*
Paris: Rapport MIRE, 1990.

Leborgne, D., and A. Lipietz. "New Technologies, New Modes of Regulation: Some
Spatial Implications." *Environment and Planning D: Society and Space* 6 (1988):
263–80.

Lebowitz, Michael. *Beyond Capital: Marx's Political Economy of the Working Class.*
New York: St. Martin's Press, 1992.

———. "Marx's Falling Rate of Profit: A Dialectical View." *Canadian Journal of Economics* 9.2 (1976): 233–54.

Lee, Butch, and Red Rover. *Night-Vision: Illuminating War and Class on the Neo-
Colonial Terrain.* New York: Vagabond Press, 1993.

Lee, Eric. *The Labour Movement and the Internet: The New Internationalism.* London: Pluto Press, 1997.

Lee, Martyn. *Consumer Culture Reborn: The Cultural Politics of Consumption.* New
York: Routledge, 1993.

Leiss, William. "The Myth of the Information Society." In *Cultural Politics in Contemporary America*, edited by Ian Angus and Sut Jhally, 282–99. New York: Routledge, 1989.

Lerner, Sally. "How Will North America Work in the Twenty-first Century?" In *Cutting Edge: Technology, Information, Capitalism and Social Revolution*, edited by
Jim Davis, Thomas Hirschl, and Michael Stack, 177–94. London: Verso, 1997.

Levidow, Les. "Foreclosing the Future." *Science and Society* 8 (1990): 59–79.

Levidow, Les, and Kevin Robins. "Towards a Military Information Society?" In *Cyborg Worlds: The Military Information Society*, edited by Levidow and Robins, 159–
77. London: Free Association Books, 1989.

Levins, Richard. "Toward the Renewal of Science." *Rethinking Marxism* 3.3/4 (1990):
102–25.

Levitt, Theodore. "The Globalization of Markets." *Harvard Business Review* 6.1
(1983): 92–102.

Linebaugh, Peter. *The London Hanged: Crime and Civil Society in the Eighteenth
Century.* London: Penguin, 1991.

Lipietz, Alain. *The Enchanted World: Inflation, Credit and the World Crisis.* London:
Verso, 1985.

———. *Mirages and Miracles: The Crisis of Global Fordism.* London: Verso, 1987.

Loeb, Paul Rogat. *Generation at the Crossroads: Apathy and Action on the American Campus.* New Brunswick, N.J.: Rutgers University Press, 1994.

London Edinburgh Weekend Return Group. *In and Against the State.* London: Pluto, 1980.

Lotringer, Sylvere, and Christian Marazzi. *Italy: Autonomia—Post-Political Politics.* New York: Semiotext(e), 1980.

Lowe, Donald. *The Body in Late Capitalist USA.* Durham, N.C.: Duke University Press, 1995.

Lucas, Martin, and Martha Wallner. "Resistance by Satellite: The Gulf Crisis and Deep Dish Satellite TV Network." In *Channels of Resistance,* edited by Tony Dowmunt, 176–94. London: British Film Institute, 1993.

Lukács, George. "Technology and Social Relations." *New Left Review* 39 (1966): 27–34.

Luke, Tim. "Informationalism and Ecology." *Telos* 56 (1983): 59–73.

Luke, Timothy. "Class Contradictions and Social Cleavage in Informationalizing Post-Industrial Societies: On the Rise of the New Social Movements." *New Political Science* 16/17 (1989): 125–53.

———. "Community and Ecology." *Telos* 88 (1991): 69–71.

———. *Screens of Power: Ideology, Domination and Resistance in an Informational Society.* Urbana: University of Illinois Press, 1989.

Lumley, Robert. *States of Emergency: Cultures of Revolt in Italy From 1968 to 1978.* London: Verso, 1990.

Lusane, Clarence. "Rap, Race, and Rebellion." *Z Magazine,* Sept. 1992, 36.

Lyon, David. *The Electronic Eye: The Rise of Surveillance Society.* Cambridge: Polity, 1994.

———. *The Information Society: Issues and Illusions.* Cambridge: Polity, 1988.

———. *Postmodernity.* Minneapolis: University of Minnesota Press, 1994.

Lyotard, Jean François. *Political Writings.* Minneapolis: University of Minnesota Press, 1993.

———. *The Postmodern Condition: A Report on Knowledge.* Minneapolis: University of Minnesota Press, 1984.

MacArthur, John R. *Second Front: Censorship and Propaganda in the Gulf War.* New York: Hill and Wang, 1992.

MacEwan, Arthur. "What's 'New' About the 'New International Economy'?" *Socialist Review* 21.3/4 (1991): 111–31.

Madhubuti, Haki, ed. *Why L.A. Happened: Implications of the '92 Los Angeles Rebellion.* Chicago: Third World, 1993.

Makdisi, Saree, Cesare Casarino, and Rebecca Karl, eds. *Marxism beyond Marxism.* London: Routledge, 1996.

Mallet, Serge. *Essays on the New Working Class.* St. Louis: Telos, 1975.

Malyon, Tim. "Might not Main." *New Statesman and Society,* 24 Mar. 1995, 24–26.

Mandel, Ernest. "In Defence of Socialist Planning." *New Left Review* 159 (1986): 5–39.

———. *An Introduction to Marxist Economic Theory.* New York: Pathfinder, 1969.

———. *Late Capitalism.* London: New Left Books, 1975.

———. *Marxist Economic Theory.* Vol. 1. New York: Monthly Review Press, 1968.

———. "The Myth of Market Socialism." *New Left Review* 169 (1988): 108–21.

Mandel, Ernest, and George Novak. *The Revolutionary Potential of the Working Class.* New York: Pathfinder, 1974.

Mann, Eric. "Labor-Community Coalitions as a Tactic for Labour Insurgency." In *Building Bridges: The Emerging Grassroots Coalition of Labor and Community,* edited by Jeremy Brecher and Tim Costello, 113–34. New York: Monthly Review Press, 1990.

———. "Labor's Environmental Agenda in the New Corporate Climate." In *Toxic Struggles: The Theory and Practice of Environmental Justice,* edited by Richard Hofrichter, 179–85. Philadelphia: New Society, 1993.

———. *Taking On General Motors: A Case Study of the UAW Campaign to Keep GM Van Nuys Open.* University of California, Center for Labor Research, 1987.

Mann, Eric, with the WATCHDOG Organizing Committee. *L.A.'s Lethal Air: New Strategies for Policy, Organizing and Action.* Los Angeles: Labor/Community Strategy Center, 1991.

Marazzi, Christian. "Money in the World Crisis: The New Basis of Capitalist Power." *Zerowork* 2 (1977): 91–111.

Marchand, Marie. *The Minitel Saga.* Paris: Larousse, 1988.

Marcuse, Herbert. *One-Dimensional Man: Studies in the Ideology of Advanced Industrial Society.* Boston: Beacon, 1964.

Martin, James. *The Wired Society.* Englewood Cliffs, N.J.: Prentice Hall, 1978.

Marx, Karl. *Capital: A Critique of Political Economy.* 3 Vols. New York: Vintage Books, 1977–81.

———. *Economic and Philosophical Manuscripts.* New York: International Publishers, 1964.

———. *Grundrisse: Foundations of a Critique of Political Economy.* Harmondsworth, Eng.: Penguin, 1973.

———. *The Poverty of Philosophy.* New York: International Publishers, 1971.

———. *Preface to the Contribution to a Critique of Political Economy.* London: International Publishers, 1971.

———. *Wage Labor and Capital: Value, Price and Profit.* New York: International Publishers, 1976.

Marx, Karl, and Friedrich Engels. *The Communist Manifesto.* New York: Washington Square, 1969.

———. *The German Ideology.* London: Lawrence and Wishart, 1938.

———. *Selected Correspondence.* Moscow: Progress Publishers, 1965.

———. *Selected Works.* 3 Vols. Moscow: Progress Publishers, 1965.

Masuda, Yoneji. *The Information Society as Post-Industrial Society.* Washington, D.C.: World Future Society, 1981.

———. *Managing in the Information Society: Releasing Synergy Japanese Style.* Oxford: Blackwell, 1990.

Mather, Celai, and Ben Lowe. *Trade Unions On-Line: The International Labour Movement and Computer Communications.* Preston, Eng.: Centre for Research on Employment and Work, Lancashire Polytechnic, 1990.

Mathews, John. *Age of Democracy: The Politics of Post-Fordism.* Melbourne: Oxford University Press, 1989.

————. *Tools of Change: New Technology and the Democratisation of Work.* Sydney: Pluto, 1989.

Mattelart, Armand. *Advertising International: The Privatization of Public Space.* New York: Routledge, 1991.

Mattelart, Armand, and Michele Mattelart. *Rethinking Media Theory.* Minneapolis: University of Minnesota Press, 1992.

Mattelart, Armand, and Seth Siegelaub, eds. *Communication and Class Struggle.* Vol. 1, *Capitalism and Imperialism.* New York: International General, 1979.

————. *Communication and Class Struggle.* Vol. 2, *Liberation and Socialism.* New York: International General, 1983.

Maylon, Tim. "Killing the Bill." *New Statesman and Society,* 8 July 1994, 12–13.

McChesney, Robert W. "The Global Struggle for Democratic Communication." *Monthly Review* 48.3 (1996): 1–20.

————. *Telecommunications, Mass Media and Democracy: The Battle for the Control of U.S. Broadcasting.* New York: Oxford University Press, 1993.

McIntosh, David. "Cyborgs in Denial: Technology and Identity in the Net." *Fuse* 18.3 (1994): 14–21.

McKenzie, Donald. "Marx and the Machine." *Technology and Culture* 25.3 (1984): 473–502.

McLuhan, Marshall. *Understanding Media: The Extensions of Man.* New York: McGraw-Hill, 1964.

McMurtry, John. "The Cancer Stage of Capitalism." *Social Justice.* On-line, Internet, PEN-L@anthrax.ecst.csuchico.edu, 24 July 1996.

————. *Unequal Freedoms: The Global Market as an Ethical System.* Toronto: Garamond, 1998.

Meehan, Eileen. "Technical Capability vs. Corporate Imperatives: Towards a Political Economy of Cable Television and Information Diversity." In *The Political Economy of Information,* edited by Vincent Mosco and Janet Wasko, 167–87. Madison: University of Wisconsin Press, 1988.

Melucci, Alberto. *Nomads of the Present: Social Movements and Individual Needs in Contemporary Society.* Philadelphia: Temple University Press, 1989.

Menzies, Heather. *Fast Forward and Out of Control: How Technology Is Changing Your Life.* Toronto: Macmillan, 1989.

————. *Whose Brave New World? The Information Highway and the New Economy.* Toronto: Between the Lines, 1996.

Mesthene, E. G., ed. *Technological Change.* Cambridge, Mass.: Harvard University Press, 1970.

Midnight Notes Collective. *Midnight Oil: Work, Energy, War 1973–1992.* New York: Autonomedia, 1992.

————. "Samir Amin, Delinking and Class Struggle: A Note." *Midnight Notes* 10 (1990): 55.

————. "The Working Class Waves Bye-Bye." *Midnight Notes* 7 (1984): 12–18.

Mies, Maria. *Patriarchy and Accumulation on a World Scale: Women in the International Division of Labour.* London: Zed Books, 1986.

————. "Why Do We Need All This? A Call against Genetic Engineering and Reproductive Technology." In *Made to Order: The Myth of Reproductive and Genetic*

Progress, edited by Patricia Spallone and Deborah Lynn Steinberg, 34–47. Oxford: Pergamon, 1987.

Mies, Maria, Veronika Bennholdt-Thomsen, and Claudia von Werlhof. *Women: The Last Colony.* London: Zed Books, 1988.

Milani, Brian. "Beyond Globalization: The Struggle to Redefine Wealth." Electronic publication. Available at http://www.web.net/~bmilani/MAI.htm.

Miller, Laura. "Women and Children First: Gender and the Settling of the Electronic Frontier." In *Resisting the Virtual Life: The Culture and Politics of Information,* edited by James Brook and Iain A. Boal, 49–59. San Francisco: City Lights, 1995.

Miller, R. Bruce, and Milton T. Wolf, eds. *Thinking Robots, an Aware Internet, and Cyberpunk Librarians.* Chicago: Library and Information Technology Association, 1992.

Mohseni, Navid. "The Labor Process and Control of Labor in the U.S. Computer Industry." In *The Labor Process and Control of Labor: The Changing Nature of Work Relations in the Late Twentieth Century,* edited by Berch Berberoglu, 59–77. Westport, Conn.: Praeger, 1993.

Montano, Mario. "Notes on the International Crisis." *Zerowork* 1 (1957): 32–60. Reprinted in Midnight Notes Collective, *Midnight Oil: Work, Energy, War 1973–1992,* 115–43. New York: Autonomedia, 1992.

Moody, Kim. *An Injury to One: The Decline of American Labour.* London: Verso, 1988.

———. "When High Wage Jobs Are Gone, Who Will Buy What We Make?" *Labor Notes* (June 1994): 8–9, 14.

———. *Workers in a Lean World: Unions in the International Economy.* London: Verso, 1997.

Moody, Kim, and Simone Sagovac. *Time Out: The Case for a Shorter Work Week.* Detroit: Labor Notes, 1995.

Mooney, Pat. *The Conservation and Development of Indigenous Knowledge in the Context of Intellectual Property Systems.* Ottawa: Rural Advancement Foundation International, 1993.

Moore, Joe. *Japanese Workers and the Struggles for Power, 1945–1947.* Madison: University of Wisconsin Press, 1983.

Moravec, Hans. *Mind Children: The Future of Robot and Human Intelligence.* Cambridge, Mass.: Harvard University Press, 1988.

———. "Pigs in Cyberspace." In *Thinking Robots, an Aware Internet, and Cyberpunk Librarians,* edited by R. Bruce Miller and Milton Wolf, 15–21. Chicago: Library and Information Technology Association, 1992.

Morley, David, and Kevin Robins. *Spaces of Identity: Global Media, Electronic Landscapes and Cultural Boundaries.* London: Routledge, 1995.

Morris-Suzuki, Tessa. *Beyond Computopia: Information, Automation and Democracy in Japan.* London: Kegan Paul International, 1988.

Mosco, Vincent. *The Pay-Per Society: Computers and Communication in the Information Age: Essays in Critical Theory and Public Policy.* Toronto: Garamond, 1989.

———. *The Political Economy of Communication.* London: Sage, 1996.

———. *Pushbutton Fantasies: Critical Perspectives on Videotex and Information Technology.* Norwood, N.J.: Ablex, 1982.

Mosco, Vincent, and Janet Wasko, eds. *The Political Economy of Information.* Madison: University of Wisconsin, 1988.

Moulier, Yves. "Introduction." *The Politics of Subversion: A Manifesto for the Twenty-first Century*, by Antonio Negri, 1–46. Cambridge: Polity, 1989.

———. "Les théories américaines de la 'segmentation du marché du travail' et italiennes de la 'composition de classe' à travers le prisme des lecteures françaises." *Babylone* 0 (1982): 175–217.

———. "L'Operaisme Italien: Organisation/représentation/idéologie: ou la composition de classe revisitée." In *L'Italie: Le Philosophe et le gendarme*, Actes du Colloque de Montreal, edited by Marie Blanche Tahon and Andre Corten, 27–63. Montreal: VLB Editeur, 1986.

Mowlana, Hamid. "Roots of War: The Long Road to Intervention." In *Triumph of the Image: The Media's War in the Persian Gulf—A Global Perspective*, edited by Hamid Mowlana, George Gerbner, and Herbert Schiller, 30–50. Boulder, Colo.: Westview, 1992.

Mowlana, Hamid, George Gerbner, and Herbert Schiller, eds. *Triumph of the Image: The Media's War in the Persian Gulf—A Global Perspective*. Boulder, Colo.: Westview, 1992.

Mujer a Mujer Collective. "Communicating Electronically: Computer Networking and Feminist Organizing." *RFR/DRF* 20.1/2 (1991): 10.

Murray, Fergus. "The Decentralisation of Production—the Decline of the Mass-Collective Worker?" *Capital and Class* 19 (1983): 74–99.

Murray, Robin. "Fordism and Post-Fordism." In *New Times: The Changing Face of Politics in the 1990s*, edited by Stuart Hall and Martin Jacques, 38–54. London: Lawrence and Wishart, 1989.

Naisbitt, John. *Megatrends*. New York: Warner, 1982.

Navarro, Vicente. *Crisis, Health and Medicine: A Social Critique*. New York: Tavistock, 1986.

———. *Dangerous to Your Health*. New York: Monthly Review Press, 1993.

Neill, Monty. "Computers, Thinking, and Schools in the New World Economic Order." In *Resisting the Virtual Life: The Culture and Politics of Information*, edited by James Brook and Iain A. Boal, 181–95. San Francisco: City Lights, 1995.

Nelson, Cary, ed. *Will Teach for Food: Academic Labor in Crisis*. Minneapolis: University of Minnesota Press, 1998.

Negri, Antonio. *La Classe ouvriere contre l'etat*. Paris: Edition Galilee, 1978.

———. "Constituent Republic." *Common Sense* 16 (1994): 88–96. Also in *Immaterial Labour, Mass Intellectuality, New Constitution, Post Fordism and All That*. London: Red Notes, 1994.

———. *Del Obrero-Masa al Obrero Social*. Barcelona: Editorial Anagrama, 1980.

———. "Domination and Sabotage." In *Working Class Autonomy and the Crisis*, edited by Red Notes Collective, 93–138. London: Red Notes, 1979.

———. "Gauche et coordinations ouvrières." *Lignes* 5 (1989): 86–97.

———. "Infinité de la communication/Finitude du désir." *Futur Antérieur* 11 (1992): 5–8.

———. "Interpretation of the Class Situation Today: Methodological Aspects." In *Open Marxism*, vol. 2, *Theory and Practice*, edited by Werner Bonefeld, Richard Gunn, and Kosmas Psychopedis, 69–105. London: Pluto, 1992.

———. "Luttes sociales et control systemique." *Futur Antérieur* 9 (1992): 15–20.

———. *Marx beyond Marx: Lessons on the Grundrisse*. New York: Bergin and Garvey, 1984.

———. *The Politics of Subversion: A Manifesto for the Twenty-first Century.* Cambridge: Polity, 1989.

———. *Revolution Retrieved: Selected Writings on Marx, Keynes, Capitalist Crisis and New Social Subjects.* London: Red Notes, 1988.

———. "Valeur-travail: Crise et problemes de reconstruction dans le post-moderne." *Futur Antérieur* 10 (1992): 30–36.

Newman, Nathan."'Third Wave Unionism' Takes to the Net." On-line, Internet, Red Rock Eater News Service [accessed 22 Aug. 1996].

Newson, Janice, and Howard Buchbinder. "Insider Trading: University Style." *Our Schools/Our Selves* 4.3 (1993): 45–52.

———. *The University Means Business: Universities, Corporations and Academic Work.* Toronto: Garamond, 1988.

Noble, David. *Forces of Production.* New York: Knopf, 1984.

———. "Insider Trading: University Style." *Our Schools/Our Selves* 4.3 (1993): 45–52.

———. "Present Tense Technology." Parts 1–2. *Democracy* (Spring, Summer, Fall 1983): 8–24, 70–82, 71–93. Reprinted in Noble, *Progress without People.*

———. *Progress without People: New Technology, Unemployment, and the Message of Resistance.* Toronto: Between the Lines, 1995.

———. "Social Choice in Machine Design: The Case of Automatically Controlled Machine Tools." In *Case Studies in the Labor Process*, edited by Andrew Zimbalist, 18–50. New York: Monthly Review Press, 1979.

———. "The Truth about the Information Highway." *CPU: Working in the Computer Industry* 13. On-line, Internet, ACTIV-L@MIZZOU1.BITNET [accessed 15 Feb. 1995].

Noble, Douglas. "High Tech Skills: The Corporate Assault on the Hearts and Minds of Union Workers." *Our Schools/Our Selves.* 1.8 (1993): 59–80.

———. "Mental Materiel: The Militarization of Learning and Intelligence in US Education." In *Cyborg Worlds*, edited by Les Levidow and Kevin Robins, 13–42. London: Free Association Books, 1989.

Nora, Simon, and Alain Minc. *The Computerization of Society.* Cambridge, Mass.: MIT Press, 1981.

Nordhaus, Hannah. "Underground by Modem." *Terminal City*, 11 Aug. 1993, 7.

Norris, Christopher. *Uncritical Theory: Postmodernism, Intellectuals and the Gulf War.* London: Lawrence and Wishart. 1992.

Nove, Alec. *The Economics of Feasible Socialism.* London: Allen and Unwin, 1985.

———. "Markets and Socialism." *New Left Review* 161 (1987): 98–104.

Novotny, Patrick. "Popular Epidemiology and the Struggle for Community Health: Alternative Perspectives from the Environmental Justice Movement." *Capitalism, Nature, Socialism.* 5.2 (1994): 29–50.

O'Connor, James. *Accumulation Crisis.* Blackwell: Oxford, 1984.

———. *The Fiscal Crisis of the State.* New York: St. Martin's, 1973.

———. *The Meaning of Crisis.* Oxford: Blackwell, 1987.

Oettinger, Anthony. "Information Resources: Knowledge and Power in the Twenty-first Century." *Science* 209 (1980): 191–98.

Ohmae, Kenichi. "Global Consumers Want Sony, Not Soil." *New Perspectives Quarterly* 8.4 (1991): 72–73.

Ovetz, Robert. "Assailing the Ivory Tower: Student Struggles and the Entrepreneur-ialization of the University." *Our Generation* 24.1 (1993): 70–95.

Panzieri, Raniero. "The Capitalist Use of Machinery: Marx versus the Objectivists." In *Outlines of a Critique of Technology*, edited by Phil Slater, 44–69. Atlantic Highlands, N.J.: Humanities Press, 1980.

———. "Surplus Value and Planning: Notes on the Reading of Capital." In *The Labour Process and Class Strategies*, edited by Conference of Socialist Economists, 4–25. London: Conference of Socialist Economists, 1976.

Parker, Mike, and Jane Slaughter. "Management by Stress." *Science as Culture* 8 (1990): 27–58.

Pelaez, Eloina, and John Holloway. "Learning to Bow: Post-Fordism and Technological Determinism." *Science as Culture* 8 (1990): 15–27.

Penley, Constance, and Andrew Ross, eds. *Technoculture*. Minneapolis: University of Minnesota Press, 1991.

Petchesky, Rosalind Polack. *Abortion and Woman's Choice: The State, Sexuality and Reproductive Freedom*. Boston: Northeastern University Press, 1990.

Piore, Michael, and Charles Sabel. *The Second Industrial Divide: Possibilities for Prosperity*. New York: Basic, 1984.

Piven, Frances Fox, and Richard Cloward. *Poor People's Movements: Why They Succeed, How They Fail*. New York: Pantheon, 1977.

p.m. "Strange Victories." In *Midnight Oil: Work, Energy, War, 1973–1992*, edited by Midnight Notes Collective, 193–215. New York: Autonomedia, 1992.

Pomeroy, Jim. "Black Box S-Thetix: Labor, Research, and Survival in the (Art) of the Beast." In *Technoculture*, edited by Constance Penley and Andrew Ross, 271–94. Minneapolis: University of Minnesota Press, 1991.

Pool, Ithiel de Sola. *Technologies of Freedom*. Cambridge, Mass.: Harvard University Press, 1983.

Poovey, Mary. 1988. "Feminism and Deconstruction." *Feminist Studies* 14.1: 51–65.

Porat, Marc Uri. "Global Implications of the Information Society." *Journal of Communication* 28.1 (1978): 70–80.

———. *The Information Economy: Definition and Measurement*. Vol. 1. Washington, D.C.: U.S. Department of Commerce, 1977.

Poster, Mark. *Existential Marxism in Postwar France: From Sartre to Althusser*. Princeton: Princeton University Press, 1975.

———. *Foucault, Marxism and History: Mode of Production versus Mode of Information*. Cambridge: Polity, 1984.

———. *The Mode of Information: Poststructuralism and Social Context*. Chicago: University of Chicago Press, 1990.

Prechel, Harland. "Transformations in Hierarchy and Control of the Labor Process in the Post-Fordist Era: The Case of the U.S. Steel Industry." In *The Labor Process and Control of Labor: The Changing Nature of Work Relations in the Late Twentieth Century*, edited by Berch Berberoglu. Westport, Conn.: Praeger, 1993. 44–58.

Progress and Freedom Foundation. "Cyberspace and the American Dream: A Magna Carta for the Knowledge Age." Washington: Progress and Freedom Foundation. On-line, Internet, 1994. Email to PFF@AOL.COM.

Przeworski, Adam. "Less Is More: In France the Future of Unemployment Lies in Leisure." *Dollars and Sense*, July/Aug. 1995, 12–15, 41.

Purdy, David. "Citizenship, Basic Income and the State." *New Left Review* 208 (1994): 30–48.

Raboy, Marc, Ivan Bernier, Florian Sauvageau, and Dave Atkinson. "Cultural Development and the Open Economy: A Democratic Issue and a Challenge to Public Policy." *Canadian Journal of Communication* 19 (1994): 291–315.

Rachleef, Peter. *Hard Pressed in the Heartland: The Hormel Strike and the Future of Labor.* Boston: South End, 1992.

———. *Marxism and Council Communism.* New York: Revisionist Press, 1976.

———. "Seeds of a Labor Resurgency." *Nation,* 21 Feb. 1994, 226–29.

Radical Chains. "'Autonomist' and 'Trotskyist' Views: Harry Cleaver Debates Hillel Ticktin." *Radical Chains* 4 (1994): 9–17.

Red Notes Collective, ed. *Immaterial Labour, Mass Intellectuality, New Constitution, Post-Fordism and All That.* Pamphlet. Red Notes: London. 1994.

———. *Working Class Autonomy and the Crisis.* London: Red Notes, 1979.

Reich, Robert. *The Work of Nations: Preparing Ourselves for Twenty-first Century Capitalism.* New York: Knopf, 1991.

Rheingold, Howard. *The Virtual Community.* Reading, Mass.: Addison-Wesley, 1993.

Rifkin, Jeremy. *The End of Work: The Decline of the Global Labor Force and the Dawn of the Post-Market Era.* New York: Putnam, 1995.

Robins, Kevin, and Frank Webster. "Athens without Slaves . . . Or Slaves without Athens? The Neurosis of Technology." *Science as Culture* 3 (1988): 7–53.

———. "Cybernetic Capitalism: Information Technology, Everyday Life." In *The Political Economy of Information,* edited by Vincent Mosco and Janet Wasko, 44–75. Madison: University of Wisconsin Press, 1988.

———. "Information as Capital: A Critique of Daniel Bell." In *The Ideology of the Information Age,* edited by Jennifer Slack and Fred Fejes, 95–117. Norwood, N.J.: Ablex, 1987.

———. "Luddism: New Technology and the Critique of Political Economy." In *Science, Technology and the Labor Process,* vol. 2, edited by Les Levidow and Bob Young, 9–48. Atlantic Highlands, N.J.: Humanities Press, 1983.

———. *The Technical Fix: Education, Computers and Industry.* London: Macmillan, 1989.

Rodriguez, Roberto. "Information Highway: Latino Student Protesters Create Nationwide Link Up." *Black Issues in Higher Education.* On-line, Internet, ACTIV-L@MIZZOU1.BITNET [accessed 16 June 1994].

Roncaglio, Rafael. "Notes on the Alternative." In *Video the Changing World,* edited by Nancy Thede and Alain Ambrosi, 206–9. Montreal: Black Rose, 1991.

Rosdolosky, Roman. *The Making of Marx's Capital.* London: Pluto, 1977.

Rose, Margaret. *The Post-Modern and the Post-Industrial: A Critical Analysis.* Cambridge: Cambridge University Press, 1991.

Ross, Andrew. "Hacking Away at the Counter Culture." In *Technoculture,* edited by Constance Penley and Andrew Ross, 107–34. Minneapolis: University of Minnesota Press, 1991.

Ross, G. "The Second Coming of Daniel Bell." In *Socialist Register,* edited by Ralph Miliband and John Saville, 56–84. London: Merlin, 1974.

Ross, Robert, and Kent Trachte. *Global Capitalism: The New Leviathan.* New York: New York University Press, 1990.

Roth, Karl Heinz. *L'Autre mouvement ouvrier en Allemagne, 1945–1978.* Paris: Christian Bourgeois, 1979. Originally *Die "Andere" Arbeiterbewegnung und die Entwicklung der Kapitalistischen Repression von 1880 bis zur Gegenwart.* Munich: Trikant Verlog, 1974.

Rothschild, Michael. *Bionomics: The Inevitability of Capitalism.* New York: Henry Holt, 1990.

Rowland, W. D., Jr., and M. Tracey. "Worldwide Challenges to Public Service Broadcasting." *Journal of Communication* 40.2 (1990): 8–27.

Ruggiero, Greg, and Stuart Sahulka, eds. *Open Fire: The Open Magazine Pamphlet Series Anthology.* New York: New Press, 1993.

Russell, Kathryn. "A Value-Theoretic Approach to Childbirth and Reproductive Engineering." *Science and Society* 58.3 (1994): 287–314.

Rustin, Michael. "Absolute Voluntarism: Critique of a Post-Marxist Concept of Hegemony." *New German Critique* 43 (1988): 146–71.

———. "The Trouble with 'New Times.'" In *New Times: The Changing Face of Politics in the 1990s,* edited by Stuart Hall and Martin Jacques, 303–20. London: Lawrence and Wishart, 1989.

Ryan, Michael. *Politics and Culture: Working Hypotheses for a Post-Revolutionary Society.* London: Macmillan, 1989.

Sachs, Wolfgang, ed. *The Development Dictionary: A Guide to Knowledge as Power.* London: Zed Books, 1992.

Sakolsky, Ron. "Zoom Black Magic Liberation Radio: The Birth of the Micro-Radio Movement in the USA." In *A Passion for Radio,* edited by Bruce Girard, 106–13. Montreal: Black Rose, 1992.

Salvaggio, Jerry, ed. *The Information Society: Economic, Social and Structural Issues.* Hillsdale, N.J.: Lawrence Erlbaum, 1989.

Santiago-Valles, William. "Memories of the Future: Maroon Intellectuals from the Caribbean and the Sources of Their Communication Strategies, 1925–1940." Ph.D. diss., Simon Fraser University, Vancouver, Canada, 1997.

Santilli, G. "Peau de leopard: L'Automatisation comme forme de controle social." *Travail* 8 (1985): 20–28.

Schaffer, Simon. "Babbage's Intelligence: Calculating Engines and the Factory System." *Critical Inquiry* 21 (1994): 203–27.

Scheer, Christopher. "The Pursuit of Techno-Happiness: Third Wavers and Tekkie Cults." *Nation,* 8 May 1995, 632–34.

Scheiner, Charles. "Electronic Resources on East Timor." On-line, Internet, ACTIV-L@MIZZOU1.BITNET [accessed 25 Oct. 1996]. Available from timor_info @igc.apc.org.

Schement, Jorge Reina, and Leah A. Lievrouw. "Introduction: The Fundamental Assumptions of Information Society Research." In *Competing Visions, Complex Realities: Social Aspects of the Information Society,* edited by Jorge Schement and Leah A. Lievrouw, 1–10. Norwood, N.J.: Ablex, 1987.

Schiller, Herbert J. *Communication and Cultural Domination* New York: International Arts and Sciences Press, 1976.

———. *Culture, Inc: The Corporate Takeover of Public Expression.* Oxford: Oxford University Press, 1989.

———. "The Global Information Highway: Project for an Ungovernable World." In *Resisting the Virtual Life: The Culture and Politics of Information*, edited by James Brook and Iain A. Boal, 17–33. San Francisco: City Lights, 1995.

———. *Information in the Crisis Economy*. Norwood, N.J.: Ablex, 1984.

———. "The Information Superhighway: Paving Over the Public." *Z Magazine*, Mar. 1994, 46–50.

———. *The Mind Managers*. Boston: Beacon, 1973.

———. *Who Knows: Information in the Age of the Fortune 500*. Norwood, N.J.: Ablex, 1981.

Schor, Juliet. *The Overworked American: The Unexpected Decline of Leisure*. New York: Harper, 1991.

Schuler, Douglas. *New Community Networks: Wired for Change*. Reading, Mass.: Addison-Wesley, 1996.

Sclove, Richard. *Democracy and Technology*. New York: Guilford, 1995.

Seabrook, Jeremy. "Biotechnology and Genetic Diversity." *Race and Class* 34.3 (1993): 15–30.

Shade, Leslie Regan. "Gender Issues in Computer Networking." Paper presented at Community Networking: The International Free-Net Conference, Carleton University, Ottawa, Canada, 17–19 Aug. 1993.

Shapiro, Andrew. "Street Corners in Cyberspace." *Nation*, 3 July 1995, 10–13.

Sherman, Barrie, and Phil Judkins. *Licensed to Work*. London: Cassell, 1995.

Shiva, Vandana. "Seeds of Struggle." In *The World Transformed: Gender, Labour and International Solidarity in the Era of Free Trade, Structural Adjustment and GATT*, edited by Cindy Duffy and Craig Benjamin, 57–70. Guelph, Ontario: RhiZone, 1994.

———. *The Violence of the Green Revolution: Third World Agriculture, Ecology and Politics*. London: Zed Books, 1991.

Shohat, Ella, and Robert Stam. *Unthinking Eurocentricism: Multiculturalism and the Media*. London: Routledge. 1994.

Shulmann, Seth. "Preventing Genetic Discrimination." *Technology Review* (July 1995): 16–18.

Siegel, Lenny. "New Chips in Old Skins: Work, Labor and Silicon Valley." *CPU: Working in the Computer Industry* 6. On-line, Internet, 1993.

Simon, Joel. "Netwar Could Make Mexico Ungovernable." Pacific News Service. On-line, Internet, ACTIV-L@MIZZOU1.BITNET [accessed 20 Mar. 1995].

Sivanandan, A. "'All that Melts into Air Is Solid': The Hokum of New Times." *Race and Class* 31.3 (1989): 1–23.

Slack, Jennifer D., and Fred Fejes, eds. *The Ideology of the Information Age*. Norwood, N.J.: Ablex, 1987.

Slaughter, Jane. "Addicted to Overtime." *The Progressive*, Apr. 1995, 31–33.

Smith, Chris. "From the 1960s Automation to Flexible Specialization: A Déjà Vu of Technological Panaceas." In *Farewell to Flexibility*, edited by A. Pollert, 138–57. Oxford: Blackwell, 1991.

Smith, Steve. "Taylorism Rules OK? Bolshevism, Taylorism and the Technical Intelligentsia in the Soviet Union 1917–41." *Radical Science Journal* 13 (1983): 3–27.

Smythe, Dallas W. "Communications: Blindspot of Western Marxism." *Canadian Journal of Political and Social Theory* 1.3 (1977): 1–27.

———. *Dependency Road: Communications, Capitalism, Consciousness and Canada.* Norwood, N.J.: Ablex, 1981.

Solomonides, Tony, and Les Levidov, eds. *Compulsive Technology: Computers as Culture.* London: Free Association Books, 1985.

Spallone, Patricia, and Deborah Lynn Steinberg, eds. *Made to Order: The Myth of Reproductive and Genetic Progress.* Oxford: Pergamon, 1987.

Spender, Dale. *Nattering on the Net: Women, Power and Cyberspace.* Melbourne: Spinifex, 1995.

Spillane, Margaret. "Unplug It!" *Nation,* 21 Nov. 1994, 600.

Stallabrass, Julian. "Empowering Technology: The Exploration of Cyberspace." *New Left Review* 211 (1995): 3–32.

Stangelaar, Fred. "An Outline of Basic Principles of Alternative Communication." Paper presented at Workshop on International Communication by Computer, Institute of Social Studies, The Hague, 27 Oct. 1985.

Sterling, Bruce. *The Hacker Crackdown: Law and Disorder on the Electronic Frontier.* New York: Bantam, 1992.

———. "A Short History of the Internet." *Magazine of Fantasy and Science Fiction,* Feb. 1993. On-line, Internet, ACTIV-L@MIZZOU1.BITNET [accessed 12 Nov. 1993].

Stewart, Edward. *The Paris Commune 1871.* London: Eyre and Spottiswoode, 1971.

Stonier, Tom. *The Wealth of Information.* London: Methuen, 1983.

"Students Fight the Contract." *Progressive,* 16 May 1995.

Surin, Kenneth. "Marxism(s) and the Withering Away of the State." *Social Text* 27 (1990): 35–54.

———. "Reinventing a Physiology of Collective Liberation: Going 'Beyond Marx' in the Marxism(s) of Toni Negri, Felix Guattari, and Gilles Deleuze." Paper presented at the Rethinking Marxism Conference, Amherst, Mass., 1993.

Swanson, Jean. "GAI: Guaranteed Disaster." *Canadian Dimension,* Dec./Jan. 1994/95, 24–27.

Tahon, Marie Blanche, and Andre Corten, eds. *L'Italie: Le Philosophe et le gendarme.* Actes du Colloque de Montreal. Montreal: VLB Editeur, 1986.

Teeple, Gary. *Globalization and the Decline of Social Reform.* Toronto: Garamond, 1995.

Templeton, Robin. "Not for Sale—Unplug Channel One." *CrossRoads* 34 (1993): 19–20.

Thede, Nancy, and Alain Ambrosi, eds. *Video the Changing World.* Montreal: Black Rose Books, 1991.

Thomas, Wes. "Hyperwebs: Pirate Radio." *Mondo 2000* 11 (1993): 26–39.

Thompson, E. P. *The Making of the English Working Class.* London: Gollancz, 1963.

———. *The Poverty of Theory.* New York: Monthly Review Press, 1978.

———. *William Morris: Romantic to Revolutionary.* London: Merlin, 1977.

Toffler, Alvin. *Future Shock.* New York: Bantam, 1970.

———. *Powershift.* New York: Bantam, 1990.

———. *Previews and Premises* London: Pan, 1984.

———. *The Third Wave.* New York: Morrow, 1980.

Tokar, Brian. "The False Promise of Biotechnology." *Z Magazine,* Feb. 1992, 27–32.

Touraine, Alain. *The Post-Industrial Society: Tomorrow's Social History: Classes, Conflicts and Culture in the Programmed Society.* New York: Random House, 1971.

Treichler, Paula A. "How to Have Theory in an Epidemic: The Evolution of AIDS Treatment Activism." In *Technoculture*, edited by Constance Penley and Andrew Ross, 57–156. Minneapolis: University of Minnesota Press, 1991.

Trend, David. "Rethinking Media Activism: Why the Left Is Losing the Culture War." *Socialist Review* 93/2 (1993): 5–34.

Tronti, Mario. "Lenin in England." In *Working Class Autonomy and the Crisis*, edited by the Red Notes Collective, 1–6. London: Red Notes, 1979.

———. *Ouvriers et capital*. Paris: Christian Bourgeois, 1977.

———. "Social Capital." *Telos* 17 (1973): 98–121.

———. "The Strategy of Refusal." In *Working Class Autonomy and the Crisis*, edited by the Red Notes Collective, 7–21. London: Red Notes, 1979.

———. "Workers and Capital." In *The Labour Process and Class Strategies*, edited by Conference of Socialist Economists, 92–125. London: Conference of Socialist Economists, 1976.

Trotsky, Leon. *History of the Russian Revolution*. London: Pluto, 1977.

Truong, Hoai-An, with Gail Williams, Judi Clark, Anna Couey, and others of Bay Area Women in Telecommunications. "Gender Issues in Online Communications." Online, Internet, ACTIV-L@MIZZOU1.BITNET [accessed 13 Jan. 1993].

Tsuzuku, Ken. "Presentation to the 1991 Labor Notes Conference." In *A Conference on Labour and Team Concepts*, Proceedings of a Conference Co-Sponsored by Capilano College Labour Studies Programme and Vancouver and District Labour Council, Vancouver, Canada, 18–19 Oct. 1991.

Tuer, Dot. "All in the Family: An Examination of Community Acess Cable in Canada." *Fuse* 17.3 (1994): 23–29.

Turner, Terisa, and Craig S. Benjamin. "Not in Our Nature: The Male Deal and Corporate Solutions to the Debt-Nature Crisis." *Review: Fernand Braudel Center* 18:2 (1995): 209–58.

Ure, Andrew. *The Philosophy of Manufacture*. London, 1835.

Valaskakis, Kimon. *The Information Society: The Issue and the Choices*. Montreal: Gamma, 1979.

Vallas, Steven. *Power in the Workplace: The Politics of Production at AT&T*. New York: State University of New York Press, 1993.

Van Parijs, Philippe. *Marxism Recycled*. Cambridge: Cambridge University Press, 1993.

———. *Real Freedom for All*. Oxford: Clarendon, 1995.

———, ed. *Arguing for Basic Income: Ethical Foundations for a Radical Reform*. London: Verso, 1992.

Van Parijs, Philippe, with Robert J. van der Veen. "A Capitalist Road to Communism." *Theory and Society* 15.5 (1986): 635–56.

Vattimo, Gianni. *The Transparent Society*. Baltimore: Johns Hopkins University Press, 1992.

Vellela, Tony. *New Voices: Student Activism in the 80s and 90s*. Boston: South End, 1988.

Viano, Maurizio, and Vincenzo Binnetti. "What Is to Be Done? Marxism and the Academy." In *Marxism beyond Marxism*, edited by Saree Makdisi, Cesare Casarino, and Rebecca E. Karl, 243–54. London: Routledge, 1996.

Vincent, Jean-Marie. "Les Automatismes sociaux et le 'general intellect.'" *Futur Antérieur* 16 (1993): 121–30.

Virilio, Paul. *Popular Defense and Ecological Struggles.* New York: Semiotext(e), 1990.

———. "Popular Defense and Popular Assault." In *Italy: Autonomia—Post-Political Politics,* edited by Sylvere Lotringer and Christian Marazzi, 266–72. New York: Semiotext(e). 1980.

Virno, Paolo. "Notes on the General Intellect." In *Marxism beyond Marxism,* edited by Saree Makdisi, Cesare Casarino, and Rebecca E. Karl, 265–72. London: Routledge, 1996.

Virno, Paolo, and Michael Hardt. *Radical Thought in Italy: A Potential Politics.* Minneapolis: University of Minnesota Press, 1996.

Vitale, Alex, and Keith McHenry. "Food Not Bombs." *Z Magazine,* Sept. 1994, 19–21.

Vogel, Steven. "New Science, New Nature: The Habermas-Marcuse Debate Revisited." *Research in Philosophy and Technology* 11 (1991): 157–78.

Wachtel, Howard. *The Money Mandarins: The Making of a Supranational Economic Order.* New York: Pantheon, 1986.

Wages for Housework Campaign. "Making Trouble, Making History: The International Wages for Housework Campaign, 1972–1994." London: Wages for Housework Campaign, 1994.

Wainwright, Hilary. *Arguments for a New Left: Answering the Free Market Right.* Oxford: Blackwell, 1994.

Wainwright, Hilary, and Dave Elliot. *The Lucas Plan: A New Trades Unionism in the Making.* London: Alison and Busby, 1982.

Waring, Marilyn. *If Women Counted: A New Feminist Economics.* San Francisco: Harper and Row, 1988.

Wark, McKenzie. *Virtual Geography: Living with Global Media Events.* Bloomington: Indiana University Press, 1994.

Waterman, Peter. "Communicating Labour Internationalism: A Review of Relevant Literature and Resources." *Communications: European Journal of Communications* 15.1/2 (1990): 85–103.

———. "From Moscow with Electronics: A Communication Internationalism for an Information Capitalism." *Democratic Communique* 11.2/3 (1994): 1, 11–15.

———. *International Labour Communication by Computer: The Fifth International?* Working Paper Series 129. The Hague: Institute of Social Studies, 1992.

———. "Reconceptualising the Democratisation of International Communication." *International Social Science Journal* 123 (1990): 78–91.

Webster, Frank. *Theories of the Information Society.* London: Routledge, 1995.

Webster, Frank, and Kevin Robins. *Information Technology: A Luddite Analysis.* Norwood, N.J.: Ablex, 1986.

Wehling, Jason. "'Netwars' and Activist Power on the Internet." On-line, Internet, ACTIV-L@MIZZOU1.BITNET [accessed 25 Mar. 1995].

Weston, Jay. "Old Freedoms and New Technologies: The Evolution of Community Networking." Paper presented at Free Speech and Privacy in the Information Age Symposium, University of Waterloo, Canada, 26 Nov. 1994.

Wilding, Adrian. Review of *Specters of Marx,* by Jacques Derrida. *Common Sense* 17 (1995): 92–95.

Williams, Raymond. *Marxism and Literature.* Oxford: Oxford University Press, 1977.

———. "The Politics of Nuclear Disarmament." In *Exterminism and Cold War,* edited by New Left Review, 65–87. London: Verso, 1982.

———. *Towards 2000*. London: Chatto and Windus, 1983.

Williamson, Thad. *What Comes Next? Proposals for a Different Society*. Washington, D.C.: National Center for Economic Security Alternatives, 1998.

Wilson, Kevin. *Technologies of Control: The New Interactive Media for the Home*. Madison: University of Wisconsin Press, 1988.

Winner, Langdon. *Autonomous Technology: Technics-Out-of-Control as a Theme in Political Thought*. Cambridge, Mass.: MIT Press, 1977.

Witheford, Nick, and Richard Gruneau. "Between the Politics of Production and the Politics of the Sign: Post-Marxism, Postmodernism, and New Times." *Current Perspectives in Social Theory* 13 (1993): 69–92.

Wood, Stephen. "The Transformation of Work?" In *The Transformation of Work? Skill, Flexibility and the Labour Process*, edited by Stephen Wood, 1–43. London: Unwin Hyman, 1989.

Wright, Steven. "Confronting the Crisis of Fordism: Italian Debates around Guaranteed Income." Forthcoming in *Capital and Class*.

———. "Forcing the Lock: The Problem of Class Composition in Italian Workerism." Ph.D diss., Monash University, Australia, 1988.

Wright, Susan. "Down on the Animal Pharm: Splicing Away Regulations." *Nation*, 11 Mar. 1996, 16–21.

Yoxen, Edward. "Life as a Productive Force: Capitalizing the Science and Technology of Molecular Biology." In *Science, Technology and the Labour Process: Marxist Studies*, vol. 1, edited by Les Levidow and Bob Young, 66–123. Atlantic Highlands, N.J.: Humanities Press, 1981.

Zeltzer, Steven. "Labor Media Communication: Voices in the Global Economy." Online, Internet, http://www.igc.org/lvpsf.

Zerowork Collective. "Introduction." *Zerowork: Political Materials* 1 (1975): 1–7.

Zimbalist, Andrew, ed. *Case Studies in the Labor Process*. New York: Monthly Review Press, 1979.

Zimmerman, Jan, ed. *The Technological Woman: Interfacing with Tomorrow*. New York: Praeger, 1983.

Zimmerman, Patricia R. "The Female Bodywars: Rethinking Feminist Media Politics." *Socialist Review* 93.2 (1993): 35–56.

Zuboff, Shoshana. *In the Age of the Smart Machine: The Future of Work and Power*. New York: Basic, 1994.

INDEX

NICK DYER-WITHEFORD teaches in the Faculty of Information and Media Studies, University of Western Ontario. He studied at the School of Communication, Simon Fraser University, Vancouver.

Typeset in 9.5/13 Trump Medieval
with Gill Sans display
Designed by Copenhaver Cumpston
Composed by Jim Proefrock
at the University of Illinois Press
Manufactured by Cushing-Malloy, Inc.

University of Illinois Press
1325 South Oak Street
Champaign, Illinois 61820-6903
www.press.uillinois.edu